ADVANCES IN
X-RAY ANALYSIS
Volume 25

ADVANCES IN X-RAY ANALYSIS

Volume 25

Edited by

John C. Russ

North Carolina State University
Raleigh, North Carolina

Charles S. Barrett
and Paul K. Predecki

University of Denver
Denver, Colorado

and

Donald E. Leyden

Colorado State University
Fort Collins, Colorado

Sponsored by
University of Denver Research Institute
and
Department of Chemistry
University of Denver

PLENUM PRESS • NEW YORK AND LONDON

The Library of Congress cataloged the first volume of this title as follows:

Conference on Application of X-ray Analysis.
Proceedings 6th-   1957-   [Denver]

    v. illus. 24-28 cm. annual.
No proceedings published for the first 5 conferences.
Vols. for 1958-   called also: Advances in X-ray analysis, v. 2-
Proceedings for 1957 issued by the conference under an earlier name: Conference on Industrial Applications of X-ray Analysis. Other slight variations in name of conference.
Vol. for 1957 published by the University of Denver, Denver Research Institute, Metallurgy Division.
Vols. for 1958-   distributed by Plenum Press, New York.
Conferences sponsored by University of Denver, Denver Research Institute.
    1. X-rays—Industrial applications—Congresses.   I. Denver. University. Denver Research Institute   II. Title: Advances in X-ray analysis.
TA406.5.C6   58-35928

ISBN 978-1-4613-9995-7   ISBN 978-1-4613-9993-3 (eBook)
DOI 10.1007/978-1-4613-9993-3

Softcover reprint of the hardcover 1st edition 1982
Proceedings of the Thirtieth Annual Conference on Applications of
X-Ray Analysis held in Denver, Colorado, August 3 – 7, 1981
© 1982 University of Denver
Denver, Colorado

Plenum Press is a division of Plenum Publishing Corporation
233 Spring Street, New York, N.Y. 10013

FOREWORD

In alternating years, the Denver X-ray Conference turns its principal attention, through the choice of subjects for the plenary lectures, to various aspects of either X-ray fluorescence or diffraction. This is a "fluorescence" year, and the three invited lecturers are experts in techniques that are at, or perhaps yet a bit beyond, the forefront of our understanding and technology in that field.

The common denominator in selecting these subjects was that each is approaching a full elucidation of theory, and active development of practical hardware, and that each presents analytical possibilities which can hardly be ignored in the next generation of commercial instrumentation. In other words, these are techniques that many of us shall likely find ourselves using by the end of the decade. The greatest difficulty in selecting the subjects was the need to overlook others, particularly 1) the increasing interest in "in-situ" or on-line analytical control, for tagging, identification or sorting; and 2) the broad subject of the computerization of instrumentation, with its powerful impact on the design of hardware, and its open invitation to the theorist to create more exact mathematical models of the analytical process, regardless of complexity, in the expectation that programmers will find ways to implement solutions in affordable, dedicated computers. Since these are subjects with which I have some personal familiarity, while the others cover fields more remote from my direct experience, it was natural (if selfish) to select the subjects that I most wanted to hear about.

Choosing the invited speakers was a much simpler task. Each is pre-eminent in his field, with impressive lists of credentials and a history of pioneering in the development of the particular technique.

The use of excitation other than the simple "continuum plus characteristic" radiation from conventional X-ray tubes has increased greatly with the introduction of energy dispersive

spectrometers, because their great detection efficiency permits
(and their limited count rate capability demands) lower power input to the specimen than diffractive ("wavelength-dispersive")
conventional spectrometers. The use of secondary targets to produce essentially monoenergetic exciting radiation is now common.
Imaginative use of filters, and even diffraction to select particular energies, have found places in many labs. Unfortunately,
these methods also restrict the range of elements which can be
simultaneously analyzed. The use of polarization to allow broad
spectrum excitation while still reducing the scattering of incident radiation from sample to detector overcomes this problem
while providing the very low spectrum background levels that permit good detection limits. Richard Ryon has pursued his interest
in this technique in both the U.S. and Europe, and brings to us a
comprehensive review of the work to date, and the possibilities
for the future, including comparisons with better known excitation
methods.

The "standard" energy dispersive X-ray detector is presently
lithium drifted silicon. With spectral resolution about an order
of magnitude poorer than diffractive spectrometers, but having
the ability to simultaneously detect X-rays from most of the
periodic table, the "ED" system has found an important niche in the
marketplace, and has both replaced the older "WD" systems in some
existing applications and created some new ones. The need to keep
the detector cooled with liquid nitrogen has been, next to the
limited count rate capability, the most important objection to the
use of ED systems in many applications. Not only is the supply
and handling of the nitrogen a bother, but the cryostat needed to
house the cooled detector limits the ability of these detectors to
be placed in many environments. Dr. Dabrowski and his coworkers
have been experimenting with other semiconductor materials for
many years, with papers reporting their progress in prior proceedings of this meeting. Their current progress with mercuric
iodide suggests that room temperature energy dispersive detectors,
in size small enough to be mounted almost anywhere, and with performance nearly as good as the cooled Si(Li) variety (and certainly
adequate for an enormous variety of analytical jobs, as witness
the success of the ED method in the first place), are now at hand.
His paper describes the performance criteria of these devices,
the physical basis for their properties, and the further stages
needed to bring them to full commercial availability.

The third paper,* by Joachim Stöhr of the Stanford Linear
Accelerator group, reminds us that X-ray analysis is neither
1) limited to instruments we can reasonably expect to find exhibited at shows, or 2) confined purely to measurement of elemental
abundances. There have been some historical applications of X-ray
emission spectra in which the profiles of lines involving

*Not published in these proceedings.

transitions from outer orbitals are interpreted to gain information on bonding configurations, so the extension of X-ray analysis
to structure is not without precedent. EXAFS (and its cousin,
EXELFS, when the absorption edge structure is studied in the spectrum of energy loss of an electron beam penetrating the sample)
offers a way to study the configuration of atoms and their immediate
neighbors. Just as the presence of an absorption edge in a spectrum of X-rays that have passed through material identifies the
elements present, so the "fine structure" near the absorption edge
can reveal the distances to neighbors and which atoms lie where.
Beyond simple comparison, the interpretation of the "squiggles" in
the curve requires both extensive mathematical treatment and good
underlying physical concepts. The information that can be gained
is particularly needed in the study of surfaces.

It is our hope that in the future, when all of these methods
have been further refined and pass into the realm of the "everyday,"
that you will refer to these printed lectures as a valuable in-
sight into the origins and growth of the techniques. Even more
important, by bringing the still young technologies to your attention, we hope to excite your enthusiasm and imagination, to hasten
their development and the day when these methods are available to
us all.

In addition, the contributed papers for the conference reflect
the usual breadth of activity of the attendees, and give a valuable
summary of the present-day state of the art. These papers, combined with the specialist workshops that have become the hallmark
of the Denver conference, provide an in-depth treatment of practically every aspect of our field. It has been a great personal
pleasure to assist in the organization of the conference, and
the review of the papers.

John C. Russ
North Carolina
December 1981

PREFACE

This volume constitutes the proceedings of the 1981 Denver Conference on the Applications of X-Ray Analysis, 30th in the series. The conference was held August 3-7, 1981 at the University of Denver and was sponsored by the University of Denver. The local conference chairmen were D. E. Leyden and P. K. Predecki, with C. S. Barrett and J. B. Newkirk as honorary chairmen.

The invited conference chairman, J. C. Russ of North Carolina State University, organized and chaired the plenary session of the conference entitled "New Techniques for the Future of X-ray Spectrometry."

The names of invited speakers on the program and the titles of their papers are listed below.

Richard W. Ryon, John D. Zahrt, Peter Wobrauschek and Hannes Aiginger, "The Use of Polarized X-Rays for Improved Detection Limits in Energy Dispersive X-Ray Spectrometry"

Andrzej J. Dabrowski, "Solid-State Room-Temperature Energy Dispersive X-Ray Detectors"

Joachim Stöhr, "Surface Crystallography by Means of Electron and Ion Yield EXAFS"

C. J. Bechtoldt, R. C. Placious, W. J. Boettinger and M. Kuriyama, "X-Ray Residual Stress Mapping in Industrial Materials by Energy Dispersive Diffractometry"

Tutorial workshops on various topics in fluorescence and diffraction were held during the first two days of the conference. These are listed below together with the names of the workshop organizers and instructors.

(1) "XRF Sample Preparation." V. E. Buhrke-The Buhrke Co. (chair), C.O. Ingamells-Amax Co., G. W. Mears-Amax Co., J. Taggart-USGS, F. Claisse-Laval Univ., J. Croke-Philips Electronic Instruments, Inc., R. Vane-Tracor X-Ray.

(2) "Energy Dispersive XRF." W. D. Stewart-Link Systems (chair), D. Gedcke-EG&G ORTEC, B. Jablonski-Shell Development Co., B. King-USGS, R. Kessler-Northrup Corp., R. Johnson-USGS.

(3) 'JCPDS Workshop on Computer Search Methods." R. Jenkins, Philips Electronic Instruments (chair), M. Holomany-JCPDS, R. Snyder-Alfred Univ., W. Schreiner-Philips Research Labs.

(4) "Fundamental Parameter XRF Software." J. W. Criss-Criss Software Inc., R. Christian-USGS, B. Jablonski-Shell Development Co., B. Artz-Ford Motor Co.

(5) "Diffractometer Alignment." J. Boslett-Rigaku USA (chair), R. P. Goehner-General Electric Co., J. Renault-New Mexico Bureau of Mines.

(6) "Wavelength Dispersive XRF." D. W. Beard-Siemens Corp. (chair), J. F. Croke, Philips Electronic Instruments, M. F. Garbauskas-General Electric Co.

(7) "XRD Sample Preparation." V. E. Buhrke-The Buhrke Co. (chair), A. J. Gude-USGS, M. Nichols-Sandia Labs.

(8) "Automated Powder Diffraction." R. L. Snyder-Alfred Univ. (chair), W. Parrish-IBM.

(9) "X-Ray Safety (XRF and XRD). R. Jenkins-Philips Electronic Instruments (chair), D. Steidley, M.D.-St. Barnabas Medical Center.

The workshop attendance continued to increase this year; 335 workshop attendees out of a total of 410 conference attendees. We are particularly indebted to the workshop organizers and instructors who gave unselfishly of their time and talent in these workshops.

We are grateful to all who co-chaired the various conference sessions. These were: C. S. Barrett, J. W. Criss, A. J. Dabrowski, C. P. Gazzara, D. A. Gedcke, C. O. Ingamells, R. Jenkins, J. J. LaBrecque, D. E. Leyden, W. Parrish, J. C. Russ, C. O. Ruud, R. W. Ryon, R. L. Snyder, W. D. Stewart, J. Taggart, and D. C. Wherry.

We are also grateful to the conference aids who worked long and unusual hours to ensure that the conference ran smoothly.

These were: Dorothy Barrett, Barb Cain, Penny Hudson, Cynthia King, Steve LaJoie, Rich Miller, Dorothy Predecki, John Shinton, Yvonne Shinton, and Linda Wallace.

A special word of thanks and appreciation to Mildred Cain, the conference secretary, for running the whole show.

Paul Predecki
For the Conference Committee

Unpublished Papers

The following papers were presented orally only and are not published here for a variety of reasons.

"Determination of Sulfur and Ash Content in Coals Using Iron[55] Excitation," T. Arai and D. G. Hempstead, Rigaku/USA Inc., Danvers, Massachusetts

"X-Ray Diffraction Analysis of Corrosion Products Formed on Iron Alloys in Contact with Molten Salts," Dale Boehme, Robert W. Bradshaw, and Monte C. Nichols, Sandia National Laboratories, Livermore, California

"Evaluation and Application of Methods for Trace Element Determinations in Water Samples Using X-Ray Spectrometry," D. E. Leyden, A. T. Ellis, University of Denver, Denver, Colorado, and W. B. Bodnar, Storage Technology, Louisville, Colorado

"A Relation of X-Ray and Mechanical Internal Stresses to Whisker Growth on Zn Electroplates," Takeshi Nagai, Kyohei Murakawa and Zenzo Henmi, Fujitsu Laboratories Ltd., Kawasaki, Japan

"Automated Measurement of Thermal Expansion Via X-Ray Diffraction," T. Nunes, P. DeHaven, and S. Lawhorn, IBM, Hopewell Junction, New York

"Molecular Orbital Spectroscopy of Thiophene," Rupert C. C. Perera, Virginia Polytechnic Institute and State University, Blacksburg, Virginia

"The Determination of Lead and Arsenic in Air Particulates Collected on Different Filter Materials by Wavelength Dispersive X-Ray Spectrometry," Christian Pupp, Joe Dlouhy and Nicole Houle, Air Pollution Technology Centre, Ottawa, Ontario, Canada

"X-Probe[TM], a Scanning X-Ray Fluorescence Spectrometer," Don Schlafke and Dale Gedcke, EG&G Ortec, Oak Ridge, Tennessee

"Application of an Automated Si(Li) Spectrometer to Prescreen Samples and Control an X-Ray Crystal Spectrometer," R. A. Semmler and Carla Plemich, IIT Research Institute, Chicago, Illinois

"Universal X-Ray Instrument Automation and Data Reduction," Rinaldo A. Spinella, Boston, Massachusetts, and James R. Lindsay, U.S. Geological Survey, Reston, Virginia

"Surface Crystallography by Means of Electron and Ion Yield EXAFS," Joachim Stöhr, Stanford Linear Accelerator Center, Stanford, California

"Toward the Optimization of EDXRF for Geological Matrices," David Wherry, Kevex Corp., Foster City, California, and Robert G. Johnson, U.S. Geological Survey, Reston, Virginia

CONTENTS

I. XRF DETECTORS AND XRF INSTRUMENTATION

Solid-State Room-Temperature Energy Dispersive X-Ray
     Detectors.   1
   Andrzej J. Dabrowski

Preliminary Study of the Behavior of HPGe Detectors with
     Ion Implanted Contacts in the Ultralow-Energy X-Ray
     Region. 23
   M. Slapa, J. Chwaszczewska, J. Jurkowski,
     A. Latuszynski, G. C. Huth, and A. J. Dabrowski

Performance of Room-Temperature X-Ray Detectors Made
     from Mercuric Iodide (HgI_2) Platelets 31
   J. B. Barton, A. J. Dabrowski, J. S. Iwanczyk,
     J. H. Kusmiss, G. Ricker, J. Vallerga, A. Warren,
     M. R. Squillante, S. Lis, and G. Entine

The Gas Proportional Scintillation Counter as a Room
     Temperature Detector for Energy Dispersive X-Ray
     Fluorescence Analysis. 39
   C. A. N. Conde, L. F. Requicha Ferreira and
     A. J. de Campos

Performance Characteristics of a High Resolution
     Si(Li) Detector Using a Time Variant Amplifier
     and a Pulsed Source of X-Rays 45
   M. A. Short and T. G. Gleason

X-Ray Tubes for Energy Dispersive XRF Spectrometry 49
   Brian Skillicorn

Toroidal Monochromators in Hybrid XRF System Improve
     Effectiveness of EDXRF Ten Fold 59
   Thomas C. Furnas, Jr., Gordon S. Kuntz, and
     Richard E. Furnas

II. XRF METHODS: PRACTICAL, MATHEMATICAL

The Use of Polarized X-Rays for Improved Detection
   Limits in Energy Dispersive X-Ray Spectrometry. . .   63
   Richard W. Ryon, John D. Zahrt, Peter Wobrauschek,
   and Hannes Aiginger

X-Ray Fluorescence of Intermediate- to High-Atomic-
   Number Elements Using Polarized X Rays   75
   R. B. Strittmatter

Examples of Analysis from an Integrated X-Ray
   Fluorescence Analysis System Using NRLXRF   81
   B. E. Artz and M. J. Rokosz

A Modular ADC/Microcomputer System for Energy Dispersive
   X-Ray Spectroscopy   85
   D. Hale, T. Satterfield, D. Blankenship, and
   J. C. Russ

Volatilization of Sulfur in Fusion Techniques for
   Preparation of Discs for X-Ray Fluorescence
   Analysis   91
   James W. Baker

Techniques for the Preparation of Lithium Tetraborate
   Fused Single and Multielement Standards   95
   Kent I. Mahan and Donald E. Leyden

III. XRF APPLICATIONS: MINERAL AND GEOLOGICAL

The Use of EDXRF for Liquids in a Uranium-Vanadium
   Solvent Extraction Process   103
   Rocky A. Smith

A Resin-Loaded Paper X-Ray Fluorescence Method for
   Determining Uranium in Phosphate Materials   107
   Benjamin W. Haynes, Jerome Zabronsky, and
   David L. Neylan

The X-Ray Analysis of Uranium Ores for Iron Sulfide
   Minerals   113
   A. J. Durbetaki, R. H. Carlson and T. F. Quail

A Statistical Comparison of Data Obtained from Pressed
   Disk and Fused Bead Preparation Techniques for
   Geological Samples   117
   Randall H. Dow

Trace and Minor Element Analysis of Obsidian from the
    San Francisco Volcanic Field Using X-Ray
    Fluorescence. 121
    Suzanne C. Sanders, John D. Zahrt, and Graydon Bell

Quantitative Determination of Ga, Zn, Cu, Ni, Mn, and
    Cr by X-Ray Fluorescence in Laterites and Bauxites
    Using Two Evaluation Methods 127
    Hasso Schorin

A Combined Dilution and Line-Overlap Coefficient
    Solution for the Determination of Rare Earths
    in Monazite Concentrates. 133
    T. K. Smith

Feasibility Study for On-Stream X-Ray Analysis of
    Barite. 139
    Yury M. Gurvich

IV.   XRF APPLICATIONS: METALS, CATALYSTS, OILS

Standard-Background" Method of X-Ray Spectral Analysis
    for Quality Control of Noble Metals in Alumina-
    Based Automobile Exhaust Catalysts 145
    Yury M. Gurvich

Some Elemental Determinations of Catalytic Materials
    Using a Thin-Film Internal Standard Technique
    by Radioisotope Excited X-Ray Fluorescence. 151
    W. C. Parker and J. J. LaBrecque

Energy Dispersive X-Ray Measurements for Cesium and
    Silver in Zeolite Ion-Exchange Columns. 157
    H. A. Vincent and M. E. Patton

Direct Analysis of Plutonium Metal for Gallium, Iron
    and Nickel by Energy Dispersive X-Ray
    Spectrometry. 163
    H. L. Bramlet and J. H. Doyle

The Analysis of Copper Alloys by Chem-X, Low Power WDX
    Multichannel Spectrometer 169
    J. Lucas-Tooth, B. W. Adamson, and Y. M. Gurvich

Energy Dispersive XRF Analysis of Lubricating Oil
    Additives with Secondary Target Excitation and
    the EXACT Fundamental Parameters Program 173
    Kenneth C. Stehr

The Analysis of Oil Additives Using Fundamental
        Influence Coefficients.   177
    Mrs. H. M. West

            V.   XRF ENVIRONMENTAL APPLICATIONS

The Measurement of Low Concentrations of Organic and
        Inorganic Gaseous Contaminants in Occupational
        Environments by X-Ray Spectrometry (XRS).   181
    N. G. West, C. J. Purnell, R. H. Brown, and
    E. Withers

The Application of X-Ray Fluorescence and Diffraction
        to the Characterization of Environmental
        Assessment Samples   189
    Albert C. Censullo and Frank E. Briden

Accurate PIXE Analysis of Thin Samples, Aerosol Loaded
        Filters and Surface Layers of Thick Samples. . . .   195
    U. Wätjen and F.-W. Richter

Energy Dispersive Analysis of Actinides, Lanthanides,
        and Other Elements in Soil and Sediment Samples . .   201
    G. R. Laurer, J. Furfaro, M. Carlos, W. Lei,
    R. Ballad, and T. J. Kneip

X-Ray Fluorescence Analysis of Welding Fume Particles . .   209
    Thomas P. Carsey

        VI.   XRD SEARCH/MATCH PROCEDURES AND AUTOMATION

A New Computer Algorithm for Qualitative X-Ray Powder
        Diffraction Analysis   213
    T. C. Huang and W. Parrish

A Versatile Minicomputer X-Ray Search/Match System . . .   221
    W. Parrish, G. L. Ayers and T. C. Huang

Automatically Correcting for Specimen Displacement
        Error During XRD Search/Match Identification . . .   231
    Walter N. Schreiner and Ronald Jenkins

X-Ray Diffraction Phase Analysis Using Microcomputers . .   237
    T. M. Hare, J. C. Russ, and M. J. Lanzo

A Second Generation Automated Powder Diffractometer
    Control System. 245
    Robert L. Snyder, Camden R. Hubbard, and
    Nicolas C. Panagiotopoulos

Application of the Modified Snyder's Program for the
    Data Processing of an Automated X-Ray Powder
    Diffractometer. 261
    G. Platbrood, J. M. Quitin, and H. Barten

INDEX, A Program to Reconcile Powder Diffractograms . . . 267
    Tommy Hom, Ron Jenkins, and Joshua Ladell

IDENT - A Versatile Microfile-Based System for Fast
    Interactive XRPD Phase Analysis 273
    Barbara A. Jobst and Herbert E. Göbel

VII. XRD METHODS AND INSTRUMENTATION

Complete Quantitative Analysis Using Both X-Ray
    Fluorescence and X-Ray Diffraction 283
    M. F. Garbauskas and R. P. Goehner

Calibration of the Diffractometer at Low Values of
    Two Theta 289
    R. Jenkins, T. Hom, C. Villamizar, and W. N.
    Schreiner

Sample Preparation and Methodology for X-Ray Quanti-
    tative Analysis of Thin Aerosol Layers Deposited
    on Glass Fiber and Membrane Filters. 295
    Briant L. Davis and L. Ronald Johnson

Differential X-Ray Diffraction by Wavelength Variation:
    A Preliminary Investigation 301
    M. C. Nichols, D. K. Smith, and Quintin Johnson

X-Ray Diffraction Quantitative Analysis Using
    Intensity Ratios and External Standards 309
    Raymond P. Goehner

A Guinier Diffractometer with a Scanning Position
    Sensitive Detector 315
    Herbert E. Göbel

Observation of an X-Ray Beam of 10 Microradian
    Divergence Without Using Any Collimator 325
    K. Das Gupta

VIII.   XRD APPLICATIONS

X-Ray Residual Stress Mapping in Industrial Materials
        by Energy Dispersive Diffractometry. 329
    C. J. Bechtoldt, R. C. Placious, W. J. Boettinger,
    and M. Kuriyama

Stress Measurement and Precision Diffraction Angles
        on Large Grained Specimens. 339
    Charles S. Barrett

Determination of Residual Stresses in Austenite and
        Martensite in Case-Hardened Steels by the
        $Sin^2\psi$ Method 343
    Chongmin Kim

X-Ray Characteristics and Applications of Layered
        Synthetic Microstructures 355
    J. V. Gilfrich, D. J. Nagel, N. G. Loter, and
    T. W. Barbee, Jr.

The Use of Energy Dispersive Diffractometry to
        Measure the Thickness of Metal and Glass Thin
        Films. 365
    Glen A. Stone

Application of Automated X-Ray Diffraction to
        Alteration Mineral Zoning Studies 373
    H. Salek, H. A. Vincent, and L. Thorpe

The Application of X-Ray Diffraction for Glass
        Batch Homogeneity Determination 379
    H. S. Kim and C. I. Cohen

X-Ray Diffraction and Fluorescence in the
        Analysis of Pharmaceutical Excipients 383
    A. J. Durbetaki and T. F. Quail

Corrections to Volume 24 389

Author Index 391

Subject Index 393

SOLID-STATE ROOM-TEMPERATURE ENERGY-DISPERSIVE X-RAY DETECTORS

Andrzej J. Dabrowski[*]

Medical Imaging Science Group
University of Southern California
4676 Admiralty Way, Suite 932, Marina del Rey, CA 90291

INTRODUCTION

Over the last several years there has been growing interest directed toward high-resolution energy-dispersive x-ray detectors which operate at room temperature. The goal has been to develop a detector which combines the advantages of room-temperature operation characteristic of scintillation and proportional counters with the excellent energy resolution of silicon and germanium cryogenically cooled spectrometers. The elimination of the cryogenic coolant and the associated vacuum cryostat will permit the design and construction of really miniature and practically convenient x-ray detection systems which will find applications in already existing instruments as well as in various new fields.

Numerous solid-state materials have been considered for radiation detection at room temperature.[1-8] Because of the noise level associated with the leakage current of the detector at room temperature, materials with large energy bandgaps have to be considered. In addition, high values of the charge transport properties are another basic requirement for these materials. In the field of x-ray detection the most extensive studies have been devoted to and the best results have been obtained with three compounds: gallium arsenide (GaAs), cadmium telluride (CdTe), and mercuric iodide (HgI_2).

The development of room-temperature solid-state x-ray detectors has progressed to the point where quantities of hundreds of detectors of different types have been fabricated and studied in various labo-

---

[*] On leave from the Institute of Nuclear Research, 05-400 Swierk, Poland.

ratories. It is now possible to draw some general conclusions regarding this new development and to define areas in which further research is needed.

CRYSTALS AND DETECTORS

Gallium Arsenide (GaAs)

GaAs belongs to the group of III-V compounds. The basic physical properties of this material are listed in Table 1. At the present time epitaxial growth from the liquid phase (LPE) or vapor phase (VPE)[2,9] is the only technique which can be used to provide low concentrations of electrically active impurities ($N_D + N_A \sim 10^{13}$ cm^{-3}). Layer thicknesses are generally limited to about 200 μm, but LPE layers of lower purity have been grown as thick as 2 mm.[10] Substrates are usually either doped n^+ or semi-insulating. An anomalous discontinuity in net carrier concentration at the interface of the epitaxial and substrate layers usually degrades the performance of the detectors. Detectors made from liquid phase epitaxial material were provided with ohmic contacts made of Ga-In alloy on n^+ substrates. Gold contacts were evaporated on the etched epitaxial surface. Ohmic contact to the vapor phase epitaxial diodes was made with special alloys of Ga-In after removal of the semi-insulating substrate by lapping. Typical leakage current for a GaAs detector is within the range 10^{-7} to 10^{-6} A. Since 1973 the development of GaAs has not progressed further.

Cadmium Telluride (CdTe)

CdTe is a II-VI compound. The stable form of CdTe has the zinc-blende structure. The most probable value for the lattice constant is 6.481 Angstroms.[11] Crystals can be cleaved quite readily along the {100} planes. The basic crystal parameters are listed in Table 1. Early crystal-growing work, which was quite successful in attaining detector-grade CdTe, used closed-system methods such as the Stockbarger[12] or modified Bridgman[13] method. Later on, different zone-passing techniques were developed.[4] Crystals of high resistivity were obtained by doping with Cl or In. For the chlorine-doped crystals a strong polarization effect was observed.[4] Good spectrometric detectors were obtained from crystals of medium resistivity (10^4 to 10^5 Ohm-cm) and of low resistivity (<10^3 Ohm-cm) grown by the travelling heater method[14] and by the sealed-ingot zone refining technique[15-19], respectively. To prepare the detectors suitable slices (0.5 to 2.0 mm) were cut from a CdTe ingot and then lapped, polished, and cleaned both mechanically and chemically. A solution of bromine in methanol was used for etching. For high-resistivity material usually Al was evaporated for detector contacts. For material of low resistivity In was adopted as a back contact and a dot of Au was evaporated for the front contact. In the case of Cl-doped

crystals special kinds of contacts (Ohmic or MIS) were developed to
suppress polarization effects. Typical leakage current of CdTe
detectors is within the range 10^{-9} to 10^{-7} A. There has been no
real progress in the development of CdTe detectors since 1977.

Mercuric Iodide (HgI$_2$)

HgI$_2$ undergoes a reversible phase transformation at 127°C from
a tetragonal structure to an orthorhombic structure. Crystals in
the room-temperature tetragonal phase are red, whereas the
high-temperature phase is characterized by yellow crystals. Since
the melting point of HgI$_2$ is at 250°C, the phase transformation
prevents the growing of crystals in the tetragonal phase by solidi-
fication from the melt. Although it is possible to grow crystals
from solution, growth from the vapor gives crystals with better
charge transport properties. The lattice constants are a = 4.3693
Angstroms and c = 12.4399 Angstroms for the tetragonal phase.
Other data are listed in Table 1. Currently crystals grown from
the vapor are the only ones which are good enough to be used as
detectors. The two vapor-growth techniques used are pulling the
crystal[20,21] and reversing the temperature gradient.[22,23] The
first is a variation of the Piper-Polich method for growing large
boules of CdS from the vapor. The second and more widely used
method was developed by Scholz and is based on the periodic reversal
of the temperature gradient between the source material and the
growing crystal.[22] Another quite successful method which has
recently been introduced is the growing of HgI$_2$ crystal platelets
by polymer-assisted vapor transport.[24,25] This new method yields
platelets of size and shape suitable for x-ray detector fabrication.
Crystals other than platelets must be cleaved or sawed into wafers
0.3 to 0.5 mm thick. Prior to the deposition of electrical contacts,
slices are etched in an aqueous solution of KI. Palladium is evaporated or a thin layer of carbon is painted onto the crystal slice
for electrical contacts. HgI$_2$ detectors exhibit leakage currents
which are typically within the range 10^{-12} to 10^{-11} A. Progress in
the development of HgI$_2$ detectors is continuously being reported.[26-38]

SOLID-STATE DETECTOR PERFORMANCE

A solid-state detector is basically analogous to an ionization
chamber in which charges generated during the absorption of radiation
constitute the signal by which the radiation is detected. The linear
coefficients for absorption of x-rays of energies from 1 to 100 keV
in HgI$_2$, CdTe, Ge, and Si are shown in Figure 1. The mean penetration depth of 10 keV x-rays is approximately equal to 12 μm in HgI$_2$,
12 μm in CdTe, 55 μm in Ge, and 120 μm in Si. The corresponding
numbers for 100 keV photons are 0.5 mm, 1.1 mm, 5.0 mm, and 115 mm.
These latter numbers illustrate what active thickness of any given

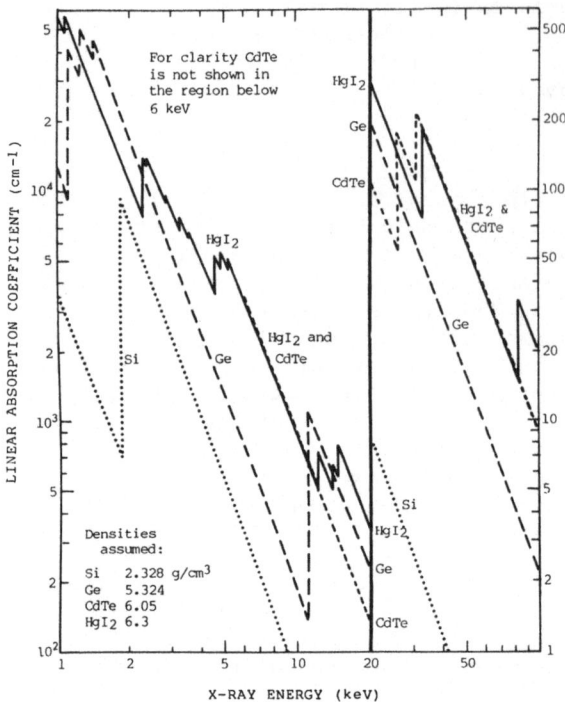

Figure 1. Linear absorption coefficients for Si, Ge, CdTe, and HgI$_2$ for x-rays with energies between 1 and 100 keV. The atomic cross-section data from Reference 39 have been used. The curve for GaAs would be very similar to the one for Ge and has not been shown.

material would be required to obtain reasonable detection efficiency for higher-energy x-rays.

The process by which x-rays are absorbed in the detector involves the production of an electron of high energy by the primary radiation. This electron in turn produces further ionization, and a cascade process ensues and continues until no electron has enough energy to cause further ionization. The mean number of electron-hole pairs produced will depend on the energy deposited by the radiation. The average energy for the creation of one electron-hole pair, ε, generally increases with increasing bandgap of the material. The values of ε versus energy bandgap E_g for various solid-state crystals have been calculated semi-empirically by Klein,[40] and fit the relation $\varepsilon = 2.67E_g + 0.87$ eV. In the case of HgI$_2$ the experimentally obtained value of ε is only 4.2 eV,[21] which does not fit Klein's relation but is very favorable from the detection point of view. The values of ε for GaAs, and HgI$_2$ are shown in Table 1.

The pulse shape from a semiconductor detector is determined by the motion of the electrons and holes created by the incident radiation. They drift under the influence of the electric field, E,

Table 1

	GaAs	CdTe	HgI$_2$
ATOMIC NUMBERS	31,33	48,52	53,80
DENSITY IN g cm^{-3}	5.3	5.85	6.4
BAND GAP IN eV	1.4	1.5	2.2
ELECTRON MOBILITY IN cm^{2}v^{-1}s^{-1}	8000	1000	100
HOLE MOBILITY IN cm^{2}v^{-1}s^{-1}	400	100	3
MEAN ENERGY FOR ELECTRON-HOLE PAIR CREATION IN eV	4.27	4.43	4.2
FANO FACTOR	0.18	?	0.27

with the velocity $V_d = \mu E$, where μ is the carrier mobility. The values of electron and hole mobilities for GaAs, CdTe, and HgI$_2$ are shown in Table 1. In compound semiconductors the values of the electron mobility are usually much higher than the values of the hole mobility. From the standpoint of x-ray spectrometry the electron mobility values found in GaAs, CdTe, and HgI$_2$ are satisfactory, but the small value of hole mobility in HgI$_2$ (only 3 cm^2/V-s) is still a limitation on the detector performance for higher-energy x-rays. To understand the way in which the signal in an external circuit builds up as the electrons and holes are swept toward the electrodes, it is necessary to consider the

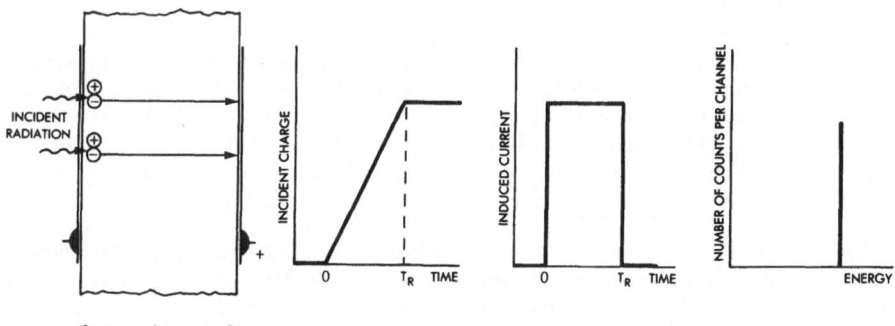

Figure 2. Basic concepts of the creation of the electron pulse in the detector. Electron-hole pairs produced by the incident radiation drift under the influence of the electric field and induce charge on the detector electrodes, induce current through the detector, and finally produce an energy spectrum.

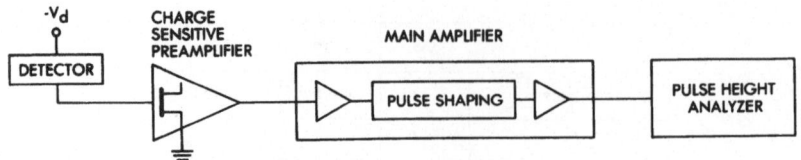

Figure 3.   General configuration of a solid-state spectrometer.

induced current due to the motion of individual charges. Carriers
produced in the photoelectric event drift under the influence of the
electric field up to the moment they reach the electrodes. According
to Ramo's Theorem[41] a charge q moving a distance Δx between parallel
plates separated by a distance L induces a charge ΔQ on the electrodes given by $\Delta Q = q(\Delta x/L)$, and current $i = dQ/dt = q(V_d/L)$, which
is illustrated in Figure 2 for the case when the electron-hole pairs
have been created only at the negative electrode of the detector.

     The general configuration of a solid-state spectrometer is shown
in Figure 3. The detector is biased directly from a high-voltage
power supply and is DC-coupled to a low-noise charge-sensitive
preamplifier. The preamplifier is followed by a voltage amplifier
containing some pulse-shaping circuits which allow optimization of
the signal-to-noise ratio. The output pulses from the amplifier are
then fed to a multichannel analyzer.

ENERGY RESOLUTION

     Most of the research on compound solid-state x-ray spectrometers
has been concentrated on obtaining progressively better energy resolution and on determining what ultimate resolution might be achieved
at room temperature. In what follows the main factors which limit
energy resolution in solid-state compound detectors will be considered.

     The shape and width of the observed spectral line corresponding
to the energy response to monochromatic x-radiation are due to
several factors. The ones which will be discussed here are statistical fluctuations, electron escape during the charge generation process, the geometrical effect of trapping and the effect of nonuniformi-
ties during charge collection, and some aspects of electronic noise.
Additionally there are several other sources of deterioration of the
spectrum such as pulse pileup effects, edge effects due to the finite
size of the detector, the microphonic effect, and polarization
effects.

Statistical Fluctuations in Charge Generation

Since the energy of incident radiation is determined from the amount of ionization which takes places in the detection medium, the accuracy of measurement depends on the fluctuation in the number of ionization events. If the events are considered to be independent, the statistical fluctuation in the number of events N will be $\epsilon = \sqrt{N}$. Fano showed[42] that the ionization events should not be treated as completely independent. In this case the standard deviation becomes $\epsilon = \sqrt{FN}$, where F is the so-called Fano factor, F<1. The contribution to the width of the energy peak from the statistical spread in the charge generated is

$$\Delta E_s (FWHM) = 2.355 \sqrt{FE\epsilon}$$

Using values of the Fano factor shown in Table 1 it can be concluded that for room-temperature solid-state x-ray spectrometers the energy resolution is not limited by statistical fluctuations during generation of the charge carriers.

Window Effect

It is known that high-purity germanium x-ray detectors have a relatively thick entrance window which renders them practically useless below 2.3 keV because of the significant tailing of the energy peaks which occurs for these energies. According to the model developed by Llacer,[43] the mean free path of the electrons produced by the incident radiation becomes longer before final thermalization and the electrons may reach the detector surface and recombine there rather than be collected by the electric field existing between the electrodes. It is not possible to make calculations according to this model for HgI_2, because at the present time there is not nearly enough data for HgI_2 to permit treatment of this subject in such a general way. If, however, we assume a linear relationship between the charge carrier mobility and the mean free path between collisions, as is the case for Ge, it can be expected that the maximum mean free path between collisions in HgI_2 will be much smaller than it is in Ge, because the mobility of electrons is two-and-one-half orders of magnitude smaller in HgI_2 than in Ge. The electrons in HgI_2 will have less chance to reach the detector surface without colliding.

Generally the window effect increases for more strongly absorbing materials. Because the absorption coefficient for Si is much smaller than for Ge, the window effect is not nearly as important in Si as it is in Ge, and Si(Li) detectors are very useful in the low-energy x-ray region. In the case of HgI_2, because of the high atomic numbers of mercury (Z = 80) and iodine (Z = 53), one might expect that the consequences of the window effect would be even more severe than for Ge detectors, but in comparing the values of the linear attenuation coefficients for Ge and HgI_2 (see Figure 1) it is seen that for low-energy x-rays these values are not so different. In fact, for the energies between 1.3 and 2.3 keV absorption in Ge is even stronger than in HgI_2.

Additionally, the window effect is more pronounced when the electric field near the detector surface is decreasing. This has been shown for HgI_2 detectors with palladium front contacts,[37] which for low-energy x-rays produce a photopeak with tailing on the low-energy side, in contrast to HgI_2 detectors with carbon front contacts, which produce symmetrical photopeaks with almost no tailing. More discussion of this subject can be found in References 43 and 37. The latest development in high-purity germanium spectrometers with implanted contacts reported at this Conference[44] can open up new prospects for application of these detectors also in the ultralow-energy x-ray region.

It can be concluded that for HgI_2 detectors with properly made contacts the window effect is not observed for x-ray energies down to 1 keV. This is one of the important (and somewhat unexpected) advantages of these detectors. For other compound solid-state detectors it has not been possible to measure x-ray spectra in the ultralow-energy x-ray region where the window effect might appear.

Trapping

One of the most critical factors influencing the performance of solid-state detectors is the possible presence of trapping centers which can cause a decrease in the amount of charge collected and hence a degradation of energy resolution. For higher energies of the incident radiation, when interactions occur throughout the whole volume of the detector, the geometric aspect of trapping usually plays a dominant role. The effect of trapping can be described by a charge collection efficiency function. If detrapping is neglected and the electric field is assumed constant, the charge collection efficiency function η for an interaction which takes place at a distance x from the negative contact in a detector with trapping centers is given[45] by

$$\eta = \frac{Q}{Q_0} = \frac{\lambda_e}{L}\left(1 - \exp\left[-\frac{L-x}{\lambda_e}\right]\right) + \frac{\lambda_h}{L}\left(1 - \exp\left[\frac{-x}{\lambda_h}\right]\right)$$

where

Q is charge collected (C)
Q_0 is charge generated (C)
λ_e is mean free drift length for electrons (cm)
λ_h is mean free drift length for holes (cm)
L is detector thickness (cm)

λ_e and λ_h are given as

$$\lambda_e = \mu_e \tau_e E, \qquad \lambda_h = \mu_h \tau_h E,$$

where

μ_e is electron mobility $(cm^2/V \cdot s)$
μ_h is hole mobility $(cm^2/V \cdot s)$
τ_e is electron trapping time (s)
τ_h is hole trapping time (s)
E is electric field (V/cm)

The function $\eta(x)$ actually determines the position and the shape of the energy peak when trapping is the major factor limiting the resolution of the spectrum. The model which predicts the observed spectrum is as follows. Consider the slice Δx of the detector located between x and x+Δx. The function $\eta(x)$ gives the relative pulse height expected from an interaction occurring within Δx, as is shown in Figure 4. The pulses from the other slices give the contributions to the other parts of the peak. The total spectrum produced is just the sum of the spectra from the individual slices. The mathematical expression is derived and the quantitative description of this model is developed in Refs. 3,46, and 47. In fact, once efficiency is plotted as a function of the position of the initial ionization event, the spread in pulse heights can be easily

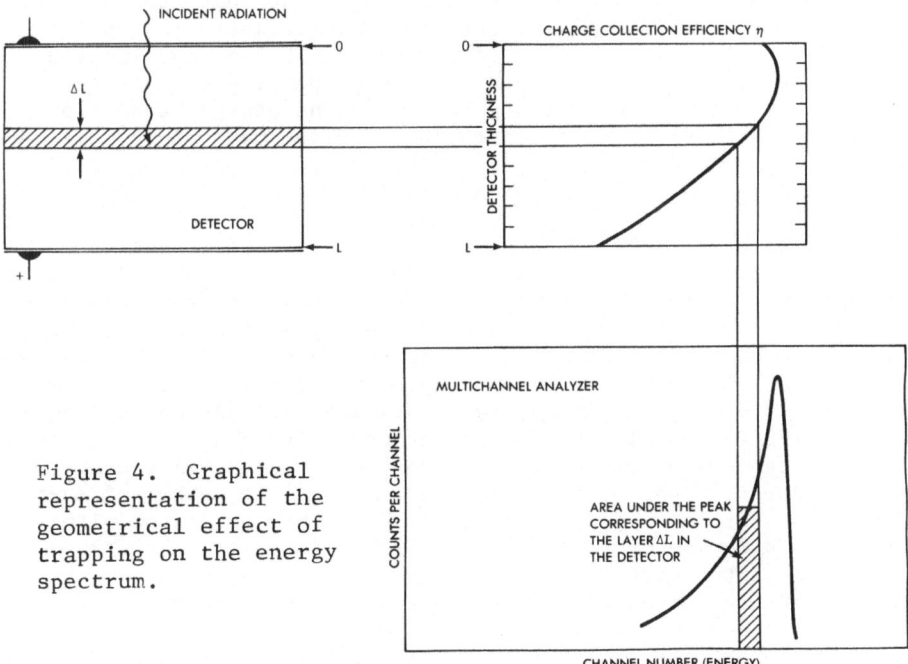

Figure 4. Graphical representation of the geometrical effect of trapping on the energy spectrum.

Table 2

For:	CdTe	HgI_2	
$\mu\tau_e$	10^{-3}	5×10^{-4}	cm^2/V
$\mu\tau_h$	10^{-4}	5×10^{-6}	cm^2/V
E	2×10^4	5×10^4	V/cm
L	0.1	0.05	cm

Table 3

λ_e/L	200	500
$\eta(0)$	0.997	0.999
λ_h/L	20	5
$\eta(L)$	0.975	0.906

ascertained to be equal to the difference between η_{max} and η_{min}. When the incident radiation interacts near the negative electrode, only electrons contribute to the charge pulse and the efficiency will be $\eta(0)$. Conversely, $\eta(L)$ is the efficiency for pure hole collection, corresponding to the case of interaction of the incident radiation close to the positive electrode. For HgI_2, η_{max} will be nearly the same as $\eta(0)$ and η_{min} will be exactly the same as $\eta(L)$. Assuming typical values of the $\mu\tau$ product and electric field for CdTe and HgI_2 (Table 2), values of $\eta(0)$ and $\eta(L)$ are calculated and listed in Table 3. The resulting pulse height spreads are approximately 9% and 2% for HgI_2 and CdTe, respectively. The smaller spread in pulse heights for CdTe reflects the smaller disparity in the values of the $\mu\tau$ product for holes and electrons in CdTe than in HgI_2. It can be concluded that the geometrical effect of trapping is a limiting factor in the energy resolution for solid-state compound detectors in the high-energy x-ray region.

Inhomogeneities in the Detector Material

Another factor which can contribute to the deterioration of energy resolution is the presence of inhomogeneities in the detector material. Material inhomogeneities give rise to variations in the electric field E, drift mobility μ, and trapping time τ. Since these three quantities are all related to the drift length λ through $\lambda = \mu\tau E$, variations in the drift length are correlated with material inhomogeneities. When low-energy x-rays are incident on the negative electrode of a plane parallel detector with a uniform electric field, charge transport due to electrons will result in a pulse whose height is given by the Hecht relation.[1] The number of events dN occuring within an energy interval dE can be correlated with fluctuations in the drift length by $dN/dE = (\partial N/\partial\lambda)(\partial\lambda/\partial E)$.[48] Here $\partial N/\partial\lambda$ is the distribution function for the number of events ∂N which result in charge transport within a range of drift length $\partial\lambda$. Graphic representation of the above is shown in Figure 5. Variations about the mean value of the drift length result in variations about the mean value of pulse height.

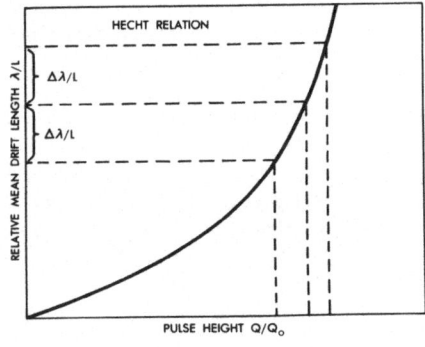

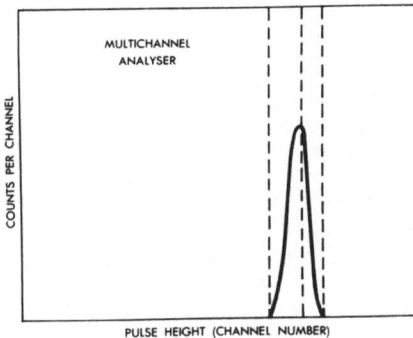

It is difficult to make any predictions about the degradation of the shape of the energy peak caused by inhomogeneities of the detector material because it is not known how λ varies over the volume of any particular detector. Experimentally, information about inhomogeneities could be obtained by scanning the detector surface with a well collimated source and comparing the results with the spectrum obtained with an uncollimated source.

Figure 5. Graphical representation of the influence of inhomogeneities of the detector material on the energy spectrum.

Electronic Noise

In order to achieve the excellent energy resolution available today with semiconductor Si and Ge spectrometers ultralow-noise amplification electronics had to be developed in addition to the work done on crystals and detectors. For these spectrometers not only is the detector cooled to liquid-nitrogen temperature, but also the first stage of preamplification electronics is cooled to about 120 K, which is optimal for operation of the input field effect transistor (FET).

In the development of a room-temperature spectrometric system special effort has been made to develop ultralow-noise preamplification electronic circuits which operate at room temperature.[36,49] The main research has been devoted to resistor feedback and pulsed-light feedback preamplifiers which operate totally at room temperature. In the case of resistor feedback preamplifiers a noise level below 300 eV (FWHM) was obtained.[50] For room-temperature pulsed-light feedback preamplifiers the noise level was further improved by about 100 eV (FWHM).[38] A discussion of the latest achievements in low-noise room-temperature preamplification systems for use with HgI_2 x-ray detectors operating at room temperature has been published by Iwanczyk et al.[36]

In general it has been found that the electronic noise related to the preamplifier input circuit is due to four main sources: 1) "white" noise in series, 2) "white" noise in parallel, 3) excess "1/f" type of noise, and 4) generation-recombination noise caused by traps present in the gate depletion region of the field-effect transistor. These four sources are represented by four parts in the equation [51]

$$\Delta E_n = 2.35 \, \frac{\bar{\varepsilon}}{q} \left\{ \left[qI_L + \frac{2kT}{R_p} \right] < N_s^2 > + 2kTR_s C_{in}^2 < N_p^2 > + A_f < N_{1/f}^2 > + BC_{in}^2 < N_{gr}^2 > \right\}$$

where

ΔE_n is the FWHM (full-width-at-half-maximum) due to the electronic noise, in eV

$\bar{\varepsilon}$ is the mean energy required to create one electron-hole pair in the detector, in eV

q is the electron charge, 1.6×10^{-19} Coulombs

I_L is the sum of the absolute values of the shunt leakage current in the input circuit, in Amperes

k is Boltzmann's constant 8.62×10^{-5} eV/K

T is absolute temperature, in K

R_p is the parallel input circuit resistance, in Ω

$< N_s^2 >, < N_p^2 >$ are the series and parallel coefficients which are functions of the pulse shaping used in the system

R_s is the equivalent series resistance, in Ω

C_{in} is the total input capacitance, in Farads

A_f is a constant expressed in $(Volts)^2$

$< N_{1/f}^2 >$ is the coefficient dependent on the pulse-shaping network

B is a constant

$< N_{gr}^2 >$ is a coefficient dependent on the pulse-shaping network

Using the above equation one can calculate the theoretical limits for different parts of the electronic noise for various detectors and amplification systems.

For an HgI_2 detector with a leakage current of 1 pA the electronic noise arising from the detector current is less than 100 eV (FWHM).[30] For CdTe and GaAs detectors the same calculation, assuming a room-temperature leakage current in excess of 1 nA, gives a contribution to the electronic noise greater than 1 keV (FWHM). High leakage currents associated with the smaller bandgaps of CdTe and GaAs impose basic limits on these materials as candidates for low-energy x-ray detectors. In fact, the best published spectrum of the Mn K-alpha photopeak (5.9 keV) measured with a CdTe detector

exhibits an energy resolution of 1.1 keV (FWHM).[52]   For HgI_2
spectrometers the energy resolution for low-energy x-rays is
limited not by the detector current but by the electronic noise
of the preamplifier.[30]

HgI_2 X-RAY DETECTORS

State of the Art

In light of the foregoing discussion it can be concluded that
as far as the actual state of development is concerned, HgI_2 is the
only solid-state crystal which combines good transport properties
of charge carriers with a large enough bandgap and thus allows the
making of detectors with a low enough current noise. Other
solid-state compound crystals lack either good enough charge carrier transport or a large enough bandgap or both. It can also be
concluded that for the future development of room-temperature
low-energy x-ray detectors the search for new materials should be
restricted to those with bandgaps above 2 eV in order to be able
to produce detectors having low enough room-temperature leakage
current.

The capabilities of HgI_2 for x-ray detection and analysis can
be illustrated by presenting the energy spectra obtained with a
room-temperature spectrometer. Figure 6 shows a typical spectrum
from Am-241 taken with a room-temperature HgI_2 detector. The
energy resolution for 59.5 keV is 2.4 keV (FWHM). The photopeak

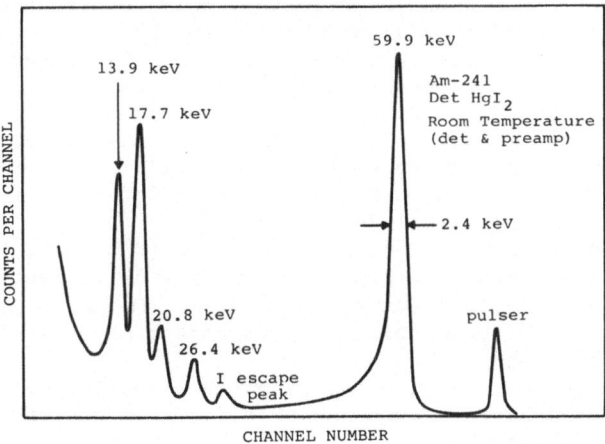

Figure 6. Typical spectrum from Am-241 source measured with
HgI_2 detector.

at 59.5 keV is noticeably asymmetric and broad compared to the
lower-energy x-ray peaks. The different types of charge transport
behavior are reflected in the shapes and widths of the photopeaks
in the spectrum. For the low-energy peaks essentially only electron
collection takes place, whereas the high-energy peak arises from
both hole and electron collection from ionization events occurring
throughout the volume of the detector.

In Figure 7 are shown the spectra from (a) a Mg target,
(b) an NaCl target, (c) a Ti target, and (d) an Fe_2O_3 target. The
photopeak measured for the Mg K_α x-ray line at 1.25 keV is symmetrical and has a full-width-at-half-maximum of 245 eV, nearly equal to
the pulser linewidth of 225 eV (FWHM).

Figure 8 shows the x-ray spectrum of Mn K x-rays using a
pulsed-light feedback preamplifier. The energy resolution obtained
was 295 eV (FWHM) for the 5.9 keV line from the Mn K_α x-ray when
the pulser linewidth was 225 eV (FWHM). The K_α and K_β lines of
5.9 keV and 6.5 keV are well separated. These resolution values
for the Mn K_α photopeak and the pulser line were used to calculate
the Fano factor for HgI_2. A value of 0.27 was obtained.

Directions of Current Investigations

HgI_2 spectrometers have been improved from an initial stage of
providing an energy resolution comparable to that of proportional
counters to the stage of providing an energy resolution only about
a factor of two worse than that of cryogenically cooled Si(Li)
spectrometers. However, before HgI_2 detectors can be widely utilized in various applications their stability of performance needs
to be improved. It has been observed that HgI_2 detectors currently
manufactured often exhibit a deterioration in their performance
with time, leading to changes in the quality of the energy spectra
they produce. Although there exist HgI_2 detectors which can operate
stably over long periods of time, still the yield of these detectors
should be considerably increased. An initial study of the problem
in HgI_2 has been undertaken recently. The preliminary work has
indicated the main directions which must be followed in progressing
toward a solution. These include:
(i) A search for procedures which can be used to prevent the
deterioration of detector performance with time or to restore good
performance after deterioration has occurred. It has been observed
that most often the deterioration of performance of a detector can
be completely reversed.
(ii) A study of the crystal growing techniques and the methods used
in fabrication of detectors with the specific goal of improving the
long-term stability of the detector characteristics. Work has been
carried out for several years on the method of growing HgI_2 crystals
by temperature gradient reversal.[23] The introduction of the new
technique of growing HgI_2 crystal platelets by polymer-assisted
vapor transport promises to greatly facilitate the effort to improve
long-term stability.[24]

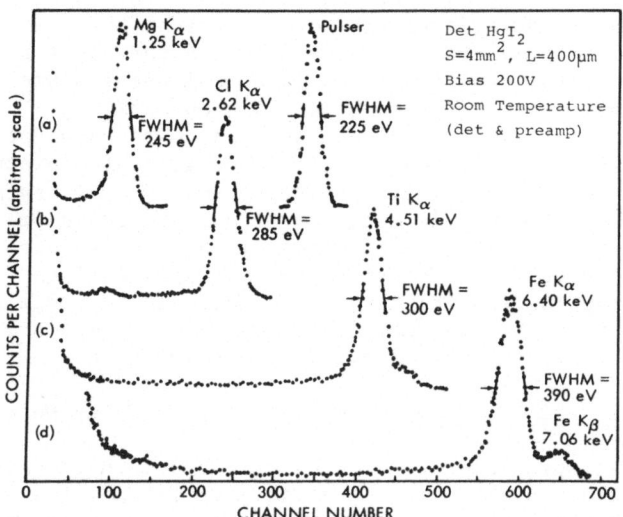

Figure 7.   X-ray fluorescence spectra from various low-Z targets taken with a room-temperature HgI_2 spectrometer.

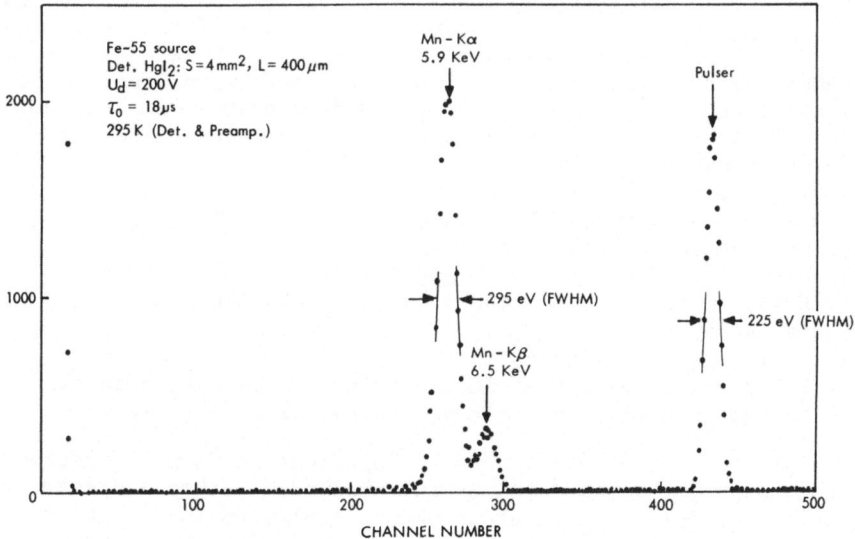

Figure 8.   Spectrum of Mn K x-rays measured with an HgI_2 room-temperature spectrometer employing a pulsed-light feedback preamplifier.

Other directions for further investigations include:
(i) The search for improvements in energy resolution for low-energy
x-rays. Further reduction of the noise level of the spectormeter can
be achieved by minimizing the excess noise in the input stage of the
preamplifier. Also, cooling of the input FET using a small Peltier
element can improve energy resolution by about 25%.
(ii) Further miniaturization. Because of room temperature operation it is possible to achieve an extreme degree of miniaturization
of the entire x-ray spectrometer by using hybrid or integrated circuit techniques. The use in addition of a microprocessor could
significantly advance the capabilities of the spectrometer system.
(iii) Special types of detectors. Specific applications will
require studying the feasibility of designing and optimizing special
kinds of detectors. These new types of detectors would include
arrays and mosaics, special geometries, and so forth.

Potential Applications of Room-Temperature HgI_2 X-Ray Detectors

HgI_2-based spectrometers offer a unique combination of good
energy resolution, high detection efficiency for x-rays, and room
temperature operation. This combination of features makes them an
attractive choice in any x-ray fluorescence (XRF) instrument which
has to be compact, lightweight, and generally convenient to use.
The absence of a liquid-nitrogen cooling system makes an advanced
degree of miniaturization of the HgI_2 spectrometer possible. As
compared with other existing room-temperature x-ray spectrometers,
the HgI_2 spectrometer allows convenient analysis of the heavy element K x-rays. These features of HgI_2 may be of great value in many
currently operated laboratory XRF systems. The main area of potential applications, however, will be the domain in which the complication of the cooled systems has been a major drawback so far.
Examples of possible applications include

(i) On-line checking of the levels of some key elements in industrial production and manufacturing processes;
(ii) Determining the amount of specific elements in ores;
(iii) Environmental protection surveys with a field-operated,
portable XRF system;
(iv) Geological field investigations;
(v) X-ray analyzing probes in electron beam devices such as
scanning and transmitting electron microscopes.

A separate, very interesting area of application is x-ray astronomy.
The unique advantages offered by the small dimensions and extremely
low weight of HgI_2 spectrometers can be exploited in designing
"flight" instruments for analyzing the x-rays from extra-solar
sources. Additional uses would include the detection and analysis
of fluorescence x-rays on planetary exploration missions.

A range of medical applications in x-ray detection and imaging systems exist. Here the portability and miniaturization of HgI_2-based instruments might be attractively convenient.

SUMMARY AND CONCLUSIONS

The desire to develop an energy-dispersive x-ray detector which combines excellent energy resolution with room-temperature operation has motivated substantial research over the last decade. Since Si and Ge, the most perfectly pure semiconductors, have been excluded as materials for room-temperature detectors because their bandgaps are not large enough, the search for other materials started. Work has been directed toward materials which have a large enough band-gap to produce detectors with low room-temperature leakage current and good enough charge transport properties to assure that the charge generated by radiation is efficiently collected in the detector. Most of the solid-state crystals which have actually gone through some stage of development as possible detectors have been found to have insufficiently good charge transport characteris-tics. Only for three compounds -- gallium arsenide, cadmium tellu-ride, and mercuric iodide -- has considerable research been done and the best results obtained.

The main objective in the development of room-temperature x-ray detectors has been to achieve progressively better energy resolution. Significant progress has been made with both CdTe and GaAs room-temperature energy resolution in the x-ray region, but the level of resolution reached is still, even in the best case, only comparable to that of gas counters which have already been developed. Thus the applications of these detectors are limited to those in which considerations other than the energy resolution are important.

HgI_2 is the only solid-state material which combines the advan-tages of good electron transport properties with a large enough energy bandgap so that detectors with very low leakage current at room temperature can be made from it. It has been possible to develop a room-temperature x-ray spectrometer that has an energy resolution appreciably better than other existing room-temperature energy-dispersive x-ray detectors. Used in combination with ultralow-noise room-temperature preamplifiers developed concurrently, HgI_2 detectors have produced energy resolution values of 245 eV (FWHM) for the Mg K-line (1.25 keV), 295 eV (FWHM) for the Mn K_α-line (5.9 keV), and 1.2 keV (FWHM) for Am-241 (59.5 keV). Furthermore, there is still additional room for improvement in energy resolution by continuing development of HgI_2 crystal growing methods, refinement of detector fabrication procedures, and further reduction in the noise level of preamplification systems.

The specific area of research which now merits the most effort in the development of HgI_2 spectrometry concerns improvement in the stability of detector performance. Although there exist HgI_2 detectors which operate stably over a long period of time, ways must be sought to increase considerably the yield of these detectors.

The development of room-temperature high-resolution x-ray spectrometers opens up a variety of potential applications in different areas such as geology, mining, environmental protection, medicine, process control, and electron microscopy. Fascinating applications can be envisioned in the fields of planetary space exploration and x-ray astronomy, ranging from the determination of the composition of cometary matter to the x-ray observation of exotic stellar objects.

ACKNOWLEDGEMENTS

The author is grateful to the Organizing Committee of the 25th Denver Conference for inviting him to present this review paper. He is further pleased to acknowledge several of his colleagues, especially Dr. Jan S. Iwanczyk, with whom he has worked almost from the beginning of his involvement with compound detectors, and Dr. Wladyslaw M. Szymczyk, for collaboration going back several years on CdTe detectors and continuing up to the present on HgI_2 detectors, and Dr. John H. Kusmiss, Jeffrey B. Barton, and Jan W. Checinski, for their vital contributions to the research on HgI_2 detectors. To Dr. Gerald C. Huth, Director of the Medical Imaging Science Group of the University of Southern California, the author would like to express his sincere thanks for encouragement and support of the HgI_2 x-ray detector research project at USC. The author gratefully acknowledges the long-standing encouragement and support of Doc. Dr. Janina Chwaszczewska, Director of the Semiconductor Detector Laboratory of the Institute of Nuclear Research in Poland, and many stimulating conversations about semiconductor physics with her and colleagues in her Laboratory. The author is also thankful to the Directorate of the INR for support of CdTe research. Collaborative associations with the EG&G Advanced Measurements Group in Santa Barbara, the Fermi Institute of the University of Chicago, and the Center for Space Research of the Massachusetts Institute of Technology are gratefully acknowledged. Finally, the author is indebted to the U.S. Department of Energy and the National Aeronautics and Space Administration for continuing financial support of much of the work on HgI_2 room-temperature x-ray spectrometry which has been carried out at USC and is summarized and reviewed in this paper.

REFERENCES

1. J.W. Mayer, Chapter 5 in "Semiconductor Detectors" ed. by
   G. Bertolini and A. Coche, North Holland Publishing Co., 1968.
2. Proc. of the Intern. Symp. on Cadmium Telluride, 1971, Strasbourg.
3. S.P. Swierkowski and G.A. Armantrout, IEEE NS-22, No. 1,
   (1975) 205.
4. Proc. of the 2nd Symp. on Cadmium Telluride, 1976, Strasbourg,
   in Rev. de Physique Applique 12, No. 2 (1977).
5. G.A. Armantrout, S.P. Swierkowski, J.W. Sherohman, and H. Yee,
   IEEE NS-24 (1977) 121.
6. Proc. of the Intern. Workshop on CdTe and HgI_2, 1977, Jerusalem,
   in Nucl. Instr. and Meth., 150 (1978).
7. R.C. Whited and M.M. Schieber, Nucl. Instr. and Meth. 162
   (1979) 113.
8. G.A. Armantrout, UCRL Preprint 86303 (also in Proc. of Fifth
   Symp. on X-Ray and Gamma-Ray Sources and Applications, 1981,
   Ann Arbor, Michigan).
9. T. Kobayashi, I. Kuru, A. Haja, and T. Sugita, IEEE NS-23, No. 1,
   (1976) 97.
10. M.B. Panish, H.J. Gueisser, L. Derick, and S. Sumski, Solid State
    Electronics 9 (1966) 311.
11. A.J. Strauss, Rev. Physique Applique 12 (1977) 176.
12. O.A. Matveev, S.V. Prokofiev, and Yu. V. Rud, Inorg. Mat. 5
    (1969) 1175.
13. N.R. Kyle, J. Electrochem. Soc. 118 (1971) 1790.
14. R. Triboulet, Y. Marfaing, A. Cornet, and P. Siffert, J. Appl.
    Phys. 45 (1974) 2759.
15. A.J. Dabrowski, J. Iwanczyk, and R. Triboulet, Nucl. Instr. and
    Meth. 118 (1974) 531.
16. A.J. Dabrowski, J. Iwanczyk, and R. Triboulet, Nucl. Instr. and
    Meth. 126 (1975) 417.
17. A.J. Dabrowski, J. Chwaszczewska, J. Iwanczyk, R. Triboulet, and
    Y. Marfaing, IEEE NS-23, No. 1 (1976) 171.
18. J. Iwanczyk, W.M. Szymczyk, M. Traczyk, and R. Triboulet, Nucl.
    Instr. and Meth. 165 (1979) 289.
19. A.J. Dabrowski, J.S. Iwanczyk, W.M. Szymczyk, P. Kokoschinegg,
    J. Stelzhammer, and R. Triboulet, Nucl. Inst. and Meth. 150,
    (1978) 25.
20. J. Saura and J.L. Regolini, J. Cryst. Growth 15 (1972) 307.
21. J.P. Ponpon, R. Stuck, P. Siffert, B. Meyer, and C. Schwab,
    IEEE NS-22, No. 1 (1975) 182.
22. H. Scholtz, Acta Electronics 17 (1974) 69.
23. M. Schieber, W.F. Schnepple, and L. van den Berg, J. Cryst.
    Growth 33 (1976) 125.
24. S.P. Faile, A.J. Dabrowski, G.C. Huth, and J.S. Iwanczyk,
    J. Cryst. Growth 50 (1980) 752.
25. J.B. Barton, A.J. Dabrowski, J.S. Iwanczyk, J.H. Kusmiss,
    G. Ricker, J. Vallerga, A. Warren, M.R. Squillante, S. Lis, and
    G. Entine, in this volume (Advances in X-Ray Analysis 25).

26. S.P. Swierkowski, G.A. Armantrout, and R. Wichner, IEEE Trans. on Nuclear Science NS-21, No. 1 (1974) 302.

27. Z.H. Cho, M.K. Watt, M. Slapa, P.A. Tove, M. Schieber, T. Davies, W. Schnepple, and P. Randtke IEEE NS-22, No. 1, (1975) 229.

28. M. Slapa, G.C. Huth, W. Seibt, M.M. Schieber, and P.T. Randtke, IEEE NS-23, No. 1 (1976) 102.

29. M. Seibt, M. Slapa, and G.C. Huth, Nucl. Instr. and Meth. 135, (1976) 573.

30. A.J. Dabrowski and G.C. Huth, IEEE NS-25, No. 1 (1978) 205.

31. A.J. Dabrowski, G.C. Huth, M. Singh, T.E. Economou, and A.L. Turkevich, Appl. Phys. Lett. 33 (1978) 211.

32. G.C. Huth, A.J. Dabrowski, M. Singh, T.E. Economou, and A.L. Turkevich, Advances in X-Ray Analysis 22 (1979) 461.

33. M. Singh, A.J. Dabrowski, G.C. Huth, J.S. Iwanczyk, B. Clark, and A.K. Baird, Advances in X-Ray Analysis 23 (1980) 249.

34. L. van den Berg and R.C. Whited, IEEE NS-25, No. 1 (1978) 395.

35. M. Schieber, I. Beinglass, G. Dishon, A. Holtzer, and G. Yaron, Nucl. Instr. and Meth. 150 (1978) 71.

36. J.S. Iwanczyk, A.J. Dabrowski, G.C. Huth, A. Del Duca, and W. Schnepple, IEEE NS-28, No. 1 (1981) 579.

37. A.J. Dabrowski, J.S. Iwanczyk, J.B. Barton, G.C. Huth, R. Whited, C. Ortale, T.E. Economou, and A.L. Turkevich, IEEE NS-28, No. 1, (1981) 536.

38. J.S. Iwanczyk, J.H. Kusmiss, A.J. Dabrowski, J.B. Barton, G.C. Huth, T.E. Economou, and A.L. Turkevich, Proc. of the 5th Symp. on X- and Gamma-Ray Sources and Applications, 1981, Ann Arbor, Michigan.

39. E. Storm, D. Gilbert, and H. Israel, Nucl. Data Tables A7, (1970) 565.

40. C.A. Klein, J. Appl. Phys. 39 (1968) 2029.

41. S. Ramo, Proc. of the IRE (1939) 584.

42. U. Fano, Phys. Rev. 70 (1946) 44 and 72 (1947) 26.

43. J. Llacer, E. Haller, R.C. Cordi, IEEE NS-24, No. 1 (1977) 53.

44. M. Slapa, J. Chwaszczewska, J. Jurkowski, A. Latuszynski, G.C. Huth, and A.J. Dabrowski, in this volume (Advances in X-Ray Analysis, Volume 25, Plenum Publishing Corp., New York and London).

45. R.B. Day, G. Dearnley, and J.M. Palms, IEEE NS-14 (1967) 487.

46. J.S. Iwanczyk and A.J. Dabrowski, Nucl. Instr. and Meth. 134, (1976) 505.

47. M. Singh, B.C. Clark, A.J. Dabrowski, J.S. Iwanczyk, D.E. Leyden, and A.K. Baird, Advances in X-Ray Analysis 24 (1981) 337.

48. K. Zanio, Nucl. Instr. and Meth. 83 (1970) 288.

49. M. Slapa, G. Hahn, J. Chwaszczewska, A.J. Dabrowski, J.S. Iwanczyk J.S. Iwanczyk, and G.C. Huth, Nucl. Instr. and Meth. 176 (1980) 567.

50. A.J. Dabrowski, M. Singh, G.C. Huth, and J.S. Iwanczyk, Proc. of the Workshop on Energy Dispersive X-Ray Spectrometry (1979), Gaithersburg, Maryland; published as NBS Special Publication 604.

51. F.S. Goulding, Nucl. Instr. and Meth. 142 (1977) 213.
52. A.J. Dabrowski, J. Chwaszczewska, J. Iwanczyk, R. Triboulet,
and Y. Marfaing, Revue de Physique Applique 12 (1977) 297.

PRELIMINARY STUDY OF THE BEHAVIOR OF HPGe DETECTORS WITH ION

IMPLANTED CONTACTS IN THE ULTRALOW-ENERGY X-RAY REGION

M. Slapa, J. Chwaszczewska, J. Jurkowski
Institute of Nuclear Research, Seminconductor Detector
Laboratory, 05-400 Swierk near Warsaw, Poland

A. Latuszynski
Institute of Physics, Univ. of Marie Curie-Sklodowska
Lublin, ul. Nowotki 10, Poland

G.C. Huth, A.J. Dabrowski
Univ. of Southern California, Medical Imaging Science
Group, Marina del Rey, California 90291, USA

ABSTRACT

   Preliminary study of performance of an HPGe detector with an
ion implanted entrance window in the spectrometry of the ultralow-
energy x-rays is presented.

   For the first time it has been shown that almost symmetric
photopeaks and absence of low energy tailing can be obtained in this
region from HPGe detectors.

1. INTRODUCTION

   High purity germanium (HPGe) detectors are widely used in spectroscopy of middle and high energy X-and gamma radiation.[1,2,3]

   The energy equivalent Noise Line Width of HPGe detectors can
be comparable with that of the Si[Li] detector,[4] since HPGe detectors exhibit very low current leakage (4×10^{-14} A)[4] and the average amount of energy required to produce a hole-electron pair is 20%
smaller in germanium than in silicon. In spite of these facts until
now, application of HPGe detectors to spectroscopy of low energy x-ray
is still limited. This limitation is mostly connected with degradation of the spectral line shape for low energy x-rays below 2.3 keV.

At energies below 2.3 keV these detectors have been thought to show an inherent low-energy tailing of the photopeak.[5,6] The fraction of counts transferred from the photopeak to the low-energy tailing region can be quite large. As it was reported in (5) for example,in the case of sulfur K_α x-ray the tailing amounts to as much as 50% of the photopeak. The degradation of spectral lineshape has been explained[5,6] as due to window effects inherent in HPGe detectors. To the present time only experimental data on HPGe detectors with Schottky barrier have been presented in the literature.

In this paper a preliminary study of performance of HPGe detectors with an ion implanted entry window in the ultralow-energy region is reported.

2. ION IMPLANTED HPGe DETECTORS

The HPGe detectors with both contacts made by ion implantation were reported first in 1971 by Harzer et al.[7] They applied a boron implantation (10–20 keV, 10^{14} ion/cm^2) layer as p^+ – type entrance contact and phosphorus implantation (4 keV, 10^{14} ion/cm^2) layer as the n^+ – type back contact. The dead layers of these detectors measured from energy loss (straggling) of α-particles were 0.2 μm and 0.8 μm for the p^+ and n^+ contact respectively. Subsequently, HPGe detectors have been developed as charged particle spectrometers.[8,9] The well established fabrication process for these detectors with boron and phosphorus ions implanted as p^+ and n^+ contacts has been worked out. The measured dead layer for these detectors are 0.2 μm – 0.4 μm.[8,9]

In the present work we have fabricated the HPGe detectors with boron or gallium implanted as an entrance electrode and a lithium diffused layer as a back contact. Boron implanted detectors behave much worse than gallium ones in the ultralow x-ray energy region. For this reason our major effort was focused on gallium implanted devices.[10]

3. EXPERIMENTAL CONSIDERATION

3.1 Detectors

The detectors we made from p-type Ge crystals exhibiting impurity concentration values from $8 \times 10^9 cm^{-3}$ to $4 \times 10^{10} cm^{-3}$. The detectors were formed in an inverted "T" structure.[11] Just before implantation the samples are etched in a 1 HF: 3HNO$_3$ solution and rinsed in deionized water. The 15 keV gallium ions with dose ranging from $5 \times 10^{13} cm^{-2}$ to $5 \times 10^{14} cm^{-2}$ were implanted. The samples were annealed in vacuum at temperature varying from

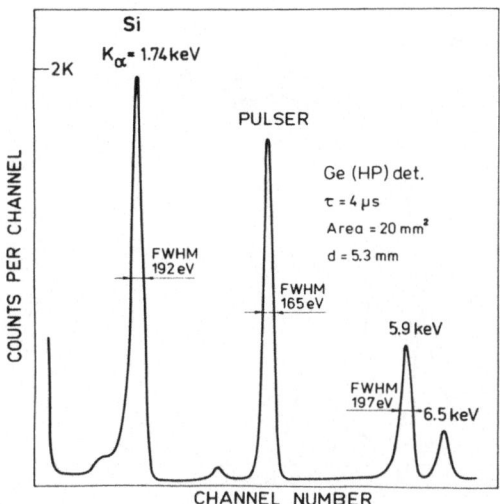

Fig. 1. Spectrum taken from our system when Si target is excited

280°C to 400°C. After the annealing process lithium diffusion at temperature about 240°C takes place.

The detectors can work with bias from three to four times larger than the full depletion voltage of the detector.

3.2. Measuring System

The common low noise, charge-sensitive amplifier connected to a pulse amplitude 4000 channel analyzer was used. To excite the x-ray fluorescence lines we used a 1 mCi Fe-55 source.

4. RESULTS

4.1. Spectroscopical Property

Fig. 1 shows the spectrum taken from our system when a silicon target is excited. The photopeak measured for the Si K_α x-ray line is almost symmetrical and the low energy part of the spectrum contains only a small fraction of total counts of the photopeak. A full width at half maximum value of 192 eV was measured when the system resolution measured (using a pulser) was 165 eV. The detector bias is 600 V when full depletion voltage of the detector is 410 V.

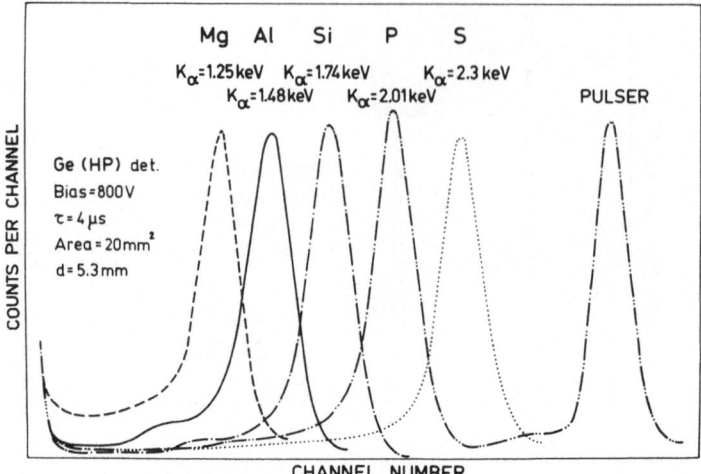

Fig. 2.   Spectra obtained for monoenergetic x-ray incident on a
           germanium detector.

     In Figure 2 the spectra obtained from x-ray fluorescence
of Mg, Al, Si, P, and S targets using the same detector as in
Figure 1 are shown.   All spectra have well pronounced photopeaks and very few counts in the low-energy part of the spectrum (i.e., for K_α-Al less than 22% of the photopeak).   The
energy resolution of photopeaks increases with decreasing the
energy of x-ray (see Figure 3).   Figure 3 shows the energy
resolution ΔE and charge collection contribution ΔE_{col} vs x-ray

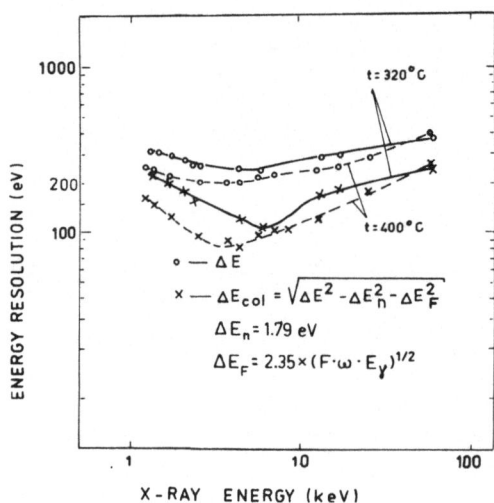

Fig. 3.   The energy resolution and charge collec-
tion contribution vs x-ray
energy.

energy.[12] These curves were obtained for two different detectors
annealed at 320°C and 400°C respectively. It is seen that the charge
collection contribution is a monotonic function of the x-ray energy
and the best results of 90 eV were obtained for a 400° annealed detector at 4 keV.

4.2. Evaluation of Window Thickness

Fig. 4 shows the fluorescence spectra of Al, Si and Cl measured
at the same conditions as in the Fig. 2. On each spectrum the generator line is also plotted. The amplitude pulses from the generator
are normalized to each photopeak assuming that at the 5.9 keV x-ray
we obtain the charge collection coefficient: $\eta = 1$.[12] One can see,
that deviation of the photopeak shape from Gaussian type (generator

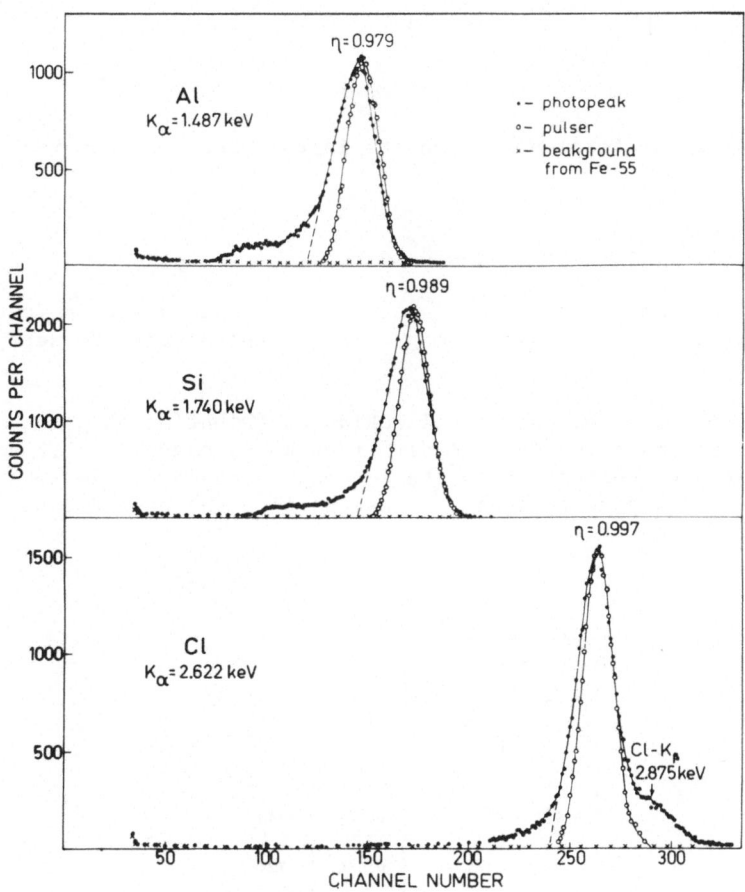

Fig. 4. Spectra obtained for selected monoenergetic x-ray.

line) increases with decreasing of x-ray energy and the charge collection coefficient decreases with decreasing x-ray energy. The origin of counts in the low-energy part of the spectra is background from the Fe-55 source as well as from window effects. The background from Fe-55 was determined by placing a Teflon foil between the detector and target. The thickness of Teflon foil was chosen so as to stop the relevant fluorescence radiation.

A very simple model has been used to determine the window thickness.[6,13] There exists a thickness L_d such that if a photoelectric interaction occurs at $x < L_d$, the result will be a count in the tail of the spectrum. If the interaction occurs at $x \geqslant L_d$, the count will fall in the photopeak. Then, the counts number at low-energy tailing N_{LET} is given by:

$$N_{LET} = N_T \left[1 - \exp(- L_d/L_p) \right] \qquad (1)$$

where:

N_T - is the total number counts in the spectrum from x-ray line

L_p - is depth penetration

L_d - is dead layer of the detector.

Our results summarized in the Table 1. The results were computed by linear extrapolation of low-energy part of the spectra to zero channel number.

Fig. 5 shows the charge collection contribution ΔE_{col} vs shaping time constant measured for two different x-ray energies. It seems that the charge collection in the low energy region (1.74 keV) is affected by trapping processes connected with the presence of non-annealed centers.

Table 1

Element	L_p[μm] Ref. 14	$\frac{N_{LET}}{N_T}$	L_d[A]
Al	0.29	0.195	650
Si	0.42	0.162	748
Cl	1.25	0.056	724

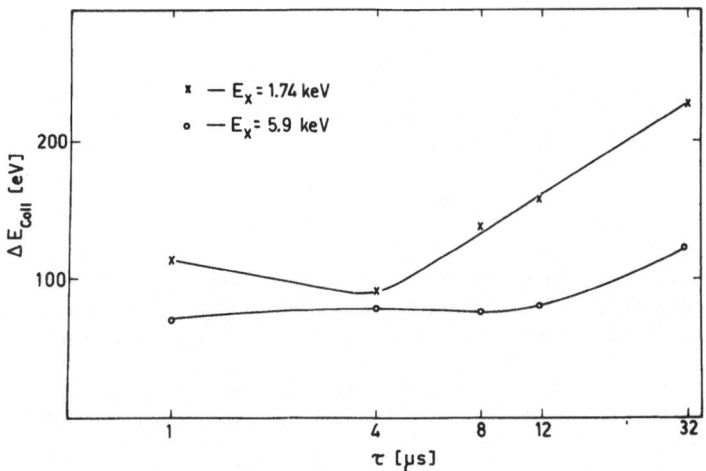

Fig. 5. The charge collection contribution vs shaping time constant.

CONCLUSION

Preliminary study of the performance of Ge[HP] detectors with ion implanted entrance contacts indicate that:

1) HPGe detectors can be useful in the ultralow energy x-ray region with almost symmetrical photopeaks.

2) Low energy tailing can be significantly reduced according to methods reported herein.

3) The dead layer for such detectors can be smaller than 750 A.

We believe that further improvement in energy resolution is possible. Further work toward optimization of the ion implantation and annealing processes and study of window effects are in progress.

REFERENCES

1.   R.H. Pehl, R.C. Cordi, and F.S. Goulding, IEEE Trans., Nuc. Sci., NS-19, No. 1, 265 (1972).
2.   I.M. McKenzie, Nuc. Inst. Meth., 162, 49 (1979).
3.   R.H. Pehl, Physics Today, 50, (1977).
4.   J. Llacer, IEEE Trans. Nuc. Sci., NS-22, No 5, 2033 (1975).
5.   F.S. Goulding, Nuc. Inst. Meth., 142, 213 (1977).
6.   J. Llacer, E.E. Haller, and R.C. Cordi, IEEE Trans. Nuc. Sci., Vol. NS-24, No 1, 53, (1977).

7. H. Herzer, S. Kalbitzer, J.P. Ponpon, R. Stuck, and P. Siffert,
    Nuc. Inst. Meth. 101, 31 (1972).
8. D. Protic and G. Riepe, IEEE Trans. Nuc. Sci., Vol. NS-24,
    No 1, 64, (1977).
9. G. Scott Hubbard, E.E. Haller, and W.L. Hansen, IEEE Trans.
    Nuc. Sci., Vol. NS-24, No 1, 161 (1977).
10. M. Slapa, J. Chwaszczewska, G.C. Huth, and J. Jurkowski, Nuc.
    Inst. Meth., (to be published).
11. J. Llacer, IEEE Trans. Nuc. Sci., NS-13, No 1, 221 (1966).
12. R. Henek, D. Gutknech, P. Siffert, L. de Lact, and W.
    W. Schwenmacher, IEEE Trans. Nuc. Sci., NS-17, No 3, 149, (1970).
13. A.J. Dabrowski, J.S. Iwanczyk, J.B. Barton, G.C. Huth, R. Whited,
    C. Ortale, T.E. Economou, and A.L. Turkevich, IEEE Trans. Nuc.
    Sci., NS-28, No 1, 536, (1981).
14. E.L. Storm, E. Gilbert, and H. Israel, Los Alamos, Report 2237.
15. J.M. Caywood, C.A. Mead, and J.M. Mayer, Nuc. Inst. Meth., 79,
    329, (1970).

PERFORMANCE OF ROOM-TEMPERATURE X-RAY DETECTORS MADE FROM

MERCURIC IODIDE (HgI_2) PLATELETS

J.B. Barton, A.J. Dabrowski, J.S. Iwanczyk, J.H. Kusmiss
Medical Imaging Science Group, USC, Marina del Rey, CA

G. Ricker, J. Vallerga, A. Warren
Center for Space Research, MIT, Cambridge, MA

M.R. Squillante, S. Lis, G. Entine
RMD Inc., Watertown, MA

ABSTRACT

Current developments in the recently introduced method of HgI_2 crystal platelet growth by polymer assisted vapor transport are described. Crystal parameters are evaluated by making electrical measurements on x-ray detectors fabricated from HgI_2 platelets. Selection for detector fabrication is on the basis of size and apparent crystalline perfection. Detectors have been fabricated with active areas averaging 2 to 3 mm^2 and thicknesses ranging from 20 to 400 μm. Values of electron mobility and mobility-lifetime product measured for HgI_2 platelet material are among the highest ever observed for HgI_2.

The combination of low leakage current and good electron transport makes HgI_2 platelets suitable for room-temperature x-ray spectrometry. An energy resolution of 370 eV (FWHM) for the 5.9 KeV Mn K_α line has been obtained, and representative low-energy x-ray fluorescence spectra are presented.

INTRODUCTION

A new method of growing HgI_2 crystals was introduced by Faile in 1980.[1] The method consists of polymer-aided vapor transport resulting in crystallization of HgI_2 in platelet form. Mercuric iodide platelets are grown from unpurified reagent grade HgI_2 in periods of a few days. This is in contrast to other methods of HgI_2 crystal growth from the vapor which require highly purified starting material and several months for purification and crystal growth.[2] Additionally, HgI_2 platelets do not need to be cleaved since they are of a suitable size and shape for x-ray detector

31

fabrication, ranging from about 10 to 100 mm² in area and 20 to
400 μm in thickness. Avoiding crystal cleavage reduces the possibility of introducing mechanical damage which may degrade detector
performance.

PLATELET GROWTH

To grow platelets, HgI_2 and a suitable polymer (~1% polyethylene or styrene by weight) are loaded into one end of a quartz ampoule
which is then evacuated (~10 μm Hg); see Fig. 1. The end with
the source material is inserted into a furnace at about 230° C.
The ampoule extends out of the furnace so that there is a temperature gradient from 230° C at the source end to room temperature at
the other end. The platelets of the red phase of HgI_2 grow at a
temperature below 127° C. A considerable quantity of black residue
consisting of polymer and impurities from the reagent grade HgI_2
starting material is left at the source end of the ampoule after
the crystal growing is completed.

Initially the procedure described by Dr. Faile was followed
exactly. After preliminary results were obtained, a systematic
study of the growth environment was undertaken in an attempt to
improve quality and reproducibility of the results. The temperature
distribution inside the ampoule has been varied in several ways.
Control over the temperature both at the source and in the growth
region has been achieved by use of a two-zone furnace. It has been
found that higher growth temperatures encourage the formation of
dendritic crystals or thin platelets, while higher source temperatures seem to result in thicker platelets. The platelet growth
zone at 127° C has been lengthened by using a more gradual temperature gradient. There is some evidence that while favoring the
production of more platelets which are less crowded, a more gradual
gradient also tends to give rise to thinner platelets. Efforts to
obtain a more favorable temperature environment for HgI_2 platelet
growth are being continued with multizone furnaces and better
temperature stability control.

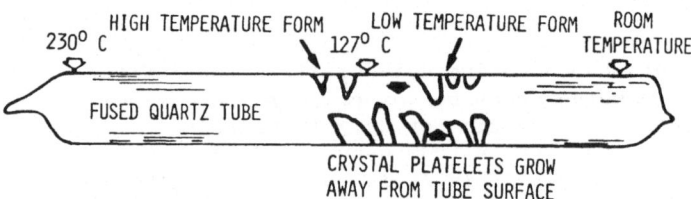

Fig. 1   Arrangement for growing HgI_2 platelets.

It has been found that reversing the ampoule in the furnace after platelet growth is complete and repeating the vapor transport results in still more residue being left behind. This procedure apparently results in purer crystals and also an improvement in both the form and quantity of the platelets grown. Such efforts toward purification have led to the re-use of HgI_2 from one growth run in a subsequent growth run. While decreased residue suggests a considerable improvement in the purity of the final crystals, obtaining good platelet formation with this purified HgI_2 starting material is more difficult.

By experimentally varying the HgI_2-to-polymer ratio (originally 100:1 by weight), crystal growth formations ranging from dendritic to platelet-like to cubic forms have been observed. Ampoule size and total HgI_2-polymer quantity are also being varied in an attempt to maximize the output of usable crystals in a given run.

DETECTOR FABRICATION

Detector fabrication from HgI_2 platelets is relatively easy since the size and shape of the crystals are suitable as grown. Crystals grown by other methods must be cleaved to obtain slices of the desired thickness for detector fabrication. The cleavage process is mechanically stressful for the crystal and may introduce structural defects which degrade detector performance. Since platelets are not cleaved, the fabrication is simplified and there is reduced chance of mechanical damage to the crystal.

The first step in platelet detector fabrication is crystal selection. There is a large variation in the size, shape, and appearance of different platelets from the same ampoule. Selection for detector fabrication is on the basis of area, thickness, and apparent crystalline perfection or optical clarity. The electrical contact must be deposited on an area of the crystal free of visible defects, and this places one limit on the contact area for any given platelet detector. The thickness of a platelet also places a limit on the contact area if optimal energy resolution is desired. Detector capacitance is determined by contact area and detector thickness:

$$C = 78 \frac{A}{L} \qquad\qquad\qquad (1)$$

where   C = detector capacitance (pf)
         A = contact area (mm^2)
         L = thickness (μm)

Detector capacitance contributes to the electronic noise level of the system.[3] If the contribution to energy resolution due to

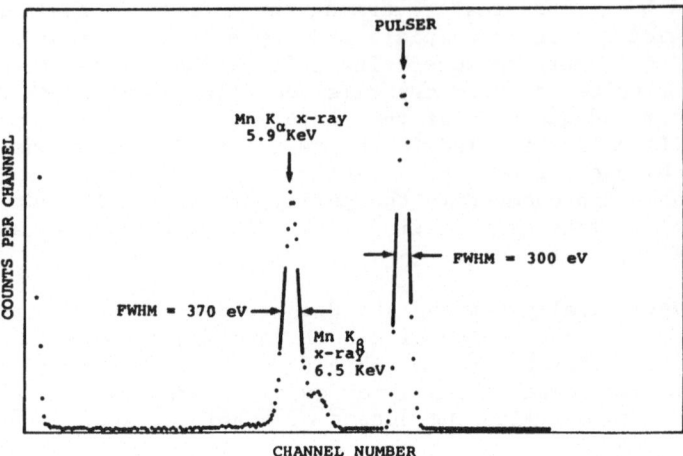

Fig. 2   Mn x-ray spectrum from ^{55}Fe source taken with HgI_2
platelet detector.

detector capacitance is not to limit low energy performance[4], its
value should be less than 3 pf. The active area which is desired
then determines which platelets may be selected for detector fabrication. For routine crystal evaluation detectors are fabricated
with active areas of about 1 to 3 mm^2.

Prior to the deposition of electrical contacts, selected HgI_2
platelets are etched in a 5% aqueous KI solution for about one
minute. This removes a small amount of the platelet surface which
may be contaminated due to chemical reaction with atmospheric
constituents. Electrical contacts are painted onto the crystal
face with an aqueous or alcoholic carbon solution. Thin leadwires
of Pd are placed in the carbon contact. The platelet is mounted
on a ceramic substrate with silicone rubber. The detector may be
protected by encapsulation in silicone rubber.

RESULTS

Since the introduction of HgI_2 platelets in 1980, we have had
the experience of fabricating and testing over 200 detectors from
this material. Considering the relative newness and simplicity of
HgI_2 platelet growth, the results have been remarkable. An energy
resolution of 370 eV (FWHM) has been obtained for the 5.9 KeV
M_nK_α x-ray line; see Fig. 2. This value is comparable to the best
energy resolution value ever obtained with an HgI_2 detector using

a resistor feedback preamplifier.

The value of the mobility-lifetime product (μτ) for electrons and holes is a measure of the charge collection efficiency of a detector.[5] By measuring the peak position from an ^{55}Fe source as a function of detector bias voltage, an estimate of μτ may be made through the Hecht relation. Some of the values measured for HgI_2 platelets are among the highest ever observed for HgI_2. Values greater than 10^{-3} cm^2/V for electrons and 5×10^{-6} cm^2/V for holes have been measured. Irradiating detectors with α-particles and measuring the rise times of pulses from the detectors permitted the values of mobility for electrons and holes in HgI_2 platelets to be measured. Detector pulse rise time is equal to the charge carrier transit time across the bulk of the detector. Knowing transit time, detector thickness, and bias voltage allows calculation of carrier drift velocity and mobility. For the best detector a mobility value of greater than 100 cm^2/V-sec for electrons has been measured.

To demonstrate the potential application of HgI_2 platelet detectors for x-ray fluorescence analysis, several targets were irradiated with Mn x-rays from an ^{55}Fe source. Ca K x-ray (3.7 KeV), Ti K x-ray (4.5 KeV), and In L x-ray (3.3 KeV) spectra are shown in Figs. 3, 4 and 5.

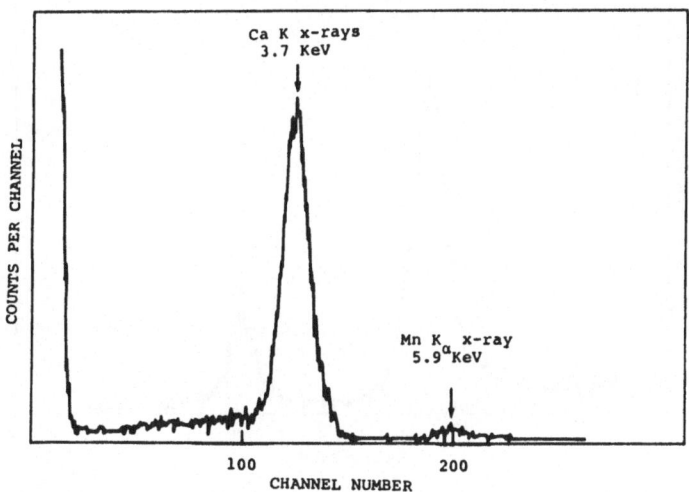

Fig. 3. Ca K x-ray spectrum taken with an HgI_2 platelet detector.

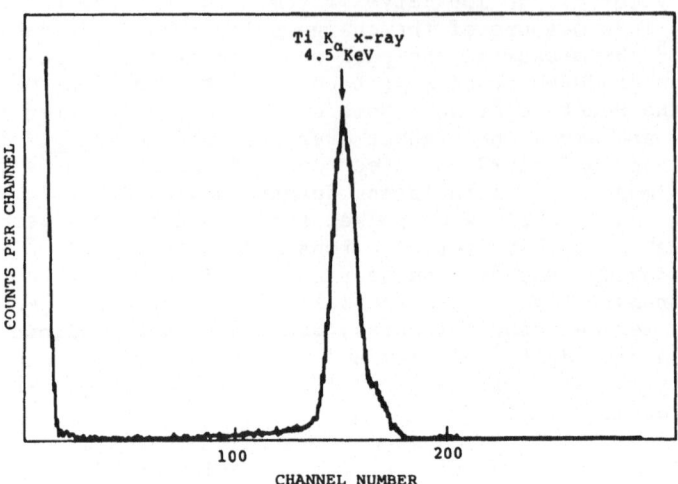

Fig. 4. Ti K x-ray spectrum taken with an HgI$_2$ platelet detector.

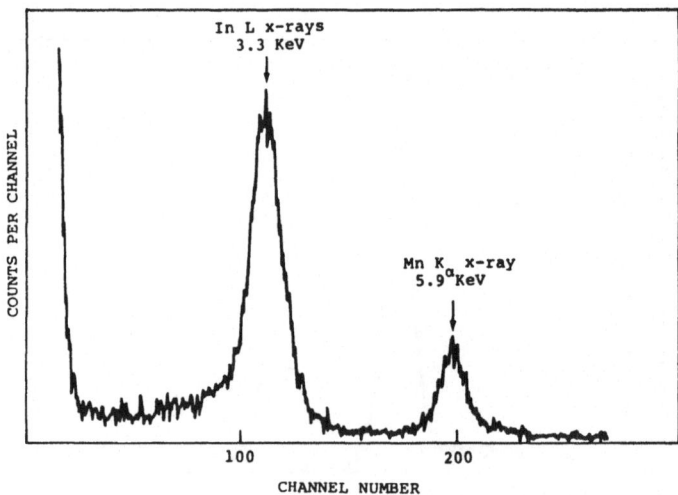

Fig. 5. In L x-ray spectrum taken with an HgI$_2$ platelet detector.

CONCLUSION

Platelets grown in a few days from unpurified starting material have given some of the best results ever obtained with HgI_2. An energy resolution value of 370 eV (FWHM) has been obtained for the 5.9 KeV Mn K_α x-ray line. Mobility-lifetime product ($\mu\tau$) values of greater than 10^{-3} cm^2/V for electrons and 5×10^{-6} cm^2/V for holes have been measured. For the best detector a mobility value of greater than 100 cm^2/V-sec for electrons has been measured.

The rapid progress that has been made with HgI_2 platelets may be attributed to two facts:
1. The speed with which platelets may be grown makes for effective and fast feedback between crystal growers and detector fabricators and testers.
2. Detector fabrication is simplified for HgI_2 platelets since cleaving of the crystal is unnecessary.

Although the results obtained with HgI_2 platelets have been encouraging, better control of the platelet growth process would be desirable. Additional work must be done to increase the yield of high-quality platelets and to obtain reproducible results.

REFERENCES

1. S. P. Faile, A. J. Dabrowski, G. C. Huth, and J.S. Iwanczyk J. Crystal Growth 50 (1980) 752.
2. M. Schieber, W.F. Schnepple, and L. Van den Berg, J. Crystal Growth 33 (1976) 125.
3. J. S. Iwanczyk, A. J. Dabrowski, G. C. Huth, A. Del Duca, and W. Schnepple, IEEE NS-28, 1 (1981) 579.
4. J. S. Iwanczyk, J. H. Kusmiss, A. J. Dabrowski, J. B. Barton, and G. C. Huth, Fifth Symposium on X- and Gamma-Ray Sources and Applications held at the University of Michigan, Ann Arbor, Michigan, June 10-12, 1981 (proceedings will appear in a special issue of Nuclear Instruments and Methods)
5. A. J. Dabrowski, in Advances in X-Ray Analysis, Volume 25

THE GAS PROPORTIONAL SCINTILLATION COUNTER AS A ROOM-TEMPERATURE

DETECTOR FOR ENERGY-DISPERSIVE X-RAY FLUORESCENCE ANALYSIS

C.A.N. Conde, L.F. Requicha Ferreira, and A.J. de Campos

Physics Department, University of Coimbra

3000 Coimbra, Portugal

ABSTRACT

A review of the basic physical principles of the gas proportional scintillation counter is presented. Its performance is discussed and compared with that of other room-temperature detectors in regard to applications to portable instruments for energy-dispersive X-ray fluorescence analysis. It is concluded that the gas proportional scintillation counter is definitely superior to all other room-temperature detectors, except the mercuric iodide (HgI_2) detector. For large areas or soft X-rays it is also superior to the HgI_2 detector.

INTRODUCTION

The continuing interest in portable instruments for energy-dispersive X-ray fluorescence analysis requires an appraisal of the characteristics of the various types of room-temperature detectors: standard proportional counters, standard scintillation counters, large-bandgap semiconductor detectors, and gas proportional scintillation counters. In the present work we present a review of the last type of detector. The basic principles of the gas proportional scintillation counter[1] are discussed, together with its present state of development, and its performance is compared with that of other room-temperature detectors.

THE SECONDARY SCINTILLATION OF NOBLE GASES

When ionizing radiation interacts with a noble gas, besides the production of electron-positive ion pairs there is production of

39

light, the so-called primary scintillation. Its intensity is rather low, and therefore the energy resolution of any scintillation counter using this effect is poor. But if the primary electrons drift under an electric field, more light can be produced if the reduced field intensity is above about 1 V cm^{-1} Torr^{-1}. This is the so-called secondary scintillation. Its intensity varies almost linearly with the field above the 1 V cm^{-1} Torr^{-1} threshold[2] and can reach values a few orders of magnitude larger than that of the primary scintillation even for electric fields below the ionization threshold. Actually one single electron drifting along the field lines can produce as many as a few hundred photons before any charge multiplication takes place. The mechanisms for light production are very efficient if the gas purity is very high; more than 70% of the energy gained by an electron from the electric field can be converted into light.[3] Indeed, if there are no impurities in the gas, the drifting electrons cannot lose energy by excitation of low-lying rotational or vibrational states (noble gas atoms have none); energy is only lost either by recoil, when electrons undergo elastic colli- sions with a noble gas atom, or, if they have enough energy, by excitation of the noble gas lower states at around 10 eV. The fact that, as shown in Monte Carlo simulations,[4] an electron can undergo tens of thousands of elastic collisions before it gets enough kinetic energy from the electric field to excite a noble gas atom shows how important gas purity can be.

Excitation of a noble gas atom, R, generally leads to the formation of an excited state, R*, which by collision forms a molecular rare gas excimer, R_2*. This breaks up as

$$R_2^* \rightarrow 2R + h\nu$$

and the photons thus produced lie in a band peaking[5] at 173 nm for Xe, 147 nm for Kr, and 128 nm for Ar, with FWHM of 14, 12, and 10 nm respectively.

This light is the secondary scintillation and can be observed only through a VUV window and detected with a VUV photomultiplier, or else converted into longer wavelengths with a film of organic wavelength-shifter deposited onto the window and detected with a standard photomultiplier. However the last solution requires con- tinuous purification (wavelength-shifter vapors poison the gas), which almost excludes its use in portable systems.

The noble gases normally used are Xe, Kr, and Ar, and the intensity of the secondary scintillation is generally larger the heavier the noble gas and does not depend on the presence of magnetic fields even for fields as strong as 4 kG.

THE GAS PROPORTIONAL SCINTILLATION COUNTER

The large intensity of the secondary scintillation produced in a noble gas can be used, in principle, to develop improved energy resolution radiation detectors. A detector based in this effect is called a "gas proportional scintillation counter" since the intensity of the secondary scintillation is proportional to the number of primary electrons and thus to the energy of the radiation detected. However, when designing a gas proportional scintillation counter, a basic principle has to be obeyed: the region of the gas volume where radiation is detected must be separated from the region where the secondary scintillation is produced; otherwise its intensity would also depend on the position of the radiation track. Detectors obeying this principle have been built with various geometries: cylindrical, spherical, and uniform field with parallel grids.

Figure 1 shows a detector using the last geometry. Electrons produced by the incident radiation drift toward grid 1 under the

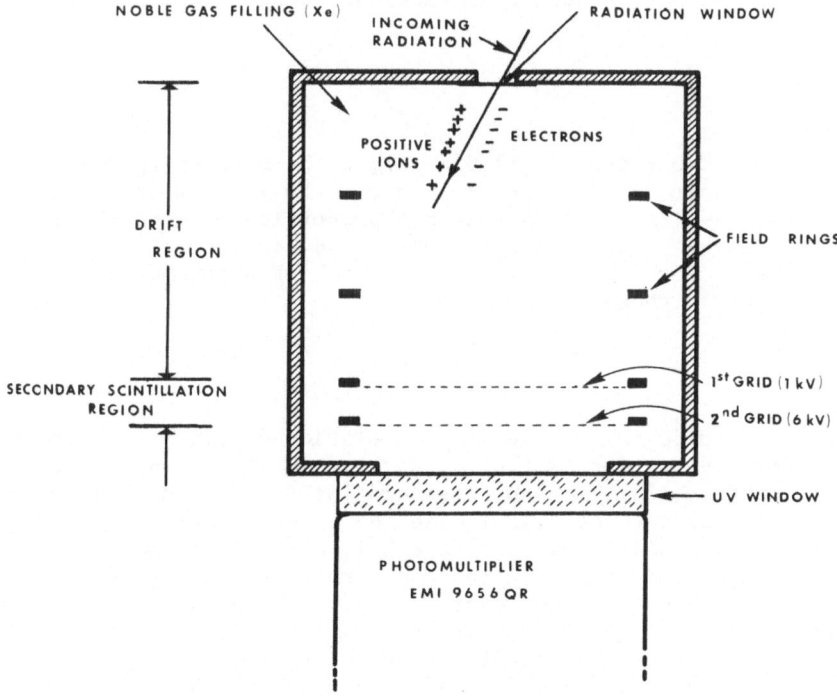

Fig. 1. The uniform field gas proportional scintillation counter.

influence of a weak electric field with intensity below the 1 V cm^{-1} Torr^{-1} threshold for light production. Thus, light is produced only when the electron swarm reaches the strong electric field region between the two grids. For radiation detected in the drift region the intensity of the secondary scintillation does therefore not depend on the position and orientation of the radiation track.

The energy resolution of a gas proportional scintillation counter results from the fluctuations in the pulse amplitude, which depend on the variances associated with the following factors: number of primary electrons, photomultiplier statistics and total number of photons produced, charge multiplication statistics, and fluctuation in the number of photons produced when an electron travels between the grids. The last effect is generally small. Because the intensity of the secondary scintillation can be made very high, the relative contribution of the photomultiplier statistics and the number of photons produced can be neglected. Since under normal working conditions there is very little or no charge multiplication, the corresponding variance can be ignored. The only contribution that cannot be neglected is the variance in the number of primary electrons. This means that the limiting energy resolution, R_{lim}(FWHM), is equal to that of a noiseless ionization chamber. Thus

$$R_{lim} = 2.35 \ (Fw/E)^{\frac{1}{2}}$$

where F is the Fano factor (0.13 for Xe), w the mean energy to produce an ion pair (21.9 eV for Xe), and E the X-ray photon energy. For 5.9 keV X-rays detected in a Xe gas proportional scintillation counter the limiting value is 5.2%, while the best experimental values[7] are 8%. For Kr and Ar fillings the best experimental values are not as good as those for Xe.

DISCUSSION

When we compare the best energy resolution obtained with these detectors for 5.9 keV X-rays (8%) with those for the standard proportional counter (12%), for the standard [NaI(Tl)] scintillation counter (45%), and the mercuric iodide (HgI_2) detector[14] (5%), we arrive at the conclusion that only the HgI_2 detector can compete with gas proportional scintillation counters. However if we consider applications to lower-energy X-rays, the HgI_2 detector loses its advantages.[7,13,15] At around 1.5 keV they have about the same energy resolution, but for lower energies the gas proportional scintillation counter is superior, offering FWHM of only 85 eV at energies as low as 149 eV. For large areas[7] the HgI_2 detector starts to lose its advantage at higher energies.

As there is little or no charge multiplication in gas proportional scintillation counters, the peak shifts characteristic of standard proportional counters at high counting rates and due to space charge effects are nonexistent.[8] The earlier technical problems that limited the use of gas proportional scintillation counters in portable systems (gas purification and wavelength-shifters) can be solved using high-vacuum techniques and VUV windows. Small-size portable detectors have been described.[9] Recent developments concern the use of vacuum photodiodes and photoionization chambers (PIPS)[10-11] in place of the photomultiplier. These chambers use a low ionization potential organic vapor (TMAE or TEA) to detect the VUV photons. These vapors are mixed with standard proportional counter gases, and the charge pulses thus obtained are proportional to the intensity of the secondary scintillation. Energy resolutions of 9% have been obtained[12] for 6 keV X-rays. These photodetectors, however promising, require more experience before they can be used in portable equipment.

We conclude that a closed-system gas proportional scintillation counter coupled to a photomultiplier through a VUV window is a room-temperature detector, using tried techniques, that can be used to advantage instead of standard proportional counters in portable X-ray fluorescence systems. Interest in them may wane, however, as a result of recently published work on HgI_2 detectors.[13-14] However a well-founded choice can only be made when more data are available concerning the relative cost, ruggedness, and stability of gas proportional scintillation counters and HgI_2 detectors. Also, in the less explored field of soft X-ray fluorescence analysis, they still offer better energy resolution than any other kind of detector, either room-temperature or cooled.

We acknowledge support from I.N.I.C. and the Gulbenkian Foundation.

REFERENCES

1 - C.A.N. Conde and A.J.P.L. Policarpo, "A Gas Proportional Scintillation Counter", Nucl. Instr. and Meth. 55,105 (1967).

2 - C.A.N. Conde, L.F. Requicha Ferreira and M. Fátima A. Ferreira, "The Secondary Scintillation Output of Xenon in a Uniform Field Gas Proportional Scintillation Counter", IEEE Trans. Nucl. Sci. NS-24, No.1, 221 (1977).

3 - A.J.P.L. Policarpo, "Light Production in Gaseous Detectors", Phys. Scripta, 23, 539 (1981).

4 - Teresa H.V.T. Dias, A.D. Stauffer and C.A.N. Conde,to be published.

5 - M. Suzuki and S. Kubota,"Mechanism of Proportional Scintillation

in Argon, Krypton and Xenon", Nucl. Instr. and Meth.164,197(1979).

6 – D.F. Anderson, "The Prospects of Using Large Vacuum Photodiodes
    with Gas Scintillation Proportional Counters",IEEE Trans. Nucl.
    Sci. NS-27, No.1, 181 (1980).

7 – M.Alice F. Alves, A.J.P.L. Policarpo and M. Salete S.C.P. Leite,
    "The Energy Resolution and Window Area Capabilities of the Gas
    Proportional Scintillation Counter", IEEE Trans. Nucl. Sci.
    NS-22,No.1, 109 (1975).

8 – A.J.P.L. Policarpo, "The Gas Proportional Scintillation Counter",
    Space Sci. Instr. 3, 77 (1977).

9 – D.A. Goganov, N.I. Komyak, V.B.El'Kind and A.A. Shul'ts,
    "Construction and Characteristics of a Gas-filled Electro-
    luminescent Detector for Soft X-rays", Prib. Tekhn. Eksper.
    21, No.4, 72 (1978).

10– A.J.P.L. Policarpo, "Coupling of the Gas Scintillation Propor-
    tional Counter to Photoionization Detectors", Nucl.Instr. and
    Meth. 153, 389 (1978).

11– A.J.P.L. Policarpo, "The Ionization Scintillation Detectors",
    Proceed. of "1981 I.N.S. International Symposium on Nuclear
    Radiation Detectors", Tokyo, March 23-26,1981 (to be published).

12– W.H.-M. Ku and C.J. Hailey, "Properties of an Imaging Gas
    Scintillation Proportional Counter", IEEE Trans. Nucl. Sci.
    NS-28,No.1, 830 (1981).

13– A.J.Dabrowski, J.S. Iwanczyk, J.B. Barton,G.C. Huth,R.Whited,
    C. Ortale, T.E. Economov and A.L. Turkevich, "Performance of
    Room Temperature Mercuric Iodide (HgI$_2$) Detectors in the
    Ultralow-energy X-ray Region",IEEE Trans. Nucl. Sci. NS-28
    No.1, 536 (1981).

14– J.S.Iwanczyk, A.J. Dabrowski, G.C. Huth, A. Del Duca and W.
    Schepple, "A Study of Low-noise Preamplifier Systems for Use
    with Room Temperature Mercuric Iodide (HgI$_2$) X-Ray Detectors",
    IEEE Trans. Nucl. Sci. NS-28, No.1, 579 (1981).

15– T.T. Hamilton, C.J.Hailey, W.H.-M. Ku and R. Novick, "A High
    Resolution Gas Scintillation Proportional Counter for Studying
    Low Energy Cosmic X-ray Sources", IEEE Trans.Nucl. Sci. NS-27,
    No.1, 190 (1980).

PERFORMANCE CHARACTERISTICS OF A HIGH RESOLUTION Si(Li) DETECTOR

USING A TIME VARIANT AMPLIFIER AND A PULSED SOURCE OF X-RAYS

M. A. Short and T. G. Gleason

Occidental Research Corporation

Irvine, CA 92713

INTRODUCTION

Two recent improvements in energy dispersive spectrometry are the development of time variant amplifiers and the introduction of pulsed X-ray sources. Either pulsed X-ray tubes or electron beams with fast beam blanking may be used.

Time variant amplifiers designed by Kandiah (1, 2) used control logic to co-ordinate charge restoration of the detector preamplifier, pole-zero cancellation and baseline restoration. In the pulse shaping circuit, time constants are switched during the processing of each pulse to optimize throughput, baseline stability and pileup rejection. Pulse processing techniques have been discussed by Statham (3).

Pulsed X-ray excitation using a grid modulated X-ray tube was suggested by Jaklevic, Goulding and Landis (4). A busy pulse processor biases the grid to interrupt X-ray excitation and reduce pulse pile-up. Real time pulse throughput is thereby increased. Pulsed X-ray tube excitation is reviewed by Stewart, Zulliger and Drummond (5) and by Jaklevic, Landis and Goulding (6).

A fast beam blanking system was devised for the electron probe microanalyzer by Statham, Long, White and Kandiah (7). The principle of operation is entirely similar to that used for the pulsed X-ray tube. Good quantitative analyses were obtained by Statham et al. at substantially improved throughput rates. Fast beam blanking is available on commercial electron microprobes.

45

     To supplement literature results and information available
from equipment manufacturers, a study has been made of the performance characteristics of a high resolution lithium-drifted silicon
detector attached to an electron probe microanalyzer equipped with
fast beam blanking and using both time variant and Gaussian shaping
pulse processors. The detector area was 10 sq. mm.

INSTRUMENTAL

     The instrumentation used for this study consisted of a Cameca
MBX electron microprobe with a Cameca fast beam blanking attach-
ment. The Si(Li) detector, Gaussian and time variant pulse processors and associated counting equipment were supplied by Tracor
Xray. Measurements were made of resolution, throughput, peak
energy shift, pulse pile-up, deadtime and throughput linearity for
input count rates, as measured on the fast discriminator, ranging
from 1K to 50K cps. All measurements were made using $MnK\alpha$ radiation, except that additional pulse pile-up measurements were made
with $MgK\alpha$ radiation. Measurements were made for several settings
of the pulse processor pulse shaping time constants: for the time
variant processor 9.5, 19 and 38 μsec (times given are times to
peak) and for the Gaussian pulse processor 9, 36 and 72 μsec.
     The results were obtained using a Tracor multichannel analyzer
system. Long counting times, up to 2000 sec live time, were
employed so that counting statistics would not be a limiting
factor. Resolutions were determined at an MCA setting of 5 ev/per
channel from X-Y plots of the spectra. The Tracor software was
used to determine $MnK\alpha$ energy centroids.

RESULTS

     A selection of the results obtained is shown in Table 1. Note
that "MCA cps Clock Time" denotes the cps processed by the MCA in
real time, and that "% Lost Counts" denotes - (Fast Discriminator
cps - MCA cps Live Time) /Fast Discrim. cps. Dead time values in
parentheses indicate that these values were obtained from the
multichannel analyzer as installed by the manufacturer, that is,
without rewiring the dead time correction circuit to take the
requirements of fast beam blanking into account. Gaps in the table
indicate that no measurements were made for these instrumental
parameters. The $MnK\alpha$ peak energy shift measurements were made
relative to the peak energy at 1K cps on the fast discriminator.
Peak resolution and peak energy shift were repeatable to within 0.5
ev. Because of their relatively low intensities and their super-
position on high backgrounds, the pulse pile-up measurements were
less reproducible.

TABLE 1: RESULTS

Amplifier Time-To-Peak μsec	Fast Beam Blanking	Fast Discriminator cps	MCA cps Clock Time	%Dead Time	%Lost Counts	Resolution eV MnKα	Peak Energy Shift eV	%Pulse Pile-Up MnKα	% Pulse Pile-Up MgKα
Time-Variant 9 1/2	yes	1K	1 K	(6)		170	—	0	0.03
		10 K	10 K	(18)		171	3	0.20	0.95
		30 K	24 K	(41)		173	11	4.7	11
9 1/2	no	1 K	970	3		170	—	0	0
		10 K	6.8 K	32		171	2	0.16	0.7
		30 K	10.9 K	64		174	3	0.40	3.2
Time-Variant 19	yes	1 K	990	(3)	less than 1.0	151	—		
		15 K	14 K	(39)		152			
19	no	1 K	870	12		148.5	—		
		15 K	5.6K	64					
Time-Variant 38	yes	1 K	1 K	(3)		139.5	—	0.02	0.09
		5 K	5 K	(21)		141		0.08	0.56
		10 K	7.5 K	(50)		142.5	2	9	10
38	no	1 K	830	17		139.5	—	0	0.29
		5 K	2.5 K	50		141		0.01	0.61
		10 K	2.8 K	73		142	2	0.12	1.1
Gaussian 9	no	1 K	950	4	less than 0.8	186	—	0	0.1
		10 K	7.5 K	25		186	2	0.05	0.82
		30 K	12 K	61		189	3	0.14	2.82
Gaussian 36	no	1 K	860	13		146.5	—	0.03	0.25
		5 K	2.5 K	50		149	0	0.05	0.25
		10 K	3.2 K	69		143	-1		
Gaussian 72	no	1 K	780	23		147	—		
		5 K	1.5 K	70			0		

DISCUSSION

   The results show that improved detector resolution is obtained using the time variant pulse processor compared with that obtained using the Gaussian shaping pulse processor at the same throughput rate. They also show that the use of fast beam blanking on an electron microprobe gives greater throughput in real time compared with that obtained using continuous electron beam excitation. The peak energy shift for both pulse processors is generally between 0 and 3 ev at the maximum throughput rate and even at 25K cps throughput the shift is only one channel. There is excellent correspondence between processed counts/unit real time and fast discriminator counts/unit real time up to and beyond the maximum throughput rate. Pulse pile-up was generally low.

   The performance of an energy dispersive spectrometer is substantially improved using a time variant pulse processor and a pulsed source of X-rays.

REFERENCES

1. K. Kandiah, Nucl. Instrum. Meth., 95, 289 (1971).

2. K. Kandiah, in: Physical Aspects of Electron Microscopy and Microbeam Analysis, B. M. Siegel and D. R. Beaman, eds., John Wiley, New York, p. 393 (1975).

3. P. J. Statham, in: Energy Dispersive X-Ray Spectrometry, NBS Special Publication 604, K. F. J. Heinrich, D. E. Newbury, R. L. Myklebust and C. E. Fiori, eds., p. 141 (1981).

4. J. M. Jaklevic, F. S. Goulding and D. A. Landis, IEEE Trans. Nucl. Sci. NS-19, 392 (1972).

5. J. E. Stewart, H. R. Zulliger and W. E. Drummond, in: Advances in X-Ray Analysis, R. W. Gould et al., eds., Kendall/Hunt, Dubuque, Iowa, 19, 153 (1976).

6. J. M. Jaklevic, D. A. Landis and F. S. Goulding, in: Advances in X-Ray Analysis, R. W. Gould et al., eds., Kendall Hunt, Dubuque, Iowa 19, 253 (1976).

7. P. J. Statham, J. V. P. Long, G. White and K. Kandiah, X-Ray Spectrom., 3, 153 (1974).

X-RAY TUBES FOR ENERGY DISPERSIVE XRF SPECTROMETRY

Brian Skillicorn

Kevex Corporation, X-ray Tube Division
P.O. Box 66860
Scotts Valley, California 95066

INTRODUCTION

X-ray tubes are used extensively in energy dispersive x-ray spectrometers in preference to radioactive sources because of their high output, variable intensity and tuneable energy spectrum. Tubes for this application have certain requirements which make them different in detail from the more common radiographic or therapy tubes and this paper sets out to present a review of some of the fundamental features of modern spectrometry grade tubes. The salient features of these tubes include:

(1) Stable output flux intensity over long periods of time.
(2) Known x-ray energy spectral characteristics.
(3) Low maximum operating voltage (<75kv).
(4) Low input power (<200 watts).
(5) Operable with full current over a wide range of voltages, typically from 4kv to the maximum tube rating.
(6) X-ray source capable of being placed very close to the specimen.

PRINCIPLES OF OPERATION OF X-RAY TUBES

An x-ray tube, Fig. 1, consists of a vacuum vessel, or envelope, containing an electron gun, a target and an x-ray window. A high voltage, positive with respect to the cathode, is applied to the target, or anode. The vacuum envelope is frequently made out of glass with provisions for making connection to the internal electrodes and for insulating the high voltage. Electrons emitted from the gun are accelerated towards the target where x-rays are produced by interaction between the electrons and the target material.

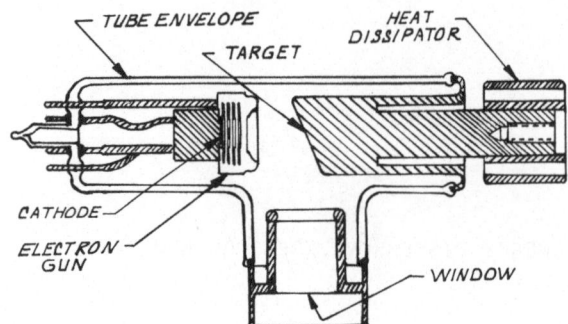

Fig. 1   50kv 2mA x-ray tube

Two types of radiation are produced at the target as shown in Fig. 2:
1. The continuous spectrum or bremsstrahlung which ranges in energy from zero to a value expressed in electron volts which is numerically equal to the voltage applied to the tube.
2. The monoenergetic rays characteristic of the elements in the target material.
The continuous spectrum is always present but any characteristic line will only be superimposed if the tube potential is higher than the energy of that particular line. The total intensity of the continuum x-rays is given by a formula published by Beatty[1].

$$I = KZE^2i$$

where

I = intensity (watts)	i = tube current (amps)
E = tube potential (volts)	Z = atomic number of target

K is a constant which has been measured by many investigators with widely differing values, but 1.3×10^{-9} per volt represents a reasonable consensus. Since the product (Ei) is the electrical input power to the tube, the overall conversion efficiency is (KZE). A typical spectrometry tube operating at 60kv with a rhodium target would therefore have a conversion efficiency of 0.35%. This value is only approximate and the conversion efficiency measured for the x-rays available through the window is much lower when the solid angle of the cone formed by the x-ray source and the window diameter is taken into account.
Intensity of the characteristic lines is given by:

$$I = Ai \ (U-1)^n \qquad U = E/E_q$$

where

E = tube voltage	i = tube current
Eq = energy of characteristic line	A = constant

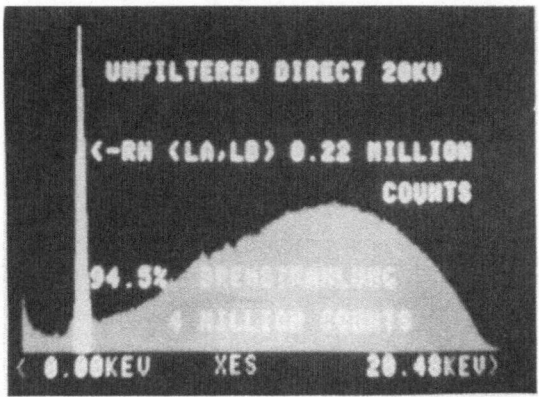

Fig. 2. X-ray spectrum from rhodium target tube operating at 20kv
        showing characteristic rhodium L lines superimposed on the
        bremsstrahlung.

    The value of the exponent n varies with the line and with the
value of U but for the k series, a value of 1.63 applies up to
U = 10.[2] For a given value of U, I is approximately proportional to
the atomic number of the target material.

TARGET DESIGN

    The low efficiency of x-ray production requires that the target
be capable of dissipating practically all of the power in the incident
electron beam therefore means must be provided for removing the heat
so generated. Modern tubes rely on conduction, convection, forced
liquid circulation, boiling liquid, radiation or any combination of
these.
    There are two important classes of targets:

    1. Transmission target. The target material thickness is
slightly greater than the range of the electrons at the highest
operating voltage. The target is then too fragile to be self
supporting so it is deposited on a beryllium foil substrate. X-rays
generated in the forward direction pass through the target and
beryllium substrate to the specimen. A transmission target tube is
shown in Fig. 3. In some applications target cooling is primarily
by convection but if the specimen chamber is evacuated the target
heat is removed by conduction through the beryllium foil and to a
lesser extent by radiation. The Fig. 3 tube is efficient when
operated near its maximum voltage rating because the intensity of
x-rays generated by a target of this type has a maximum value forward

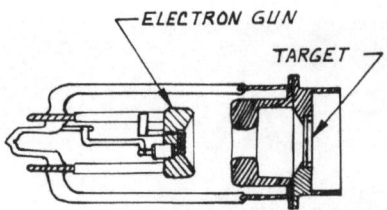

Fig. 3.   30kv 0.3mA transmission target tube

of the target and the target can be positioned very close to the
specimen without the need for an additional window. An undesirable
feature of transmission target tubes is the fall off in intensity as
the tube voltage is lowered and the x-rays are generated nearer to
the target surface so that they are attenuated as they pass through
the remainder of the target and the beryllium foil. The low end of
the output energy spectrum is more strongly attenuated in comparison
with higher energies consequently the shape of the spectral curve
changes. At voltages where the electron range is small compared with
the target thickness, the x-ray intensity varies as the ninth or
tenth power of the voltage instead of the normal square law dependency.
Transmission target tubes are limited in their power capabilities to
about 100 watts dissipation because of the difficulty of cooling the
target foil.

      2. Solid target. The vast majority of spectrometry tubes use
the x-rays generated at the surface of a target which is thick in
comparison with the range of the incident electrons and the tube
shown in Fig. 1 is an example of this type.

      The target material is brazed to a copper support structure
which conducts the target heat to a dissipator. The tube is mounted
in a liquid filled housing which provides electrical insulation and
also cools the heat dissipator. At low power levels the cooling is
by conduction through the liquid but at higher power levels the
liquid boils in the vicinity of the dissipator which therefore does
not rise above the boiling point of the liquid over a wide range of
values of target power dissipation. The vapour bubbles condense
back to liquid and the heat is removed by convection cooling the
outside of the tube package.

      Fig. 4 shows a tube designed to make maximum use of the x-ray
flux generated at the target by positioning it very near to the
window. Close coupling between target and specimen is possible as
a consequence and the inverse square law intensity fall-off with
distance to the specimen is minimized. This tube is particularly
suitable for use with spectrometers using secondary target excitation
where close proximity of the primary and secondary targets is desired

to maximize the intensity of the characteristic radiation from the
secondary target. The target is water cooled and mounted in close
proximity to the window at the end of a drift tube which can be
inserted into the specimen chamber through a hole with an o-ring seal.

The choice of target material depends on the intended application
of the tube. For maximum intensity of the continuum radiation high
atomic number materials such as gold, platinum, rhenium, tungsten or
tantalum are used. For specific characteristic radiation silver,
palladium, rhodium, molybdenum, copper, nickel, cobalt, iron or
chromium have been used. Compatibility with normal tube construction
techniques and materials plus a low vapour pressure at target operating and fabricating temperatures are essential criteria used to
judge the suitability of any given material. Relatively thick target
inserts are used instead of plated coatings to avoid contamination by
fluorescent x-rays from the copper support structure.

Tubes necessarily use many different materials in their construction and to maintain spectral purity of the x-ray output the window/
target geometry must be carefully designed otherwise contamination by
the characteristic lines of these materials may result when they are
excited either directly by scattered electrons or by the x-rays
generated at the target. Internal shields made out of the same
material as the target are sometimes necessary to avoid such contamination.

ELECTRON GUN

All modern tubes use an independent electron gun as introduced
by Coolidge in 1913.[3] Fig. 4 is an example of such a tube in which a
hot tungsten filament produces thermionic electrons. The filament is
wound in a spiral and mounted in an electrode shaped to give an
electric field configuration which will focus the electrons on to the
target in an image of the filament. Because the filament spiral is
long in comparison with its diameter the source of x-rays will be
rectangular with the long dimension usually three or four times the
short dimension. The angle of the target surface to the electron
beam axis is chosen so that the projected image of the x-ray source
as seen from the window appears approximately square. Filament
temperature and therefore tube current is controlled by passing a
current through the filament from an independent power supply usually
connected in a feedback loop which keeps the electron beam current
constant. The filament operating temperature is low to ensure long
life and to prevent evaporation of the tungsten which could
contaminate the target material.

In another type of gun commonly used in spectrometry tubes,
Fig. 1, electron emission is from a space charge limited indirectly
heated cathode with current control by adjustment of the voltage

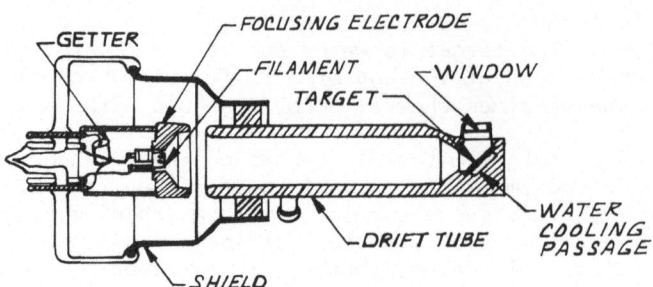

Fig. 4.   60kv watercooled tube

applied to an adjacent control electrode. This type of cathode can
be poisoned by the backstreaming positive ions which always exist due
to residual gas in the tube envelope and to avoid poisoning, an annular cathode is used to form a hollow cylindrical electron beam. The
positive ion beam then passes down the centre of this hollow cylinder
and is intercepted by a metal disc placed in the centre of the
cathode. The x-ray source in tubes of this type is the ring shaped
projection to the hollow electron beam on the target. These tubes
are used extensively in pulsed systems where the electron beam is
gated on and off in response to a voltage applied to the control
electrode. The gating voltage is derived from the x-ray detector so
that x-rays are produced only when the detector is in a receptive
condition.

X-RAY WINDOW

     The window is usually made out of beryllium foil chosen because
it is a metal which can be brazed to give a true hermetic seal, will
withstand tube processing temperatures, has a low x-ray absorbtion
coefficient and has characteristic lines at energies too low to
interfere with measurements. Fig. 5 shows the transmission through
beryllium foils of various thicknesses as a function of x-ray energy.
Problems with fragility and porosity of thin foils make it difficult
to construct tubes reliably with window thinner than 0.003 inches
(0.076mm) consequently x-rays with energies below 3KeV are heavily
attenuated. Most spectrometry tubes in use have 0.005 inch (0.127mm)
thick windows this representing a compromise between the conflicting
requirements of low energy transmission and manufacturing yield. High
power tubes, particularly those with high atomic number target materials sometimes require the use of thicker window foils to facilitate
removal of heat caused by bombardment with electrons scattered from
the target.

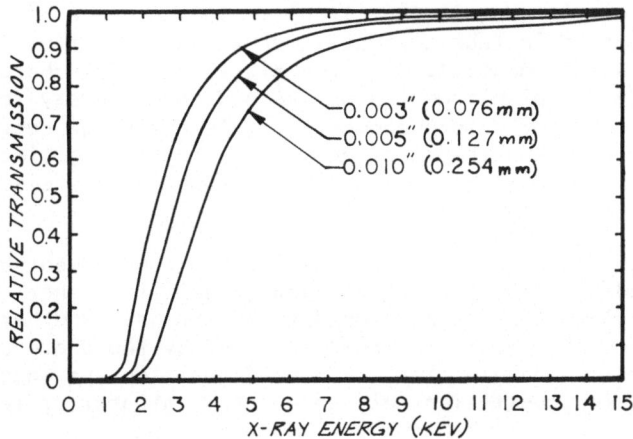

Fig. 5. X-ray transmission through beryllium foil

OUTPUT STABILITY

Stability of the x-ray flux intensity depends upon several factors. The two most obious relate to the stability of the target voltage and the electron beam current as controlled by the performance of the power supply. Modern semi-conductor power supplies operating with high gain feedback loops are capable of keeping tube voltage and current constant to with $\pm$ 0.01% of the nominal operating values.

A major cause of flux intensity variation is the unwanted acquisition of electrical charge by insulating surfaces in the tube envelope. There are several mechanisms by which this can happen:-

1. X-rays generated at the target irradiate the glass walls of the vacuum vessel generating photo-electrons which travel towards the anode structure leaving the glass with a positive charge which leaks away due to conduction in the glass. The rate of charge leakage is a function of the glass electrical conductivity which in turn is a function of temperature and x-ray flux intensity. If the charge is created at a rate greater than that at which it can leak it will build up until the electric gradient along the glass exceeds its breakdown strength causing a local discharge as the accumulated charge flows to the electron gun connections along the surface of the glass. This charge/discharge cycle can continue indefinitely and the discharge path in the glass becomes visible as characteristic dendritic track marks etched in the glass which eventually may perforate the vacuum envelope. In many systems the discharge causes a current to flow in

electron beam current regulating circuits causing these to call for a
momentary reduction in tube current so that the x-ray flux intensity
fluctuates at a rate controlled by the glass charge/discharge cycle.
Repetition rates ranging from ten per second to one every few hours
have been observed for this phenomenon. This discharge current can be
prevented from reaching the electron current regulating circuits by
constructing the tube with a guard ring surrounding the electron gun
connections.

2. The tube envelope is also subjected to bombardment by electrons scattered back from the target and those with higher energies
may reach the glass generating secondary electrons. Whether this
effect increases or reduces the charge caused by the photo-emission
process depends on the electric field configuration and whether the
secondary emission coefficient of the glass is greater or less than
unity.

3. The vacuum envelope is usually not symmetrically disposed with
respect to the electron beam axis and also the glass is not homogeneous in its electrical properties so that charge build-up is not uniform about the beam axis. This asymmetrical charge configuration will
cause the electron beam to deflect so that the position of the x-ray
source moves with the charge/discharge cycle. In equipment using an
x-ray tube in conjunction with a small diameter collimator, this position modulation will cause an apparent variation in x-ray flux intensity if the source moves sufficiently far out of the line of sight of
the collimator.

4. Source position movement resulting from glass charging is also
manifest after a period of operation at high voltage when the glass ac-
quires a charge corresponding to operation at that voltage level.
When the tube voltage is lowered, the charge does not leak away instan-
taneously so that the lower energy electron beam may be deflected by a
significant amount. The beam will drift back to its correct position
over a period of time ranging from minutes to hours.

From the preceding discussion on glass charging it is obvious
that the phenomena involved are impossible to quantify and the tube
designer must use techniques to minimize their effect on tube performance. The tube shown in Fig. 4 incorporates several features
specifically aimed at minimizing glass charging. The glass is
shielded from primary x-rays emitted at the target by the electron
gun assembly while fluorescent x-rays emitted from the gun assembly
cannot reach the glass. Scattered electrons from the target are
captured in the metal drift tube and electrons produced by field
emission in the high gradient electron-gun/anode gap are captured by
the metal shield. This tube is completely free from any of the
previously mentioned effects associated with charging of the glass
and has demonstrated very high stability of x-ray output even when
cycled rapidly from 60kv down to 5kv.

TUBE PROCESSING

Every effort is made during the various stages of tube manufacture to ensure cleanliness of those parts exposed to the internal vacuum. Final pump-out is done at an elevated temperature to outgas the internal components as completely as possible. While still attached to the vacuum pump, the tube is then conditioned up to a voltage and current somewhat in excess of the maximum ratings. The mechanism of the conditioning process is not completely understood but the tube will arc between the cathode and anode structures as the voltage is slowly increased.[4] Discharges seem to be necessary in order for the tube to be made capable of operating at full voltage. Two possible explanations are:

1. The high electric field existing at the edge of sharp microscopic protrusions from the electron gun structure produces field emission currents which heat the protrusions and vapourize the material. Similarly electrons striking the anode tend to hit the ends of protrusions on the anode assembly. In each case local outgassing occurs in addition to the material vapourization.

2. Electrostatic forces tear away loose clumps of material from the electrode surface which are accelerated towards the opposite electrode where they impact and cause heating and consequent vapourization. Outgassing can also occur both at the clump source and point of impact.

Neither of these mechanisms explain the slow loss of withstand voltage which occurs when a tube is not operated for a period of time. One suggestion is that the electrode surfaces adsorb residual gas from within the tube during idle periods and this gas is evolved when the tube is operated again. The tube must then be put through another conditioning process to re-distribute the evolved gas so that it can be adsorbed by other surfaces within the envelope or by the getter.

REFERENCES

1. R. T. Beatty, Proc. Roy. Soc. 89, 314 (1913).
2. R. L. Myklebust, D. E. Newbury and H. Yakowitz. NBS Publication 460, 116 (1976).
3. W. D. Coolidge, Phys. Rev. 2, 409 (1913).
4. R. Hawley, Vacuum, Vol. 10, No. 4. 310-318, Pergamon Press Ltd. (1961).

TOROIDAL MONOCHROMATORS IN HYBRID XRF SYSTEM

IMPROVE EFFECTIVENESS OF EDXRF TENFOLD

Thomas C. Furnas, Jr.
Molecular Data Corporation
2869 Scarborough Road, Cleveland, Ohio 44118

Gordon S. Kuntz
Sherwin-Williams Company
P. O. Box 6027, Cleveland, Ohio 44101

Richard E. Furnas
Cornell University, Dept. of Mathematics
Ithaca, New York 14850

INTRODUCTION

Our objectives in instrument design & development are to improve the QUALITY OF THE DATA which can be obtained and presented to a computer system for analysis and to explore the corresponding choices which such improved data quality allows.

In usual EDXRF systems, the detector views the entire scattered and fluorescence radiation from a sample illuminated either by a direct beam or by a secondary target. In either case, the beam may be filtered before it strikes the specimen. Most efforts to monochromatize the exciting radiation or to reduce the background under the fluorescent peaks from the sample result in (a) decreases in X-ray flux available for excitation, (b) significant increases in power expended in the X-ray tube, or (c) extremely close coupling of the system components. These are NOT THE ONLY CHOICES AVAILABLE.

A new HYBRID XRF system[1] (Fig. 1) which includes both energy and wavelength dispersive components (Fig. 2) has yielded an improvement (Fig. 3 and Table I) of at least ten fold in the effectiveness ratio over the usual EDXRF systems described above. Along with providing higher monochromatic flux for FUNDAMENTAL PARAMETERS TECHNIQUES[2,3], the HYBRID XRF system offers unique characteristics and advantages in situations requiring very high accuracy or very low detection limits in reasonable data acquisition times.

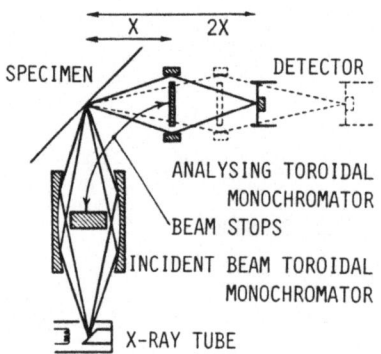

FIGURE 1. Geometry of HYBRID XRF SPECTROMETER. Note that the beam stops play a critical role in achieving the desired performance.

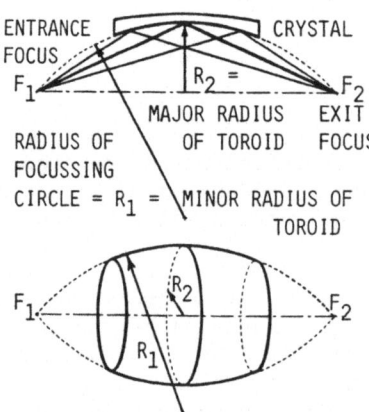

FIGURE 2. (a) Basic geometry of focussing X-ray diffraction optics showing relationship to a toroidal arrangement for increasing the efficiency of flux transfer from entrance to exit focus. (b) A toroid of overlapped circles ($R_1 > R_2$) yields the barrel shape of the final toroidal arrangement.

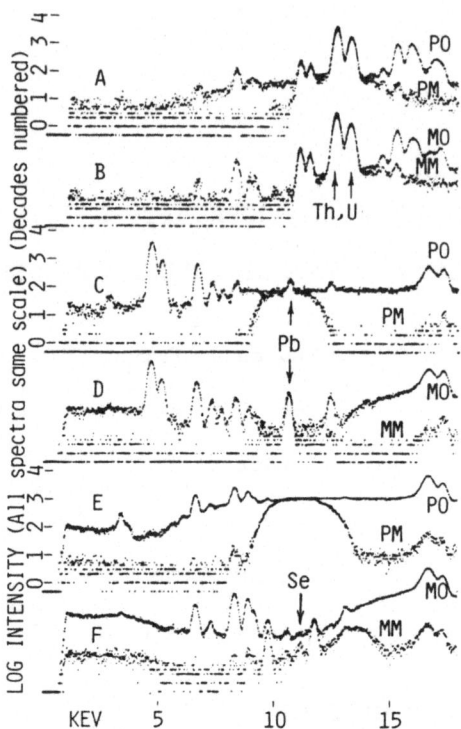

FIGURE 3. Typical spectra illustrating the four different operational modes of the HYBRID XRF SPECTROMETER. PO = direct illumination of the specimen by the X-ray tube target and specimen viewed by an open detector. PM = direct illumination of the specimen viewed with an analysing monochromator in front of the detector. MO = incident beam monochromator to illuminate the specimen which is viewed by an open detector. MM = incident beam and analysing monochromators both were used. Curves A and B are a thorium–uranium mixture to simulate fuel pellets. Curves C and D are a paint containing ≈600 ppm of lead. Curves E and F are SRM 1577 NBS Bovine Liver which contains 1.1 ppm selenium.

TABLE I.   A COMPARISON OF TYPICAL DATA

INSTRUMENT	MDC HYBRID XRF SPECTROMETER				KEVEX 810	SPECTRACE 440
CONFIGURATION	PO	PM	MO	MM	AgMoMoO	KMoO

A 50 sec. MAJOR COMPONENT ANALYSIS: Thorium plus Uranium pellet

	PO	PM	MO	MM	AgMoMoO	KMoO
KV x MA	30x2	30x2	30x2	30x2	30x5	30x0.2
Det. to Sample	217 mm	217 mm	217 mm	217 mm	39 mm	31 mm
cps (collected)	4112	3272	4989	4495	4379	3473
Peak Ht. (Th)	4286	6073	6272	8740	12531	12791
Peak Ht. (U)	2016	2398	2741	3367	5299	4465
P/B (Ht) (Th)	71	99	156	238	266	229
P/B (Ht) (U)	33	39	68	92	112	87
Integ. net (Th)	72351	101765	106201	147525	296152	71188
Integ. net (U)	33352	40452	46464	56591	134772	27811
MDL ppm (Th)	616	442	343	236	247	390
MDL ppm (U)	496	413	291	228	203	365
Eff. ROI	1.16	8.91	1.66	11.55	1.52	1.39
Eff. net	1.05	6.55	1.57	9.71	1.46	1.32
Fig. 3 Spectra	A:PO	A:PM	B:MO	B:MM	--	--

A 100 sec. MINOR COMPONENT ANALYSIS: Lead (≈600 ppm) in PAINT

	PO	PM	MO	MM	AgMoMoO	KMoO
KV x MA	38x2	38x2	38x2	38x2	35x10	38x1
Det. to Sample	176 mm	176 mm	176 mm	176 mm	39 mm	31 mm
cps (collected)	1519	115	1213	38	4625	8165
Peak Ht.	183	168	134	144	459	739
P/B (Ht)	3.1	2.6	17	>30	7	16
Integ. net	1576	1174	1771	2082	4495	5709
MDL ppm	60	66	19	5.8	20	12
Eff. ROI	0.029	0.64	0.018	1.28	0.016	0.009
Eff. net	0.010	0.15	0.015	1.22	0.010	0.007
Fig. 3 Spectra	C:PO	C:PM	D:MO	D:MM	--	--

A 1000 sec. TRACE ANALYSIS: Se (1.1 ppm) NBS Bovine Liver SRM 1577

	PO	PM	MO	MM	AgMoMoO	KMoO
KV x MA	38x2	38x2	38x2	38x2	30x5	30x0.25
Det. to Sample	187 mm	187 mm	187 mm	187 mm	39 mm	31 mm
cps (collected)	972	173	1116	21	670	4086
Peak Ht.	1023	1075	58	27	22	1237
P/B (Ht)	1.1	**	1.4	5	**	**
Integ. net	525	**	212	288	**	**
MDL ppm	1.1	**	0.6	0.15	**	**
Eff. ROI	0.03	0.24	0.0015	0.023	--	0.008
Eff. net	0.0005	**	0.0002	0.014	**	**
Fig. 3 Spectra	E:PO	E:PM	F:MO	F:MM	--	--

** = calculated MDL greater than 1.1 ppm present or net integral was negative. -- = complete data not available. Eff. = Effectiveness = ratio of counts net or in ROI divided by all other counts collected by the detector.

PERFORMANCE

The central theme of the HYBRID XRF SPECTROMETER is to collect a large solid angle of radiation, monochromatize it by diffraction from a suitable crystal and bring it to focus at a desired location. This is done by using toroidal monochromators which collect a hollow cone of radiation and bring it to focus further along its axis. It provides both good monochromatization and an INCREASE in FLUX at the focus over that obtainable by removing the monochromator assembly entirely. The arrangement even "works better" for small sources and small foci so the illumination of small specimen with high flux mono-chromatic radiation is almost an inherent feature of the input por-tion of the HYBRID XRF system.

To complete the system, a second toroidal monochromator is placed between the specimen and the detector. The beam stop inter-cepts direct view of the specimen so only the band pass of X-rays diffracted by the toroidal monochromator crystal can reach the de-tector. This band pass is determined by special design of the mono-chromator and/or is readily varied by 2:1 translation of the detector and analysing monochromator.

Some comparisons using major, minor and trace constituent sam-ples were run on two standard commercial instruments (a KEVEX 0810 at Mogul Corp. in Chagrin Falls, Ohio, and a SPECTRACE 440 at Union Carbide Parma Technical Center, Parma, Ohio) and on the HYBRID XRF SPECTROMETER (built by Molecular Data Corporation with UCAR highly oriented graphite in the monochromators and located at Cleveland State University) in each of its four different modes of operation. Typical spectra are shown in Figure 3 and data regarding those par-ticular spectra are given in Table I.

BIBLIOGRAPHY

1. T. C. Furnas, Jr., M. C. Lambert, and R. E. Furnas, "Use of toroidal monochromators in a hybrid XRF system to obtain increased effectiveness ratios", Transactions of Fifth Symposium on X- and Gamma Ray Sources and Appli-cations", in special issue of Nuclear Instruments and Methods (in press) (1982) (Includes an additional 25 references).
2. G. S. Kuntz and R. L. R. Towns, "Determination of lead in paint by EDXRF spectrometry", to be published.
3. Gordon S. Kuntz, Ph. D. Thesis, Cleveland State University, Cleveland, Ohio (1981).

THE USE OF POLARIZED X-RAYS FOR IMPROVED DETECTION LIMITS IN ENERGY DISPERSIVE X-RAY SPECTROMETRY*

Richard W. Ryon

Lawrence Livermore National Laboratory
P.O. Box 808, L-310
Livermore, California 94550

John D. Zahrt

University of Northern Arizona
Flagstaff, Arizona

Peter Wobrauschek and Hannes Aiginger

Atomic Institute of the Austrian Universities
Vienna, Austria

ABSTRACT

The use of polarized x-rays to excite fluorescence spectra with decreased backgrounds and improved detection limits is reaching a mature state of development. With bulk, low-Z specimens, polarized x-ray sources have produced detection limits which are ∿1 to 3 times lower than are obtained with the best unpolarized photon sources. Based upon experience and the known properties of larger solid angle geometries, further significant to dramatic improvements are anticipated.

*Work performed under the auspices of the U.S. Department of Energy by the Lawrence Livermore National Laboratory under contract number W-7405-ENG-48.

INTRODUCTION

   While x-ray fluorescence is a powerful analytical tool, it
does have limited sensitivity. In general, x-ray fluorescence
has detection limits of about 1 µg/g for trace elements in bulk
low-Z specimens, and about 10 ng/cm^2 for thin films. In order
to extend the range of applications of photon-excited x-ray
fluorescence, electronic and geometric approaches have been
suggested, investigated, or developed.

   The subject of this paper is limited to a geometric improvement, namely, polarized x-rays.[1,2] The use of polarized x-rays
applied to energy dispersive x-ray fluorescence was first
discussed in 1973.[3] Continued progress and additional
applications have been reported up to the present time.[1-12]

   The basic geometry for a polarized x-ray spectrometer is
shown in Figure 1. When the beams are highly collimated, there is
virtually no background due to source radiation when the fluorescence spectra are observed in the direction perpendicular to
the initial scatter plane (that is, at x). This polarization
phenomenon has been known ever since the classic experiments of
Barkla in 1906.[13] X-ray polarization is a general phenomenon,
and is observed in such interactions as Compton, Bragg, and Raman
scattering, and in Borrmann transmission. Some of the possible
phenomena have been suggested as sources of polarized x-rays for
use in spectrometry.[14] Very intense and highly polarized
synchrotron radiation has been used for those applications
requiring very high sensitivity.[15] Polarized radiation from
exotic sources such as channeling radiation[16] or the elusive
x-ray lasers[17-19] may one day be available in the laboratory.
Of the possible sources of polarized x-rays, Bragg and Barkla
scattering have been developed to a high degree.

   For the energy region being discussed here (<30 KeV),
relativistic effects can be ignored. (At high energies, the more
exact Klein-Nishina formulation should be used.[20]) The
approximate cross section for scattering unpolarized x-rays
follows a cosine^2 function of the scattering angle. Upon
scattering once, one of the electric vectors in Figure 1 is
eliminated; when scattered twice, the other vector is eliminated,
and a dramatic reduction in background takes place. The
reduction of scattered radiation is beneficial because the
signal-to-noise ratio is improved and count-rate limitations are
overcome.

   There is always some divergence about the angles in Figure 1,
and scattering will occur at all angles other than 90^0. The
result is that some source radiation will be detected. In the

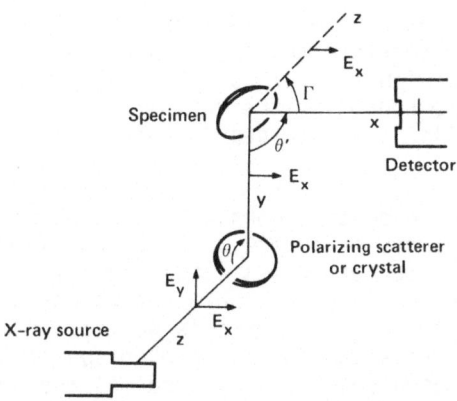

Figure 1. Sketch of system geometry. After scattering twice, the
           scatter intensity measured along the X-axis is much
           less than that measured along the Z-axis

limit of very highly collimated beams, the polarization would
approach 100%, the detected source radiation would approach zero,
and the spectrum would consist only of fluorescent peaks (if one
ignores the contributions from multiple scatter, bremsstrahlung
production, inelastic resonance scattering[15] in the specimen,
and detector background). Unfortunately, the geometric
attenuation factor in this case is very large and very long
counting times would be required*, and the detector does not
ignore the aforementioned contributions.

It has been demonstrated that there is a compromise between
the source intensity and low background which gives optimum
performance.[6] With Barkla (amorphous) polarizers, and with
each beam defined by a single tubular collimator, the minimum
detection limits are given by the approximate expression

$$MDL \propto \bar{\omega}^{-2},$$

where $\bar{\omega}$ is the average half-range of the angles θ, θ', and Γ
about $\pi/2$ (Figure 1). The lowest detection limits are obtained
by using the orthogonal geometry, with inter-component distances
minimized (<2.5 cm), and with collimator apertures opened to
give maximum counting rates (even though the polarization
$\approx 1-2\bar{\omega}^2$ is degraded).

---

*That is why peak-to-background ratios are misleading. Minimum
 detection limits include time in the definition, and are, therefore a better measure of performance.

With Bragg (crystalline) polarizers, the functional
relationship for the detection limits is different because of the
concentrated, directional nature of the scattering, and because
of the line-focus produced by the crystal. Even so, the best
detection limits also are obtained in this case with open
collimators.

There are some geometric considerations which are peculiar
to the Bragg polarizers (Figure 2). Since no significant
radiation is diffracted at angles larger than the rocking angle
of the crystal, any emission on the anode which is outside the
area subtended by the rocking curve is wasted. Fine focus
diffraction x-ray tubes can sustain higher specific power loads
than diffusely focused spectroscopy tubes so the former will
produce higher brightness at the specimen.

When Barkla polarizers are used, a much larger anode area
may be intercepted by the collimator, so a spectroscopy x-ray tube
would be the best choice. While the specific power is lower, the
usable integrated photon flux becomes about the same.

The same considerations apply when comparing flat crystals
with amorphous materials. The greater reflectivity per unit solid
angle of a crystal is off-set to some degree by the larger area
which scatters from the amorphous material.

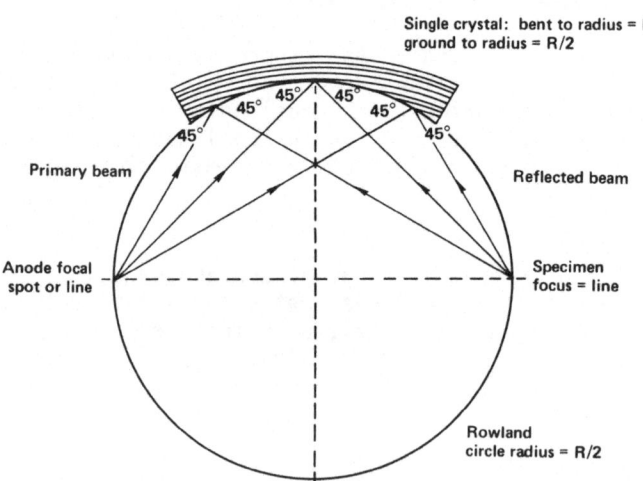

Figure 2.   Johansson (or Johann) crystal polarizer shown in the
             YZ plane. For a flat crystal, only the central ray is
             diffracted.

COMPARISON OF EXCITATION METHODS

In the comparisons which follow, measurements were made using the same experimental apparatus. Only the secondary source was changed to produce Bragg polarization, Barkla polarization, or secondary fluorescence. Thus, while the absolute counting rates may not be the best that can be achieved, the data obtained do serve to demonstrate the relative effectiveness of the methods of excitation.

Table 1 lists the characteristics of the three types of excitation being compared in this report. As may be surmised, there is no clear-cut best method. The choice is determined by the analytical application. If an existing apparatus is to be used, the choice may depend upon the x-ray tube (focus, power, anode), the collimation (if not variable), and the steps taken to reduce multiple scatter.

Bragg polarizers are best for lower energy radiation, such as $CrK\alpha$ and $CuK\alpha$, where suitable crystal planes can be found. This behavior is shown in Table 2. A copper anode fine focus x-ray tube was used in an apparatus which has long inter-component distances. The apparatus was designed for studying the polarization phenomenon, rather than being optimized for analyses. The intensities are, therefore, low. In the comparison, the Bragg-polarized $CuK\alpha$ yields higher counting rates and gives detection limits ~ 1.5 times lower for K and Ca in the NBS Orchard Leaves sample than are obtained with $FeK\alpha$ secondary radiation. The Barkla B_4C scattering source gives very poor results in this case because of the very restrictive geometry.

A close-coupled system was used to obtain the data in Table 3. The close coupling as well as the change from $CuK\alpha$ to $CrK\alpha$ primary radiation yields higher counting rates and lower detection limits. The relative performance of the three methods of excitation narrows and even crosses, as shown in Figures 3 and 4. As the collimator aperatures are increased, excitation intensity from the crystal increases less rapidly and the secondary target becomes as good as, or better than, the Bragg-polarized source.

At higher energies, Barkla polarizers are favored over Bragg-polarizers because of the lack of good crystal reflections. In fact, if a narrow range of elements is to be measured, a secondary fluorescer with its nearly monochromatic x-rays is preferred. The data in Table 4 were taken in Livermore. These data were taken with a Mo anode x-ray tube and boron carbide polarizer. It is seen that an extended range of elements can be measured simultaneously with good sensitivity using a Barkla polarizer and bremsstrahlung radiation.

Table 1. Characteristics of Excitation Methods

	Secondary Fluorescers	Bragg-Polarized	Barkla-Polarized
Scatter Efficiency per unit solid angle	Moderate	Moderate to High	Low
Intensity increase with solid angle	High	Low	High
Monochromatic Source	Shifted by fluorescence	No change	Compton-shifted (Broad band at wide solid angle)
Polychromatic Source	Nearly Monochromatic	Monochromatic	Polychromatic
Self collimating	No	Yes	No
Useful energy ranges	Useful for most elements	Cannot use for high energies due to non-availability of crystal d-spacings $(d=n\lambda/\sqrt{2})$	Less useful at low energies due to absorption $(\mu_{scatter}/\mu_{total})$
Multi-element capability in single spectrum	Narrow	Narrow	Broad

Table 2. Comparison Between Barkla-Polarized, Bragg-Polarized, and Secondary Target Excitation*

SPECIMEN: NBS ORCHARD LEAVES

ALL : 40 kV, 36 mA

BRAGG-POLARIZED: Cu anode tube (fine focus); Cu(113) crystal

BARKLA-POLARIZED: Cu anode tube (fine focus); B_4C scatterer

SECONDARY TARGET: Cu anode tube (fine focus); Fe sec. target

	Bragg-Polarized		Sec. Target		Barkla-Polarized	
	c/s**	MDL(3σ), ppm	c/s	MDL(3σ), ppm	c/s	MDL(3σ), ppm
K	16	47	11	65	.95	800
Ca	40	29	26	42	2.4	500
Mn	1.5	2.6			.046	90
Fe	7.4	1.9			.53	26
xyz Geom. Target Kα	12.2		5.0		.38	

*Long-collimator system, Atomic Institute, Vienna
**c/s = counts/sec.

$$MDL = 3 \sqrt{\frac{Background\ counting\ rate}{1000}} \times \frac{\mu g/g}{c/s}$$

MDL = Minimum detection limit in ppm with a 1000 second counting time. 3σ means signal has statistical significance 3 times that of background.

FUTURE DEVELOPMENTS TOWARD OPTIMIZATION

There is considerable room for improvement of the Bragg-polarizers. Foremost among these improvements is the use of curved crystals.[8] The aperture at the crystal is essentially the rocking angle, which is usually on the order of 0.2°. If the crystal is bent rather than flat (Figure 2), the angular acceptance can reach about 10°. The increase in intensity would therefore be ~10/.2 = 40-50 times. A further factor of up to 16 can be obtained using a doubly curved toroidal crystal.[21] We may, therefore, anticipate further reductions in detection limits by a factor of more than 10X.

Table 3.   Comparison Between Barkla-Polarized, Bragg-Polarized,
           and Secondary Target Excitation*

SPECIMEN: NBS ORCHARD LEAVES

ALL   :   25 kV, 70 mA

BARKLA-POLARIZED: Cr anode tube, B_4C Scatterer

BRAGG-POLARIZED :   Cr anode tube; Ta(002) crystal

SECONDARY TARGET: Cr anode tube; Ti sec. target

	Bragg		Sec. Target		Barkla	
	c/s	1000 sec MDL(3σ), ppm	c/s	1000 sec MDL(3σ), ppm	c/s	1000 sec MDL(3σ), ppm
S	6.1	120	26	50	2.5	170
Cl	4.4	50	21	18	2.0	65
K	528	7.3	1960	6.2	183	14
Ca	955	5.4	3380	5.8	309	11
xyz Geom. Total	5980	-	13,600	-	1810	-

*Close coupled system, coarse collimators, Vienna

    Experiments are underway in Vienna* using a bent copper
crystal to diffract CuKα radiation. While the dramatic increase
in intensity anticipated has not yet been observed, the detection
limit for Mn in solution is 0.3 ppm and the minimum measurable
quantity from an evaporated solution is 0.1×10^{-9} grams.

    There are other refinements to the geometry of the apparatus
which can be made by carefully considering the effects of various
arrangements of the components.[4,8]

---

*J. Laurensen, B. Schmidt Nielsen and K. Maach Bisgård at the
 Royal Veterinary and Agricultural University in Copenhagen,
 Denmark are also doing similar experiments.

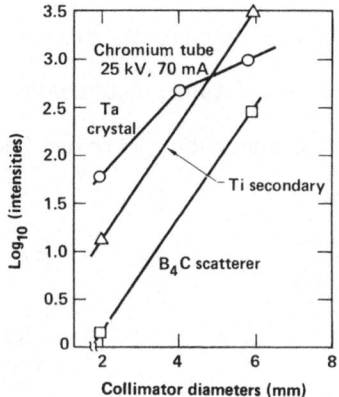

Figure 3. Calcium intensities, NBS Orchard Leaves

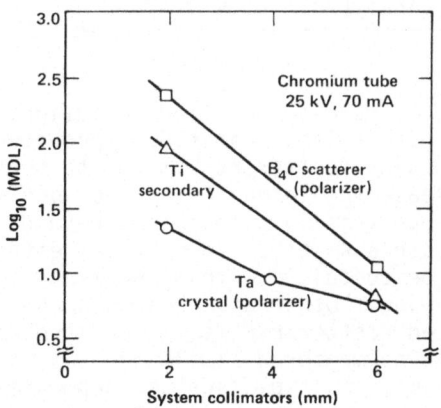

Figure 4. Calcium detection limits, NBS Orchard Leaves

Table 4. Comparision Between Barkla-Polarized and Secondary
          Target Excitation*

SPECIMEN: NBS ORCHARD LEAVES PELLET

BOTH   : 40 kV; current to give 50% dead time.

BARKLA-POLARIZED: Mo anode tube; B_4C scatterer

SECONDARY TARGET: Mo anode tube; Yttrium Sec. Target

	Polarized			Secondary Target	
	c/s	1000 sec. MDL(3σ),ppm		c/s	1000 sec. MDL(3σ),ppm
K	320	18		59	54
Ca	590	13		120	37
Fe	110	.84		26	1.9
Br	15	.3		11	0.3
Rb	22	.3		-	-
Sr	78	.3		-	-
PbLβ	25	1.		-	-

*Ryon, Adv. X-Ray Anal.(20), p. 588 1977.

In the case of Barkla-polarized radiation, there are important, but probably less dramatic, improvements to be anticipated. Circular scatterers similar to bent crystals (Fig. 2) can increase the intensity at the specimen by a factor of about 10. A report is in preparation regarding the optimization of such a system. To be fully effective, the scatterer should be contained within the x-ray tube, which is not a trivial undertaking. Other advances which have been suggested are: multi-layered scatterers[9] to extend the range of elements which can be measured similtaneously; constructing the polarizers to minimize multiple scatter depolarization;[10] and the use of a bundle of tubular collimators to maintain reasonably high intensity with a high degree of polarization.[1]

SUMMARY AND CONCLUSION

Data indicating the usefulness of polarizers have been presented. Polarized x-rays as discussed here are not esoteric or exotic. The materials used to produce polarization can be considered and used as easily as secondary fluorescers and used to advantage in many applications.

Further developments and improvements as discussed above are anticipated.

REFERENCES

1. T. G. Dzubay, B. V. Jarrett, and J. M. Jaklevic, "Background Reduction in X-Ray Fluorescence Spectra Using Polarization," Nuclear Instruments and Methods 115, pp. 197-299 (1974).
2. H. Aiginger, P. Wobrauschek, and C. Brauner, "Energy Dispersive Fluorescence Analysis using Bragg-Reflected Polarized X-Rays," in "Measurement, Detection, and Control of Environmental Pollutants," International Atomic Energy Agency, Vienna, pp. 197-212 (1976).
3. J. C. Young, R. A. Vane, and J. P. Lenehan, "Background Reduction by Polarization in Energy Dispersive X-Ray Spectrometry," Western Regional Meeting of the American Chemical Society, San Diego, California (Oct. 1973).
4. R. H. Howell, W. L. Pickles, and J. L. Cate, Jr., "X-ray Fluorescence Experiments with Polarized X-Rays," Advances in X-Ray Analysis 18, 265-77 (1974).
5. L. Kaufman and D. C. Camp, "Polarized Radiation for X-Ray Fluorescence Analysis," ibid, pp. 247-58.
6. R. W. Ryon, "Polarized Radiation Produced by Scatter for Energy Dispersive X-Ray Fluorescence Trace Analysis," Advances in X-Ray Analysis 20, pp. 575-90 (1977).
7. L. Kaufman, D. C. Price, M. A. Holliday, B. Payne, D. C. Camp, J. A. Nelson, F. Deconninck, "Fluorescent Excitation Analysis in Medicine," J. Radioanalytical Chemistry, 43, pp. 321-46 (1978).
8. P. S. Ong and J. N. Randall, "A Focusing X-Ray Polarizer for Energy-Dispersive Analysis," X-Ray Spectrometry, Vol. 7(4), pp. 241-8 (1978).
9. R. W. Ryon and J. D. Zahrt, "Improved X-Ray Fluorescence Capabilities by Excitation with High Intensity Polarized X-Rays," Advances in X-Ray Analysis 22, pp. 453-460 (1979).
10. J. D. Zahrt and R. W. Ryon, "Multiple Scattering and the Polarization of X-Rays," Advances in X-Ray Analysis 24, pp. 345-50 (1981).
11. H. Aiginger and P. Wobrauschek, "X-Ray Fluorescence Analysis in the Nanogram Region with a Totally Reflected and a Bragg Polarized Primary Beam," Journal of Radioanalytical Chemistry 61 (1-2), pp. 281-293 (1981).

12. R. B. Strittmatter, "X-Ray Fluorescance of Intermediate to High Atomic Number Elements Using Polarized X-Rays," Thirtieth Annual Denver X-Ray Conference (August 1981). To be published in Advances in X-Ray Analysis, Vol. 25.

13. C. G. Barkla, "Polarisation in Secondary Röntgen Radiation," Proceedings of the Royal Society (London), A77, pp. 247-255 (1906).

14. R. H. Howell and W. L. Pickles, "Possible Sources of Polarized X-Rays for X-Ray Fluorescence Spectra with Reduced Backgrounds," Nuclear Instruments and Methods, 120, pp. 187-188 (1974).

15. C. J. Sparks, Jr., "X-Ray Fluorescence Microprobe for Chemical Analysis," in "Synchrotron Radiation Research," Herman Winich and S. Doniach, eds., pp. 459-512, Plenum Publishing Corp. (1980).

16. M. J. Alguard, R. L. Swent, R. H. Pantell, B. L. Berman and S. D. Bloom, "Observation of Radiation from Channeled Positrons," Physical Review Letters, 42(17), pp. 1148-51 (23 April 1979).

17. R. A. Fisher, "Possibility of a Distributed-feedback X-Ray Laser," Applied Physics Letters, 24(12) pp. 598-599 (15 June 1974).

18. S. A. Akhmanov and B. A. Grishanin, "Coherent Emission of Characteristic Lines on Passage of Charged Particles through a Single Crystal," JETP Letters 23(10, pp. 515-8 (20 May 1976).

19. K. Das Gupta, "Observation of Non-Divergent X-Ray Beam from Germanium Mono-Crystal," submitted for publication (June 1981).

20. M. A. Blokin, "The Physics of X-Rays," Second revised edition, Moscow, 1957. English translation from U.S. Atomic Energy Commission Office of Technical Information, AEC-tr-4502, p. 247.

21. E. P. Bertin, "Principles & Practice of X-Ray Spectrometric Analysis," 2nd edition, Plenum Press, N.Y. 1975, p. 204.

X-RAY FLUORESCENCE OF INTERMEDIATE- TO HIGH-ATOMIC-
NUMBER ELEMENTS USING POLARIZED X RAYS

R. B. Strittmatter

Los Alamos National Laboratory
Los Alamos, NM 87545

INTRODUCTION

The use of polarized x rays as the excitation source for
x-ray fluorescence (XRF) measurements has been shown to significantly improve signal-to-background ratios.[1-4] However,
previous studies on polarized x rays applied to XRF techniques
have concentrated on low-energy fluoresced x rays (<30 keV).
In many cases strong matrix effects exist or the analyte is
encased by a material that strongly attenuates low-energy
x rays. These situations may preclude accurate assays based on
L x-ray detection, and techniques based on the detection of
higher energy K x rays may be more suitable because of the
increased penetrability of higher energy x rays. The
measurements and calculations reported in this work were made
to assess the improvement in signal-to-background ratios and
the increase in accuracy and detection sensitivity achievable
by using polarized x rays as the excitation source for
fluoresced x rays having energies between 25 and 110 keV.

MEASUREMENTS AND ANALYSIS

The measurements were made using a 200-kV, 15-mA constant
potential x-ray generator. The x-ray polarization was achieved by
90° scattering from a polarizing target in a manner similar to the
method used for low-energy polarized x-ray measurements.[1-4] The
polarized x rays were used to fluoresce x rays in various samples
containing intermediate- to high-atomic-number (50-92) analyte
elements, and the fluoresced x rays emitted perpendicular to the
initial scattering plane were detected using a high-resolution

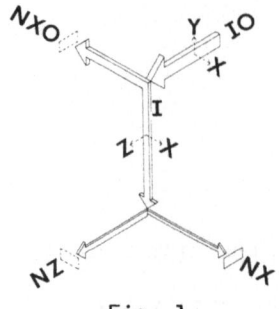

Fig. 1.

Schematic of geometry used for polarized x-ray fluorescence measurements.

germanium detector. The primary beam from the x-ray generator and the 90°-scattered x rays that hit the sample defined the initial scattering plane. Figure 1 is a schematic of the measurement geometry where IO is the primary beam, I is the polarized beam resulting from the 90° scatter, and NX is the detector position that minimized the scattered flux. The polarizers were aluminum and iron disks 2.54 cm in diameter and located 30.5 cm from the x-ray generator target. The polarizer-to-sample distance was 4.75 cm, and a 1.6-cm-diam by 1.9-cm-long tungsten collimator defined the polarized beam divergence. The sample-to-detector distance was 10 cm, and a 1.6-cm-diam by 1.9-cm-long tungsten collimator located 1.5 cm from the sample defined the sample area viewed by the detector. Two samples were measured: a 0.16-cm-thick brass plate containing approximately 0.05-at.% tin and 0.5-at.% lead and a 0.003-cm-thick uranium foil sandwiched between two 0.09-cm-thick iron sheets.

    To quantify the improvement in signal-to-background ratios obtained using polarized x rays as compared to conventional direct XRF, a quantity called the scatter fraction was measured. The scatter fraction, R, is defined as

$$R \equiv \frac{(N_{scatter}/N_{fluoresced}) \text{ polarized}}{(N_{scatter}/N_{fluoresced}) \text{ unpolarized}} \quad , \tag{1}$$

where $N_{scatter}$ is the measured count rate of scattered x rays in a narrow energy interval and $N_{fluoresced}$ are the fluoresced x rays used for beam intensity normalization. Referring to Fig. 1, the scatter fraction is determined by measurements of the polarized XRF at position NX and the unpolarized XRF at position NXO. The measured scatter fraction depends on the average polarization of the photons impinging on the surface of the sample and the asymmetry ratio of the sample-detector configuration. The change due to the 90° scatter in the energy spectra of the x rays striking the sample was compensated for by lowering the x-ray generator potential when measuring the unpolarized XRF. The polarization, P, for the beam I in Fig. 1 is defined as $P = (I_x - I_z)/(I_x + I_z)$, where I_x and I_z are the components of the beam intensity having electric vectors in the x and z directions, respectively. The asymmetry ratio is defined as $A = \epsilon(\pi/2)/\epsilon(0)$, where $\epsilon(\pi/2)$ and $\epsilon(0)$ are the efficiencies for detecting photons whose electric vectors are perpendicular and parallel, respectively, to the scatter plane defined by the polarized beam and the detected photons. Using the measured scatter fraction and a calculated or measured asymmetry ratio, the polarization of the beam I can be determined using the relationship

$$P = (1 - R) \frac{1 + A}{1 - A} \quad . \tag{2}$$

The above discussion assumes an initial beam that is unpolarized, a condition that is not always met when using x-ray generators as a primary source.

RESULTS AND DISCUSSION

The maximum improvements in signal-to-background ratios obtained for low-energy polarized x rays have been limited by geometrical and multiple scattering effects.[1-4] However, the Klein-Nishina formula for Compton scattering contains a component of the scattering that is polarization independent. For the energies of the K x rays of high-atomic-number elements, this component limits the practical achievable polarization. The Klein-Nishina formula was used to calculate the maximum polarization and the minimum scatter fraction that can be achieved from pure 90° scattering. The results of this calculation are presented in Fig. 2. It should be noted that the maximum polarization occurs at scattering angles slightly less than 90°. The results of this calculation indicate that the minimum scatter fractions achievable at final energies of 25 and 100 keV are 0.005 and 0.113, respectively.

The effect of the polarizer material and thickness on the scatter fraction and polarized beam intensity was measured at 80 keV using aluminum and iron polarizers and the brass sample. The measured scatter fractions are presented in Fig. 3. The decrease in the measured scatter fraction with decreasing thickness is accompanied by a decrease in the polarized beam intensity. The results of these measurements and calculations of scattering efficiencies indicate that polarized beam intensities comparable to those obtained for low-energy x rays should be achievable.

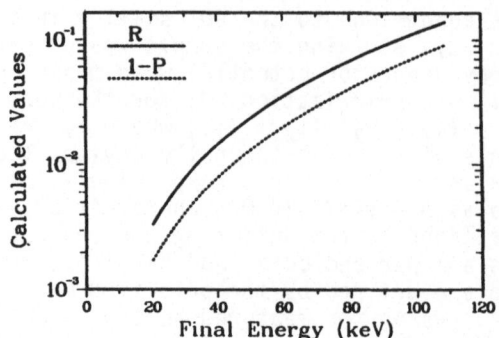

Fig. 2.
Calculated scatter fractions, R, and polarizations, P, based on
the Klein-Nishina formula for pure 90° scattering. The results
are plotted as a function of the detected photon energies.

The measured scatter fractions, R_M, at 27, 80, and 100 keV
using the K_α lines of tin and the K_{α_1} lines of lead and
uranium for normalization in Eq. 1 are presented in Table 1.
Also given in Table 1 are the scatter fractions, R_{KN},
calculated for pure 90° scattering using the Klein-Nishina
formula, and the ratio of $1-R_M$ to $1-R_{KN}$, which is an
estimate of the value of $1-R_M$ due to geometry and multiple
scattering effects. The signal-to-background ratios obtained
in this work at 27, 80, and 100 keV were increased by factors
of 10.5, 5.5, and 5.4, respectively, compared to the direct XRF
measurements.

To realize improvements in detection sensitivity and assay
accuracy using polarized x rays, the observed increases in

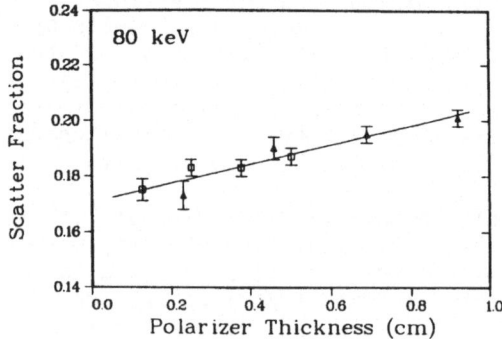

Fig. 3.
Scatter fractions measured at 80 keV using aluminum (Δ) and
iron (□) polarizers.

Table 1. Measured and Calculated Scatter Fractions Near K X-Ray
Lines of Tin, Lead, and Uranium

Fluoresced Element (Z)	Energy of Scatter (keV)	R_M	R_{KN}	$1-R_M/1-R_{KN}$
Tin (50)	27	0.095	0.006	0.91
Lead (82)	80	0.183	0.068	0.89
Uranium (92)	100	0.184	0.115	0.92

signal-to-background ratios must be accompanied by fluoresced
analyte x-ray intensities comparable to those measured using
conventional direct XRF. For the present experimental configuration, the ratios of the signal intensity for the K_{α_1}
lead and uranium lines for the polarized XRF measurements compared to the unpolarized XRF measurements at a through rate of
15×10^3 counts/s were 0.01 and 0.03, respectively. Using
the same x-ray generator with closer coupling to the polarizer
and closer sample-to-detector coupling, a factor of 50-100 increase in signal should be realizable. Improvements in polarizer geometry such as a curved polarizer should also increase
the polarized beam intensity.[2,4] Assuming a factor of 50
improvement in signal intensity, the detection sensitivity
would be three times better than the sensitivity attainable
using conventional XRF for uranium. With even greater signal
intensities, polarized XRF methods could provide detection sensitivities for high-atomic-number elements an order of magnitude lower than conventional direct XRF methods.

REFERENCES

1. T. G. Dzubay, B. V. Jarrett, and J. M. Jaklevic, "Back-
    ground Reduction in X-Ray Fluorescence Spectra Using
    Polarization," Nucl. Inst. Methods 115:297-299 (1974).
2. R. H. Howell, W. L. Pickles, and J. L. Cate, Jr., "X-Ray
    Fluorescence Experiments with Polarized X Rays," in: "Ad-
    vances in X-Ray Analysis," W. L. Pickles, C. S. Barrett,
    J. B. Newkirk, and C. O. Ruud, eds., 18, Plenum Pub. Co.,
    New York (1974).
3. L. Kaufman and D. C. Camp, "Polarized Radiation for X-Ray
    Fluorescence Analysis," in: "Advances in X-Ray Analysis,"
    W. L. Pickles, C. S. Barrett, J. B. Newkirk, and C. O.
    Ruud, eds., 18, Plenum Pub. Co., New York (1974).
4. R. W. Ryon, "Polarized Radiation Produced by Scatter for
    Energy Dispersive X-Ray Fluorescence Trace Analysis," in:
    "Advances in X-Ray Analysis," H. F. McMurdie, C. S.
    Barrett, J. B. Newkirk, and C. O. Ruud, eds., 20, Plenum
    Pub. Co., New York (1977).

EXAMPLES OF ANALYSIS FROM AN INTEGRATED

X-RAY FLUORESCENCE ANALYSIS SYSTEM USING NRLXRF

B. E. Artz and M. J. Rokosz

Engineering and Research Staff
Scientific Research Laboratory
Ford Motor Company
Dearborn, Michigan

Methods of correction for matrix differences are required in X-ray Fluorescence (XRF) Analysis when the overall composition of the unknowns is substantially different from the available standards. Sample preparation techniques used to minimize matrix differences often require development time and can consume irreplaceable sample material. Alternatively, the increasing computer power available to the analyst and the refinement of computer programs using fundamental parameter calculations has made this approach more attractive.

A system consisting of a Siemens SRS-1 wavelength dispersive spectrometer (WDS)[1,2], a KEVEX 0810-A/NS880 energy dispersive spectrometer (EDS), software for data collection and manipulation and a 40 element version of the NRLXRF[3] fundamental-parameters analysis program has been put together to simplify XRF analysis of samples lacking standards of a similar composition. This configuration is shown schematically in Figure 1.
    In order to take full advantage of this configuration generalized software was written for collection, preprocessing and merging of x-ray data that would output a file compatible with NRLXRF input file format requirements. The PDP11/34 computer system associated with the WDS spectrometer was used as the central site for data merging, formatting and storage. The WDS system runs any number of samples for any combination of elements and acquisition parameters and stores the necessary sample information such as sample type, known compositions, particle sizes, sample thickness, homogeneity, number and names of elements for analysis, chemical form of analyte being reported as well as X-ray generator and spectrometer parameters. This information is also stored and can be accessed later, thus simplifying operator inputs for analyte lines already used. The acquisition program also determines the optimal counting times for analyte peaks & backgrounds and orders the data acquisition sequence to minimize spectrometer 2θ driving.

Software developed for the EDS system accepts the output from a modified version of the Tracor Super ML program[4] and also allows the operator to specify sample and analysis conditions.

A merge and sort program accesses the output from both the WDS, and EDS software to provide an output file that can serve as an input to NRLXRF. This program must merge data from the same sample, obtain any required stored data for standards, and develop an organized LINES group for NRLXRF.

Ancillary programs are available for maintaining the various parameter files. These programs automatically update the line parameter file with the most current value of the peak position for a particular analyte line-spectrometer setting combination. They also update the standard intensity and composition files.

Three examples involving the analyses of soils, catalytic converters and experimental electrode materials using the previously described system are presented.

Several soil samples from various locations were submitted for complete XRF analysis. Initial examination indicated that these samples fell into two categories characterized by high and low calcium content respectively and that there were 29 elements present that could be measured by XRF. The only standards which were available were four NBS clay materials (97, 98, 97a, 98a) which included 18 of the required 29 elements. Of these 18 elements, only 9 covered the range expected in the soils. Additional standards were made by mixing an oxide or compound of an analyte element at about the 1 to 2% level with alumina. Two additional mixtures of higher calcium levels were also prepared. The final set of standards used consisted of the multi-element clays and 28 of the simple binary mixtures. Of the 29 elements analyzed, 25 were also analyzed by other methods on samples taken from the same original larger sample. Table 1 compares the results of the analyses by the different methods. In general NRLXRF seems to agree to within $\pm 20\%$ for elements with concentrations greater than 1%. Discrepancies in trace elements are probably due mainly to sampling errors. Figure 2 shows the CPU time required to analyze the first soil sample as a function of the number of elements analyzed. The 4 clay standards were used throughout and binary standards added as the various elements were added. Figure 2 indicates that the use of NRLXRF for more than 15-20 elements under these conditions becomes extremely time consuming. Analyses of similar samples in the same run are accomplished much more rapidly usually at less than 10% of the time for the first sample.

Minor and trace elemental analyses of catalytic converters have been done routinely at this facility using a single element standard/effective wavelength method[5] which has been verified many times by comparison with other laboratory results. Two catalyst samples were analyzed by this method and by NRLXRF using the same set of standards. The standards consisted of 38 binary mixtures containing various levels of the single analyte elements mixed with catalyst substrates. The analysis was for eleven elements on each sample. Table 2 compares the results from the two methods. With the exception of iron, NRLXRF agrees with the other method to within

Table 1. Comparison of results for soil analysis by NRLXRF, inductively coupled plasma–atomic absorption spectroscopy (ICP), and neutron activation analysis (NAA).

	NRLXRF	ICP	NAA			NRLXRF	ICP	NAA
Mg	.6	.43	1.4		Co	.006	.003	.002
Al	6.0	4.0	5.4		Ni	.008	.003	
Si	33.8	30.4			Cu	.002	.004	
P	.11	.052			Zn	.007	.019	
Cl	.007	<.1			Rb	.01	.01	
K	2.2	2.4			Sr	.005	.004	
Ca	.24	.15	.29		Zr	.032	.019	
Ti	.54	.34	.50		Ba	.049	.030	.041
V	.009	.009	.007		La	.007		.006
Cr	.005	.007			Ce	.018		.011
Mn	.18	.22			Pb	.005	.008	
Fe	2.43	2.93	2.95					

Table 2. Analysis of converters. Calib refers to the analyte amount read directly from the binary calibration curves. Cur/Met is the current analysis method.[2]

		CONVERTER-1			CONVERTER-2	
Elem	Calib	NRLXRF	Cur/Met	Calib	NRLXRF	Cur/Met
P	1.35	1.35	1.36	.88	.90	.90
S	.02	.02	.02	.03	.02	.03
Mn	.01	.01	.01	<.01	<.01	<.01
Fe	.24	.24	.27	.31	.31	.37
Ni	.02	.01	.02	1.65	1.74	1.95
Zn	.22	.25	.26	.19	.26	.27
Rh	----	----	----	.012	.022	.021
Pd	.034	.050	.047	----	----	----
Ce	.68	.70	.72	.67	.71	.76
Pt	.09	.10	.10	.16	.21	.22
Pb	.56	.68	.68	.66	.91	.91

Table 3. Weight fractions for electrode materials. I/I_{100} is the relative intensity and Expd is the stoichiometric composition.

Elm	I/I_{100}	NRL XRF	Expd	Cmpd	Elm	I/I_{100}	NRL XRF	Expd	Cmpd
Nb	.671	.60	.63	$NbSe_2$	Ta	.755	.48	.43	$TaSe_3$
Se	.279	.40	.37		Se	.185	.52	.57	
Nb	.595	.73	.72	$NbSe_3$	Ti	.380	.43	.43	TiS_2
Se	.161	.27	.28		S	.413	.57	.57	
Ta	.778	.45	.43	$TaSe_3$	Ti	.203	.24	.33	TiS_3
Se	.202	.55	.57		S	.574	.76	.67	

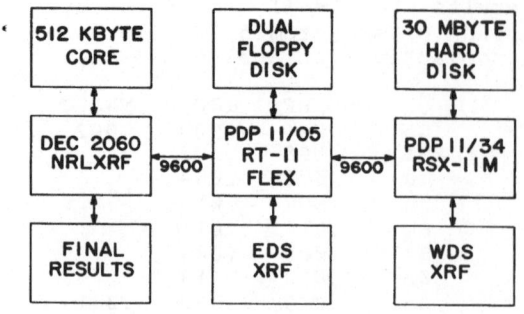

Figure 1. System configuration

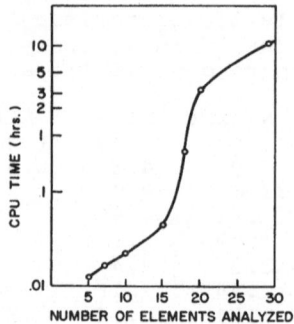

Figure 2. CPU time DEC 2060

the expected accuracy (+10%) of that method. The result for iron is probably better using NRLXRF since it more effectively corrects for the small amount of iron in the substrate material.

Experimental electrode materials, consisting of a few grams of metallic type "whiskers" were also analyzed using NRLXRF. The samples were pressed into small irregular flat disks and analyzed using the EDS system. The samples were binary mixtures of Ti-S, Nb-Se, and Ta-Se. The standards used were the pure elements. The XRF analyses agreed to within 10% of the expected stoichiometric composition in 5 of 6 compounds analyzed. Table 3 summarizes these results.

A software system has been developed which ties together a WDS system, an EDS system and NRLXRF to simplify XRF analysis of samples for which closely similar standards do not exist. Analysis of several different types of materials has demonstrated the effectiveness of this approach and analysis accuracies are in the +10 to +20% relative range.

References

1. B. E. Artz, C. J. Kelly, M. A. Short, "A Computer Control for an X-ray Fluorescence Analysis Unit, Advances in X-ray Analysis, (18), 309-316, (1975).
2. B. E. Artz, E. C. Kao, M. A. Short, "Using DEC RSX-11M Operating System for X-ray Diffraction and Xray Fluorescence Analysis", Advances in X-ray Analysis, 22, 425-431, (1979).
3. J. W. Criss, NRLXRF, A Fortran Program for X-ray Fluorescence Analysis, COSMIC Program #DOD-00065, Univ. of Georgia Computer Center, Athens, GA. (1977)
4. F. Schamber, N. Wodke, J. McCarthy, 12th Annual Conf. of M.A.S., Boston, Mass., paper 85, (1977).
5. B. E. Artz, "X-ray Fluorescence Analysis of Catalytic Converters using Single-element Standards and Theoretical Corrections for Interelement Effect", X-ray Spectrometry, Vol. 6, No. 3, 165-170, (1977).

A MODULAR ADC / MICROCOMPUTER SYSTEM FOR ENERGY DISPERSIVE X-RAY SPECTROSCOPY

D. Hale, T. Satterfield, D. Blankenship, J. C. Russ **

The Nucleus, Oak Ridge, TN
** Engineering Research, N.C State University, Raleigh NC

INTRODUCTION

Multichannel analyzers, as used in energy dispersive X-ray spectrometers using Si(Li), High Purity Ge, or other types of detectors, receive a series of pulses, generally in the 5 to 10 volt range full scale, which they measure and count in memory. Some are constructed using hardwired digital logic circuits to perform these memory control operations, while others incorporate standard computer memories and central processing units, with programmed logic for storing, displaying and (generally) processing the spectrum. With the advent of ever more powerful and lower cost microcomputers, it becomes increasingly attractive to adapt one to this purpose. We have done so, using an Apple II, and will describe the hardware of the interface, the logic of the data acquisition programs, and the utilization of memory in the computer.

ANALOG TO DIGITAL CONVERTER

Pulse measuring, as contrasted to the simpler task of measuring a steady or slowly varying DC voltage using a single-chip "ADC" (analog to digital converter) requires that the pulse trigger a circuit to detect the peak and retain that voltage for measurement. The peak voltage is then transferred to a small capacitor and discharged using a constant current supply. During the time required to discharge the voltage, a counter records the number of oscillations of a high frequency clock, and it is this number which subsequently represents the magnitude of the pulse.

This description applies in general terms to virtually all of the ADC's used for pulse measurement, which are of the Wilkinson or run-down type. They contrast sharply to the rather more widely used

85

successive approximation type used for measuring steady voltages
(several of which are available in interfaces for microcomputers
such as the Apple). The successive approximation type uses a binary
search tree to generate a comparison voltage. The initial 'guess' is
half of the full scale voltage. If this is less than the input
voltage level, it is increased (and vice versa) with each successive
step being one-half the preceding one. A register keeps track of
whether each step was up or down, as a string of ones and zeroes,
and the final value forms a number which gives the magnitude of the
voltage. This method is much faster than the run-down type and has
digitisation times essentially independent of the voltage magnitude,
but suffers from noise and linearity problems so that it is
generally necessary to make several readings to obtain good
precision. This erodes the speed advantage and is awkward to adapt
to pulse measuring.

The run-down ADC we have designed accepts voltage pulses with 0-8
volts magnitude into a 2 Kohm input impedance, with a shaping time
contant not less than 0.25 usec. (typical spectroscopy amplifiers
used in energy dispersive systems use time constants from 2-12
usec.). The input is DC coupled, with zero stability better than
0.01% per deg. C from 15 to 40 C, and a +-2% adjustable range. There
is, additionally, a logic input for coincidence counting not
normally used in spectroscopy applications. The run-down clock
operates at 10 MHz, for a total conversion gain of 1024, that is, an
8 v. pulse (maximum size) will be converted to the number 1024 while
a smaller pulse will give a proportionately smaller value. The gain
shift due to variation in the current supply is less than 0.01% per
deg. C.

The entire circuit board (7x18 cm.) plugs into one of the eight
peripheral slots within the microcomputer chassis. This provides
power (the low power Schottky devices used allow the card to consume
less than 1 watt total, of +-12 and +5 volts), and incidentally
assures a stable thermal environment and high immunity to external
noise. When the pulse conversion is finished, the number is held in
a 10 bit binary register and an interrupt flag raised on the
computer bus to signal the processor. One of the programs to be
described uses this interrupt capacity, while others poll the busy
flag on the card to await availability of information.

The card also contains a clock which counts pulses from the same
high frequency oscillator used for the run down timer. This can be
gated off by a variety of signals, including amplifier pileup,
preamplifier reset, and computer busy signals, to accurately reflect
system live time. An up/down counter is also used to keep track of
pileup events signalled by the amplifier, and to periodically halt
the live time clock while lost pulses are made up with accepted
ones. The clock can also raise an interrupt flag in the computer at
a 1 Hz rate, to allow for preset time counting.

PROGRAM LOGIC FOR DATA ACQUISITION

The microcomputer sets aside a 1024x2 byte area to record counts. Two bytes (16 bits) give a maximum count capacity of 65535 per channel, so in applications requiring larger numbers, three bytes (24 bits) would be used to allow over 16 million counts. When the ADC produces a value on the bus, it appears to the processor as two successive bytes of memory (all peripherals in the Apple are memory mapped). This number must be added to the base address of the area used for counting, and the memory location at the byte thus addressed must be incremented. If it 'rolls over' (reaches 256), the next byte (1024 locations higher in memory) must be incremented, and so on for the third byte. Since the microprocessor (6502) runs at 1 MHz, this takes approximately 20 µsec., which is shorter than the typical amplifier pulse width and ADC run down times. Consequently, it is not necessary to have circuitry and logic on the ADC board to directly increment memory (DMA – direct memory access), which would both complicate the design and reduce the flexibility of allocating memory usage, to be discussed below.

One way to acquire data is to have the program run in a small waiting loop, watching the ADC's busy flag and using the resulting value to increment memory whenever it is ready. The program then clears the ADC (by writing to a memory address) and restarts it. Usually, however, we want to see the spectrum as it is acquired, and this scheme does not provide for that capability. A program to access the spectrum memory and display the 1024 channels (or any 256 channels) on the computer's video display with some alphanumeric information showing full scale and limits, requires approximately 50 msec. Figure 1 shows the typical display appearance. One way to provide display and storage capability while maintaining fast ADC service is to update the display once per second (on the clock ticks). Using this method, we have a semi-live display.

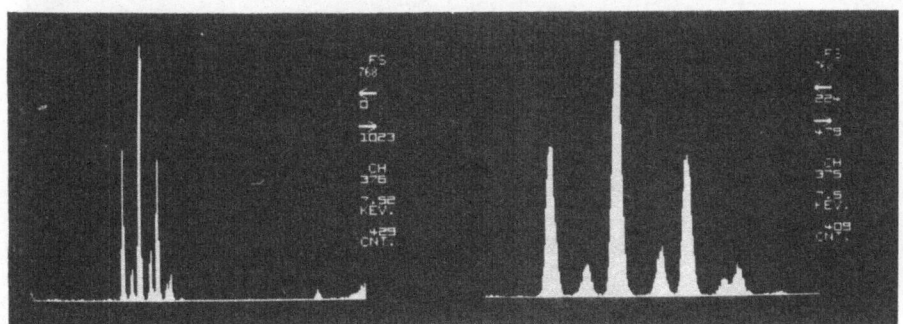

Figure 1. Spectrum display on Apple video monitor. Left: 1024 channels. Right: 256-channel segment. Vertical full scale, horizontal limits, and locations of 'bug' or cursor movable by user are shown.

It is also possible to have the program normally dedicated to drawing and re-drawing the display, responding to any user commands to change vertical scale, horizontal limits, etc. (this is accomplished by pressing a key on the computer keyboard). Then when the ADC interrupt request occurs, this process is halted very briefly while the memory incrementing takes place. In this case, the interrupt service routine must preserve the status of all processor registers and restore them afterward so that the normal or "foreground" program can continue smoothly,and this adds to the time required per pulse (it becomes over 40 μsec.). While still invisible to the human watching the screen, who sees a 'live' display of the spectrum as it is acquired, the increased time is enough that there is a greater chance that at high count rates (eg. 10Kcps and greater) some incoming pulses will be lost due to the processor being busy, and so the dead time of the system increases.

Consequently, the true live display is best suited to lower counting rate applications, and the intermittent display update mode to preset time or higher counting rate modes of data acquisition. Other special purpose modes in which certain address ranges (energy windows or regions of interest corresponding to preselected elements) are totalled and displayed as time histograms (multichannel scaling mode) or a multielement ratemeter display on the video screen, or are simply counted to provide data for alloy sorting or some other dedicated application not requiring a full spectrum display, place no great demand on the system and can be handled by polling the ADC with no significant dead time. Figure 2 shows two of these modes of operation. During spectrum acquisition, a region of interest can be established in the program to obtain total counts for one element while pulses are produced for X-ray mapping on an SEM (by writing to an address on the circuit board which triggers a one-shot to generate the pulse).

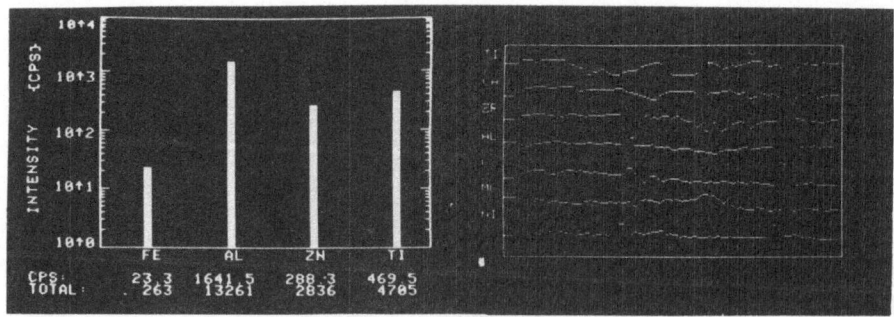

Figure 2. Display during counting for preselected elements. Left: Multielement logarithmic ratemeter format. Right: Time histogram (multichannel scaling) for seven elements corresponding to distribution across surface of specimen in SEM.

MEMORY MANAGEMENT

The programs described above for data acquisition and display are written in machine language for speed, and occupy altogether about 4K bytes of memory. The storage array for the spectrum takes another 2K bytes, and the high-resolution video display of the Apple is in fact a bit-mapped 8K segment of memory. Together with a minimal monitor or system program, this can fit into a stand-alone 16K computer. However, as memory has dropped precipitously in price, it makes sense to incorporate the full 64K memory which the Apple is capable of addressing. This allows a full disk operating system for storage of programs and data (a 5" floppy disk will hold 140K bytes).

With this amount of memory available, we have apportioned it among the various functions as shown in the map below:

address range	purpose/programs
0- 2 K	system monitor
2- 4 K	display labelling & graphics
4- 6 K	spectrum memory
6- 8 K	ADC service routines
8- 16 K	bit-mapped video display
16- 38 K	application program (see below)
38- 48 K	disk operating system
48- 52 K	memory mapped peripherals (ADC printer, keyboard, etc.)
52- 64 K	BASIC interpreter or math libraries for other languages (eg. FORTRAN)

This leaves a substantial space for data processing programs, of which we have written quite a few in BASIC. These include:
a) Peak identification using KLM markers and elemental symbols
b) Spectrum stripping by least squares fitting of stored or generated peaks
c) "Z-A-F" and other quantitative programs for SEM, microprobe and TEM (including thin section, particle, inclined surface, and even Monte-Carlo models)
d) XRF programs including regression by LaChance-Traill, Lucas-Tooth and Rasberry Heinrich models, linear and quadratic calibration curve fitting, and a fundamental parameters (effective wavelength) method

Most of these make further use if the computer's graphics display capability as shown in Figure 3.

Athough the microcomputer is somewhat slower than a more costly minicomputer, the factor of 2-3 in computing speed seems not to be a serious limitation for typical applications computations (regression with 6 correction factors and 20 standards takes less than 30 seconds and could be speeded up by compilation if needed), especially when compared to data acquisition time. Data acquisition

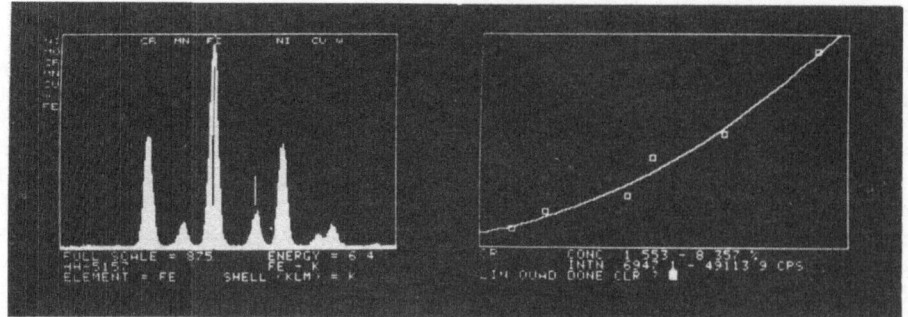

Figure 3. Applications program displays. Left: Elemental peak identification
using superimposed KLM lines from extensive tables. Right: Calibration
curve fit to chromium data from standards. The screen can also be used
for alphanumeric or tabular printout of results.

during computation is possible, but not simultaneously with display
of the spectrum, and not at high count rates (which simply take all
of the available time so that no computing is done). If simultaneous
counting and computing is not needed, then the full memory can be
used for data processing (recalling stored spectra from disk), in
which case about 36K is available in addition to the operating
system and language. With these limitations, the combination seems
well suited and cost effective for laboratories that do not have
extremely high sample throughput. It is also practical for dedicated
monitoring operations, or where special experiments may require
unique programming.

CONCLUSION

Adapting a conventional ADC design   to work with a standard
microcmputer requires careful attention to the interface design,
logic of software to provide data acquisition and display, and
memory utilization to allow applications programs to access the
data. We have used a memory-mapped peripheral design with ability to
generate interrupts from either the modular ADC or the live time
clock. The data can be acquired by polling for very high count rate
or preset time applications and for counting of specific elements
without a full spectrum display, or serviced on interrupt for a true
live display at lower count rates (eg 5 Kcps and below). The data
acquisition software, graphics display, full computer disk operating
system, and language such as a BASIC interpreter, leave about 22K
bytes for applications progams in a 64K byte Apple II computer, or
larger programs can be run to process previously stored data. Broad
applications possibilities exist for such a system.

VOLATILIZATION OF SULFUR IN FUSION TECHNIQUES FOR PREPARATION OF DISCS FOR X-RAY FLUORESCENCE ANALYSIS

James W. Baker

U.S. Geological Survey
Box 25046, MS 928, Denver, Colorado 80225

The fusion technique for sample preparation for x-ray fluorescence analysis has been proven very successful for a wide variety of sample types. An inherent problem in the fusion technique has been the loss of sulfur during the fusion process. To extend our quantitative major element analysis method for geologic materials (1,2,3) we have examined the problem of sulfur volatilization and have tested a variety of fluxes to determine their suitability to quantitatively retain sulfur during fusion, to produce a homogeneous glass disc that is suitable for presentation to an x-ray spectrometer with no additional sample preparation, and to help protect platinumware from attack by sulfide minerals.

Highly oxidizing conditions are necessary both to retain sulfur and to protect platinumware from sulfide attack. Previous researchers have shown that sulfur may be lost under reducing conditions (4). All of the fusions in this study were performed in an oxidizing environment, using a muffle furnace. Our studies indicate that sulfur can be retained only in its highest oxidation state, as a sulfate, during the fusion process. To quantitatively retain all of the sulfur in a sample, a highly oxidizing flux in an oxidizing environment must be used to insure that the sulfur is all oxidized to and remains in the sulfate oxidation state. Table 1 lists various fluxes rejected because they did not quantitatively retain all forms of sulfur or did not form a homogeneous glass bead. Table 2 summarizes the most suitable flux mixtures examined, all of which conform to the criteria given above. Caution is recommended whenever high concentrations of sulfide ores are present; some metals may still alloy with the platinum despite the oxidizing environment.

Table 1. Various Rejected Fluxes[a]

Flux	Quantitative Sulfur Retention	Glass Disc[b]
$Na_2B_4O_7$	No	Yes
$Li_2B_4O_7$	No	Yes
$Li_2B_4O_7$ (90%) Li_2CO_3 (10%)	No	Yes
$Li_2B_4O_7$ (56%) Li_2CO_3 (44%)	No	No
$Li_2B_4O_7$ (47%) Li_2CO_3 (37%) La_2O_3 (16%)	No	No
$Li_2B_4O_7$ (62%) Li_2CO_3 (7%) B_2O_3 (15.5%) $LiNO_3$ (15.5%)	No	Yes
$Li_2B_4O_7$ (72%) B_2O_3 (28%)	No	Yes
$Li_2B_4O_7$ (73%) $BaCO_3$ (27%)	No	Yes

[a]All samples fused with addition of approximately 0.45% LiBr with respect to flux weight.
[b]Glass disc is homogeneous and suitable for analysis with no additional sample preparation.

The temperature and time of fusion are both important factors in the retention of sulfur. If good control of temperature and fusion time are exercised by the analyst, the amount of sulfur lost from any non-oxidizing flux may appear to be quantitative, as the standards and samples may lose proportional amounts of sulfur. To insure that the fluxes tested were quantitatively retaining sulfur, all of the sulfur analyses were verified by analyzing ground portions of the fused discs with a LECO sulfur analyzer*.

*Any trade names used are for identification only and do not constitute endorsement by the U.S. Geological Survey.

Table 2. Suitable flux mixtures for sulfur retention[a]

$Na_2B_4O_7$ (80%)
$NaNO_3$ (20%)

$Li_2B_4O_7$[b]   (80%)
$LiNO_3$ (20%)

$Li_2B_4O_7$ (80%)
$CsNO_3$ (20%)

[a]All fluxes contain approximately 0.45% LiBr as a non-wetting
agent, and were fused at 1000° C.
[b]Cs_2CO_3 may be added to make the $Li_2B_4O_7$: Cs_2CO_3 ratio as great
as 4:1, depending on amount of heavy absorber desired.

Although sodium tetraborate-sodium nitrate flux retains
sulfur and protects the platinum, it does eliminate the capability
of determining sodium in the sample. Therefore, its usefulness is
limited for geologic materials and some other applications.

Temperature of the fusion is very important. Close
temperature control is necessary if sulfur is to be quantitatively
retained. For example, the lithium tetraborate-lithium nitrate
flux does not quantitatively retain sulfur if the fusion is
performed at 1100° C instead of 1000° C. This loss may be due to
the volatilization of lithium sulfate at the higher temperatures.
Pure lithium tetraborate has been successfully used to produce
glass discs (3), but, because of its high melting point, even
sulfate sulfur is not quantitatively retained. The addition of
the lithium nitrate increases the alkalinity and oxidizing power
of the flux. It also effectively shifts the composition of the
flux from lithium tetraborate towards the eutectic, located near
lithium metaborate, thus significantly lowering the melting point
of the flux (5). Although we use extensive interelement
correction routines to obtain quantitative analyses using the
lithium tetraborate-lithium nitrate fusion discs, some analysts
may prefer to use heavy absorbers, higher dilution ratios, etc.
Previous investigators (6) have indicated that the addition of
lanthanum oxide causes difficulties due to an interference of
lanthanum with magnesium and that the severe hygroscopicity of
lathanum oxide is objectionable. Furthermore, our studies
indicate that the addition of lathanum oxide to the flux
frequently induces crystallization of the fusion discs. We found
that the addition of cesium carbonate to the lithium tetraborate-
lithium nitrate flux or the replacement of lithium nitrate with
cesium nitrate are effective alternatives to the lathanum oxide.
Good glass discs are obtained over a wide range of cesium

concentration, allowing the analyst much freedom in selecting the amount of heavy absorber to be used in the final disc. Although the cesium carbonate, cesium nitrate, and lithium nitrate are all hydroscopic by themselves, the final fused discs are only mildly hydroscopic and can be preserved for long periods with virtually no change if kept in a dessicator.

The addition of small amounts of LiBr (approximately 0.45% of the flux weight) has a two-fold advantage. The LiBr serves as a non-wetting agent, preventing the molten flux from sticking to the platinum crucibles and molds. The addition of a small amount of LiBr also reduces the melting point of the flux, increasing fluidity of the molten material during pouring into the mold.

I gratefully acknowledge the assistance of Van E. Shaw, who performed the verification analysis of the sulfur in the fused discs.

References

1.   J. E. Taggart, Jr. and J. S. Wahlberg, New Mold Design for Casting Fused Samples, in: "Advances in X-ray Analysis," Volume 23, p. 257-261, (1980).

2.   J. E. Taggart, Jr. and J. S. Wahlberg, A new in-muffle automatic fluxer design for casting glass discs for x-ray fluorescence analysis, Abs. 327a, Federation of Analytical Chemists and Spectroscopy Societies Conference, Sept. 19, 1980.

3.   J. E. Taggart, Jr., F. E. Lichte, and J. S. Wahlberg, Analysis of Samples from Mount St. Helens Using X-ray Fluorescence and Induction Coupled Plasma Spectroscopy, in: "The 1980 Eruptions of Mount Saint Helens, Washington," USGS Professional Paper 1250, In press.

4.   K. Norrish and J. T. Hutton, An Accurate X-ray Spectrographic Method for the Analysis of a Wide Range of Geological Samples, Geochim. et Cosmochim. Acta, 33, p. 431-453, (1969).

5.   B. S. R. Sastry and F. A. Hummel, Studies in Lithium Oxide Systems: V, Li_2O--Li_2O-B_2O_3, Jour. Am. Ceramic Soc., Vol. 42, p. 216-218, (1959).

6.   B. P. Fabbi, A Refined Fusion X-ray Fluorescence Technique and Determination of Major and Minor Elements in Silicate Standards, Am. Min., 57, p. 232-245, (1972).

TECHNIQUES FOR THE PREPARATION OF LITHIUM TETRABORATE
FUSED SINGLE AND MULTIELEMENT STANDARDS

Kent I. Mahan

University of Southern Colorado

Pueblo, Colorado

Donald E. Leyden

University of Denver

Denver, Colorado

INTRODUCTION

Since first introduced by Claisse[1,2] twenty-five years ago, borate fusions have been used in XRF analysis of a wide variety of materials because fusion eliminates many of the problems associated with particle size and matrix effects encountered in the analysis of powders. Geological and environmental materials such as rocks, ores, sediments, slags, air-borne particulates, etc., lend themselves nicely to lithium or sodium tetraborate or lithium metaborate fusion because the inherent presence of silica in these materials stabilizes the resultant glass from crystallization.

In our laboratory, it was desired to apply the lithium tetraborate fusion technique to the simultaneous determination of major and minor elements in river and lake sediments by energy dispersive X-ray fluorescence spectrometry (EDXRF). Calibration and reference standards for these determinations would have to be prepared in concentrations ranging from a few ppm to a few percent depending on the concentration ranges anticipated in the sediment samples taking into account dilution of the sample by flux. This offered what we thought would be an excellent opportunity to test some different techniques in preparing fused synthetic multielement calibration standards. The three techniques used in this study include the addition of high purity compounds to flux before fusion, dilution of standards previously fused with additional flux, and the addition of aqueous solutions of the metals to flux followed by evaporation of the excess water before fusion. Pella, Lorber and Heinrich[3] described the preparation of synthetic standards by adding finely ground powders of pure compounds to flux in the analysis of standard reference fly ash and particulate materials. In the determination of trace levels of RB, Sr, Zr and Nb in geological samples, Jagoutz and Palme[4] micropipetted aqueous solutions of the metals directly to the Pt-Au crucible used in the fusion, then evaporated away the water before adding flux and fusing the mixture.

In this study, data have been collected on a total of five sets of four calibration standard discs fused from a $Li_2B_4O_7$, SiO_2, $LiNO_3$ flux, each set containing 2-11 elements. In addition, a set of five blanks were prepared. Combinations of these six sets including blanks produced 4-15 calibration points for 16 elements - K, Ca, Ti, V, Cr, Mn, Fe, Co, Ni, Cu, Zn, Rb, Sr, Y, Zr and Pb. Calibration curves generated from EDXRF data were for the most part linear. Scatter about the calibration curves appears to be more a function of the concentration range used for a particular element rather than the methods used in introducing the element.

METHODS AND MATERIALS

Reagents

Solid compounds used to make the synthetic reference and calibration standards were analytical reagent grade potassium acid phthalate and $CA(NO_3)_2 \cdot 4H_2O$ (Baker Chemical Co.); ultrahigh purity (99.999% or greater) TiO_2, Cr_2O_3, Fe_2O_3, ZrO_2 and Y_2O_3 (Aldrich Chemical Co.). Aqueous solutions were prepared from ultrahigh purity compounds or purchased as commercially prepared atomic absorption standards. These included 1000 µg/mL V (V_2O_5 in dil. HCl), 1000 µg/mL Cr (K_2CrO_4 in dil. HNO_3), 1000 µg/mL and 10,000 µg/mL Mn ($Mn(NO_3)_2 \cdot 6H_2O$ in dil. HNO_3), 1000 µg/mL Ni ($Ni(NO_3)_2$ in dil. HNO_3), 1000 µg/mL Cu (Cu in dil. HNO_3), 10,000 µg/mL Pb ($Pb(NO_3)_2$ in dil. HNO_3), 10,000 µg/mL Rb ($RbNO_3$ in dil. HNO_3), and 10,000 µg/mL Sr ($SrCO_3$ in dil. HNO_3). Compounds used in the flux were $Li_2B_4O_7$ (Spex Industires, Lot #178), $LiNO_3$ (Spex Industries, Lot #6791) and analytical reagent grade silica gel (Baker Chemical Co.).

Preparation of Synthetic Standards for Fusion

Flux was prepared in 10-12 g portions for fusion by weighing $Li_2B_4O_7$, SiO_2 (silica gel) and $LiNO_3$ to carefully maintain a proportion of nine parts $Li_2B_4O_7$, one part SiO_2, and enough $LiNO_3$ to make the final mixture 1.5% $LiNO_3$. After weighing the flux into a clean glass container, it was thoroughly mixed and the desired elemental components added. Solid compounds were simply weighed and mixed thoroughly with the flux before fusion.

Aqueous solutions in volumes of 30-1000 µL per element added were micropipetted directly into the flux. Care was taken to insure that the total volume of liquid added was not so great that it would not all be absorbed by the flux and thus break through to the surface of the glass container. The maximum volume that could be absorbed in 10-12 g of flux was approximately 3 mL. After drying the flux for several hours at about 80°C, it was thoroughly remixed before fusion.

Serial glass dilutions were performed by diluting portions of glass previously fused with known elemental concentrations with additional flux. Table 1 summarizes the particular technique used with respect to each element included in the study.

Table 1: Introduction of Elements Into
$Li_2B_4O_7$, SiO_2, $LiNO_3$ Flux

Solid Compounds	Aqueous Solutions	Glass Dilution
K, Ca, Ti, Cr, Fe,	V, Cr, Mn, Co, Ni, Cu,	Ti, Y, Zr
Y, Zr	Zn, Rb, Sr, Pb	

Fusion

Fusion was carried out in a 95% Pt, 5% Au crucible (Matthey-Bishop, Malvern, PA) over a compressed air/natural gas flame for 30-60 minutes or until clear. The fused product was poured into a round graphite mold (31 mm diameter) maintained at 450°C on a hot plate. Stress cracking of the discs was minimized by leaving the discs in the graphite mold at 450°C for 20-30 minutes after the product had solidified to allow the glass to anneal. This procedure produced a 31 mm diameter disc weighing 8-10 g which could be analyzed by XRF assuming infinite thickness.

Before analysis, the discs were surfaced by wet grinding on a 120 grit belt grinder, polishing with alumina compound and finally finsing in sequence with tap water, dilute HNO₃, deionized water and methanol. Reference standards required for obtaining a pure spectra of each element were prepared using the same techniques as were used for the calibration standards.

Five sets of four calibration standards along with a set of five blanks were ultimately prepared. The first two sets appearing in Table 2 are referred to as "major" element standards since they range generally from a few tenths of a percent by weight to a maximum of 1-2%. Calibration sets 3-5 in Table 2 are referred to as "minor" element standards since the individual element concentrations range from a few μg/g to several hundred μg/g. Iron is also present in two of the three minor element standards in concentrations ranging from a few tenths of a percent to just over 1%. This was done to maintain iron in those standards in the same range as the anticipated iron concentration in the samples to be analyzed.

Table 2: Elemental Combinations Within
Calibration Standard Set

Calibration Set	Elements Present
1	Ca, K, Fe
2	Cr, Fe
3	Ti, V, Mn, Fe, Cu, Zn, Rb, Pb, Sr, Zr, Y
4	V, Ni, Cr, Co, Fe
5	Ti, V, Cr, Mn, Ni, Cu, Zn, Pb, Zr, Y

Apparatus

Spectra of the reference and calibration standards were obtained using a United Scientific (Nuclear Semiconductor Division) Spectrace 440 energy dispersive X-ray fluorescence spectrometer with Tracor Northern 880 analyzer equipped with a PDP-11/05 computer. A pulsed Ag tube operating at 30 kV and 0.03 mA was used along with a 0.025 mm Ag filter and vacuum. The counting time was 1000 seconds. These conditions were selected to give a reasonable sensitivity for all the elements included in the study. The spectra obtained from the fused multielement calibration standards were processed using a multiple least squares fitting routine (XML) that deconvolutes overlapped peaks with the help of a standard reference library of spectra. The library spectra were obtained from single element reference standards fused in the same manner as the calibration standards.

RESULTS AND DISCUSSION

Calibration curves constructed from the data appear in Figures 1-5 and give 17 calibration curves plotted for 16 elements, while Table 3 summarizes the sensitivities calculated from the slope of the calibration curves, the theoretical detection limit (TDL) calculated from $3\sigma_B$, the range of standards used for the calibration as well as the correlation coefficient for the calculated least squares line.

Since ultimately the calibration standards were designed to be used in the determination of major and minor elements in river and lake sediments, concentration ranges were chosen to correspond to those anticipated in those samples. In some cases, as can be seen in Table 3, the lower end of the concentration range for some of the elements fell below or near the TDL accounting for some of the lack of precision observed for the calibration curves of V, Cr, Co, Ni and Cu. All the imprecision in the Cu and Ni curves cannot, however, be accounted for by the nearness of the concentrations to the detection limit. It was observed in fusing the reference standards for Cu and Ni that a dull gray residue was formed on the surface of the crucible. Subsequent fusion in the crucible in which the nickel reference was fused gave high Ni counts decreasing with each subsequent fusion of blank flux. This would seem to imply that in the case of Ni, reduction is taking place at the walls of the Pt-Au crucible. Use of higher proportions of the $LiNO_3$ oxidizer may help to alleviate this problem.

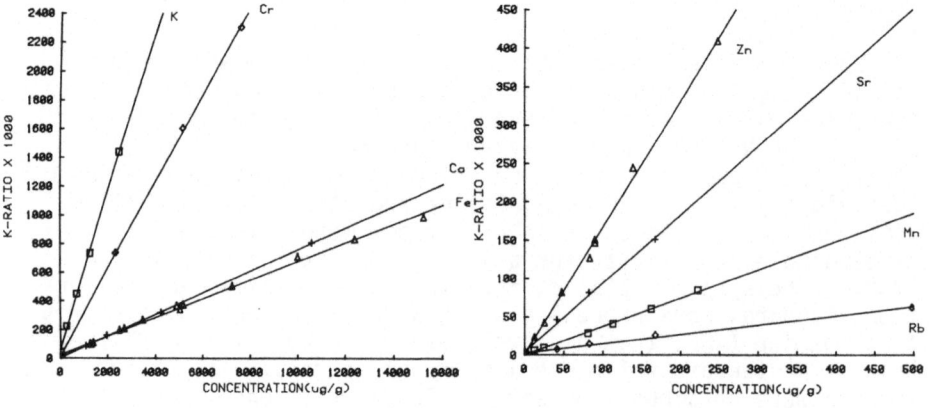

Figure 1:   Calibration curves for Fe, Ca, K, and Cr.   Figure 2:   Calibration curves for Zn, Sr, Rb, and Mn.

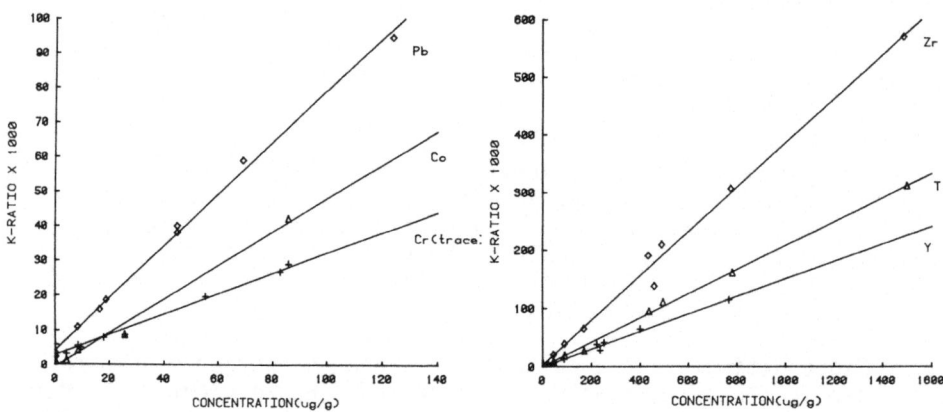

Figure 3:   Calibration curves for Co, Pb, and Cr (trace).   Figure 4:   Calibration curves for Ti, Zr, and Y.

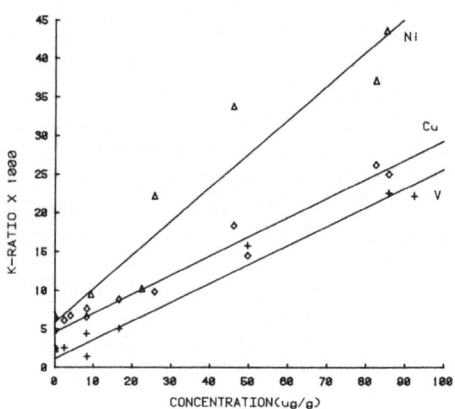

Figure 5:   Calibration curves for Ni, Cu, and V.

Table 3: Sensitivity and Detection Limit Values Computed for 1000 sec Counts at 30 kV and 0.03 mA Using a Thin Ag Filter

ELEMENT	SENSITIVITY (CPS/PPM)	TDL ($3\sigma_B$)	STDS RANGE (PPM)	CORRELATION COEFFICIENT (R)
K	0.002	77	257-2502	.9994
Ca	0.003	46	1100-13000	.9999
Ti	0.009	14	44-1500	.9990
V	0.016	10	2-166	.9896
Cr	0.022	6	4-86	.9977
Cr	0.022	6	2400-10000	.9998
Mn	0.043	3	12-250	.9989
Fe	0.055	4	1200-12000	.9983
Co	0.071	2	4-86	.9939
Ni	0.129	2	4-83	.9552
Cu	0.102	2	4-86	.9859
Zn	0.262	2	12-250	.9987
Rb	0.808	3	41-500	.9956
Sr	1.100	2	41-500	.9958
Y	1.320	3	22-766	.9942
Zr	1.340	2	44-1500	.9944
Pb	0.092	2	8-124	.9983

Comparison of the Cr curves in Figures 1 and 3 gives a good test of the relative merits of using solid compounds versus aqueous solutions. While the curve produced from the solid Cr_2O_3 (Fig. 1) has a slightly better fit, the sensitivities computed from the slopes of the two curves are precisely (within 0.67%) the same, illustrating the wide range of linearity obtainable. In general, linearity was excellent regardless of the method applied. The curves for Ti, Y and Zr (Fig. 4) each have points derived from discs in which the oxide was added directly and in which glass fused with about 1% Zr and Ti and 0.5% Y was diluted with additional flux to their final concentrations.

CONCLUSIONS

Addition of solid compounds and aqueous solutions to flux along with serial dilutions of glass all appear to be effective techniques for producing synthetic calibration standards fused with $Li_2B_4O_7$. Selection of a technique would appear to be most dependent upon the concentration range desired in the standards. In retrospect, it might be more advantageous to use 10,000 µg/mL aqueous standards and/or multielement aqueous standards to reduce the volume of solution transferred to the flux, thus reducing the danger that the flux would not absorb all the liquid and also reducing drying time. Use of multielement aqueous standards would reduce the time and tedium required in pipetting individual solutions as well as reduce the volume of liquid transferred.

ACKNOWLEDGEMENT

The authors acknowledge the AMAX Foundation and the Camille and Henry Dreyfus Foundation for partial support of this work. Kent I. Mahan acknowledges the University of Southern Colorado for support of a sabbatical leave.

REFERENCES

1. F. Claisse, Quebec Dept. Mines, P.R. 327, 1956.

2. F. Claisse, Norelco Reporter, 4, 1957.

3. P.A. Pella, K.E. Lorber and K.F.J. Heinrich, Anal. Chem., 50, 1268, 1978.

4. E. Jagoutz and C. Palme, Anal. Chem., 50, 1555, 1978.

THE USE OF EDXRF FOR LIQUIDS IN A URANIUM-VANADIUM SOLVENT

EXTRACTION PROCESS

Rocky A. Smith

Energy Fuels Nuclear, Inc.
Post Office Box 787
Blanding, Utah   84511

Elements of importance in a uranium-vanadium solvent extraction process can be monitored using EDXRF. Fifteen independent circuits were monitored for six different elements on a hourly basis. The precision of the analysis varies from element to element and from circuit to circuit, but all precision was within acceptable limits set for mill control by more conventional means of analysis. The elements of primary interest were uranium and vanadium.

INTRODUCTION

The circuit matrix types include acidic leach solutions, caustic strips, sodium chloride strips, and extractive kerosene. The analysis of the elements in each circuit using conventional methods is not economically feasible. EDXRF is well suited for this problem because it can analyze a large number of different elements in a solution without any major sample modifications. The analysis is much faster and as accurate in most cases as conventional control analysis. This report describes the use of EDXRF in coordination with a linear regression program to handle the data. The elements of primary interest in the circuit are uranium and vanadium, but a group of secondary elements including sulfur, chloride, iron, and molybdenum are also important. The goal of this program is to produce mill control assays faster and more efficient.

EXPERIMENTAL

Apparatus

A United Scientific Spectrace TM 440 energy dispersive XRF spectrophotometer, with a Mo sidewindow, 50 KV maximum, x-ray tube was used to measure the intensities of the samples. A Tracor Northern TN2000 microprocessor was used to control the x-ray and manipulate the data generated. The x-ray tube was operated at four different power and current settings. Other instrumental specifications and conditions were as follows: automatic twenty position sample changer for 1.25 inch samples, externally controlled transmission filters, 150 ev, 30 mm^2 detector with a 1.0 mil Be window, pulsed tube subsystem with remote current and voltage capability. A collimator insert was used to reduce the view to a 3/8 inch diameter.

Reagents

All chemicals used were of reagent grade quality. Sulfuric acid was used to increase the solubility of vanadium in several of the vanadium circuits. Commercial grade nitrogen gas was used to purge the x-ray chamber of argon and other interferring gases.

Procedure

The vanadium feed and raffinate solutions were acidified with concentrated sulfuric to the 2% level, to redissolve the vanadium precipitate. These samples are then filtered to remove any solids which might be entrained. Filtering is not always necessary. Five ml of sample is put in the cups that have been covered with film, no cap is used. The samples are placed on the carousel in specific order to coordinate with the order of the calibration curves on file in the computer. Nitrogen gas is injected into the x-ray chamber for 45 to 60 seconds to purge any argon present before the x-ray measurements are made. Each of the four acquisition conditions were run for twenty seconds live time. Dead time was fifty percent or less in all acquisitions.

Discussion

Matrix effects beome minimal when you are dealing with a circuit that changes very little over a long period of time. Each circuit has a linear regression curve established for its particular matrix and concentration ranges. The overall result is a program that works for a large variety of sample types on an individual basis.

TABLE 1 - ACQUISITION CONDITIONS

ELEMENTS	FILTERS	KV	MA	KeV RANGE
S, Cl	None	15	.2	0-10
V, Fe	Cellulose	15	.5	0-10
U	Moly	30	.3	0-20
Mo	Copper	50	.5	15-25

TABLE 2 - CIRCUIT CONCENTRATION RANGES

CIRCUIT	U_3O_8 g/l	V_2O_5 g/l
V_2O_5 Loaded Strip	*	180-250
V_2O_5 Feed	*	0.5-10.0
V_2O_5 Loaded Organic	*	1.0-15.0
U_3O_8 Feed	0.5-2.0	0.2-10.0
U_3O_8 Raffinate	*	0.2-10.0
U_3O_8 #1 Strip	15.0-45.0	*
U_3O_8 Loaded Organic	0.75-2.0	*
U_3O_8 Stripped Organic	0.01-0.10	*

* See below for Threshold Levels

TABLE 3 - THRESHOLD LEVELS

CIRCUIT	U_3O_8 g/l	V_2O_5 g/l
V_2O_5 Raffinate	----	0.012
V_2O_5 Stripped Organic	----	0.020
U_3O_8 Raffinate	.020	-----
U_3O_8 Stripped Organic	.015	-----

TABLE 4 - CIRCUIT PRECISION FOR TYPICAL SAMPLES

CIRCUIT	MEAN g/l		RSD %
V_2O_5 Loaded Strip	265.10	3.90	1.47
V_2O_5 Feed	4.12	.132	3.21
V_2O_5 Raffinate	.259	.025	9.80
V_2O_5 Loaded Organic	10.15	.086	0.85
U_3O_8 Loaded Organic	1.34	.038	2.84
U_3O_8 Stripped Organic	.053	.006	10.09
U_3O_8 Feed	.590	.049	8.28
U_3O_8 #1 Strip	28.41	.470	1.68

The acquisition conditions were chosen after a wide variety of matrix and elemental differences were considered. Acquisition conditions can be seen on Table 1. A remote sample and power control is essential for this many different acquisitions and samples. The solutions that were analyzed were of a volatile nature, pre-empting the use of the vacuum. Nitrogen was used as the purge gas in place of helium because of cost and availability.

The solvent extraction process involves a wide range of elemental concentrations. For example, vanadium ranges from 0.10 g/l to 250.0 g/l. Table 2 shows probable concentration ranges for uranium and vanadium in several circuits. Reasonable threshold levels are typically three times the calculated detection limit.[1] Table 3 shows some typical threshold levels for uranium and vanadium for a few major circuits. Precision of analysis for uranium and vanadium has been calculated (see Table 4) for each matrix type and concentration extremes. The precision is the lowest for sample types with lower uranium and vanadium concentrations. Count times were held to twenty seconds, even though higher count times gave better precision.

This program assumes that the samples from a circuit remain constant in element ratios. This need not be absolutely true in every case for every element, but no inter-elemental effects are incorporated in this program. Large concentration changes can cause problems. Matrix interferences are minimal because each matrix has its own calibration curve. Intensities were collected as K-ratios and calculated using a multi-element linear regression program. There is a curve for each circuit and a specific order must be maintained when putting the samples in the carousel.

CONCLUSION

Application could be widespread in the uranium-vanadium industry or any industry that monitors large liquid circuits of a similar nature. The limit of this method is reached when concentration changes become large, or when inter-elemental effects are more prominent. The ideal situation would be a mill processing ore from a single ore deposit of consistent makeup. Solvent extraction and leach circuits are somewhat sensitive to changes in the ores basic matrix composition. Different ores generally have different elements which might cause some interferences that the system is not calibrated for. Several sets of curves can be put in the computer for the different ore bodies.

REFERENCES

(1) J. Russ, R. Shenand, R. Jenkins, "Exam: Principles & Experiments," Edax International, Inc., 1978, p. 66.

A RESIN-LOADED PAPER X-RAY FLUORESCENCE METHOD FOR

DETERMINING URANIUM IN PHOSPHATE MATERIALS

Benjamin W. Haynes, Jerome Zabronsky, and David L. Neylan

U.S. Department of the Interior, Bureau of Mines
Avondale Research Center
4900 LaSalle Road
Avondale, Maryland 20782

INTRODUCTION

The assessment of the environmental impact of mining and mineral processing is an important part of the research program of the U.S.D.I. Bureau of Mines. Under the Minerals Environmental Technology program, the Avondale Research Center evaluated the accessory minerals and elements associated with the mining, beneficiation, and processing of phosphate rock in the central Florida phosphate district (1).[1] In support of this project, a resin-loaded paper (RLP) X-ray fluorescence (XRF) method was developed for determining uranium in these processing and waste materials.

Several methods for determining uranium in a variety of matrices including phosphate rock have been published (2-5). Sill et al. reported methods for determining radionuclides from Pb (Z = 82) to Cf (Z - 98) in uranium mill tailings, soils, dusts, and biological samples (6-8). The uranium is determined by alpha spectrometry after a $BaSO_4$ precipitation and an electrodeposition step.

To avoid some of the tedious chemistry and the electrodeposition step, development of a method of using RLP and XRF analysis was undertaken. The objective was to develop a method capable of determining uranium over a concentration range of 1 to 200 µg/g in a matrix very high in calcium. Since the original work on resin-loaded papers, done at this research center (9), the use of such papers followed by XRF or neutron activation analysis has resulted

---

[1]Underlined numbers in parentheses refer to items in the list of references at the end of this report.

in methods for the determination of trace elements in various matrices (10). Various types of uranium collection were tried, all involving U^{+6}. Anion RLP collection of uranium as $UO_2(SO_4)_2^=$ was chosen as the best method for preconcentrating uranium from phosphate rock processing and waste materials.

In applying this technique to determining uranium in phosphate materials, the high concentration of calcium interferes with uranium collection as a cation. Eliminating this interference while developing a reliable method for determining uranium which allowed determining other radionuclides was the research goal.

PROCEDURE

The procedure developed uses the fusion-dissolution and precipitation of interfering radionuclides with BaSO$_4$ as described by Sill et al. (6-8), followed by uranium extraction from the filtrate by TOPO (tri-n-octyl phosphine oxide) (11). The TOPO is stripped with $(NH_4)_2SO_4$ solution, and the uranium collected as $SO_4^=$ anion on $SO_4^=$ saturated SB-2 papers (Whatman, Inc., Clifton, N.J.)[2]. This procedure allows effective uranium separation from the other radionuclides, removal of the calcium interference by extraction with TOPO, uranium collection on RLP, and uranium determination by XRF analysis. Though not discussed in this manuscript, the other radionuclides could be determined by conventional radioanalytical techniques (6-8).

Fusion-Dissolution

The fusion dissolution procedure of Sill et al. (6-8) was used with a 0.5 g sample and a final sample volume of approximately 20 ml to simplify the stripping procedure. According to Sill, the uranium is quantitatively retained in the filtrate, and the remaining radionuclides are retained in the BaSO$_4$ precipitate.

Extraction-Stripping

The uranium is readily extracted from the resulting high chloride solution with TOPO (11) using the following procedure.

1. One ml of conc. HCl is added for each 10 ml of solution present in the separatory funnel to ensure a minimum of 1N HCl solution.
2. To the separatory funnel is added 5 drops of 2 M hydroxylamine sulfate to reduce any Fe^{3+} to Fe^{2+} and keep the uranium as U^{6+}.
3. 0.1 M TOPO in cyclohexane is added until the organic to aqueous (O:A) ratio is 1:2 (i.e., 10 ml organic/20ml aqueous).

---

[2]Reference to specific suppliers or products is for reference purposes only and does not imply endorsement by the Bureau of Mines.

4. The separatory funnel is shaken for 2 min, and the mixture allowed to settle for 10 min. The aqueous phase is discarded.
5. A 2.0 M pH 1 $SO_4^=$ solution is used as a stripping agent with the O:A ratio of 1:2. The phases are shaken for 2 minutes and allowed to equilibrate for 10 minutes. The aqueous phase is removed into a clean beaker, and the organic is discarded.
6. The stripping solution is adjusted to a pH of 2.0 to 2.5 with concentrated aqueous NH_3. The solution is then diluted with H_2O so that the volume of this dilution is about three times the amount of stripping solution used (60 ml total volume).
7. The pH is adjusted with 1:100 v/v% solution of NH_3 to pH 3.0. Final dilution to 0.5 M $SO_4^=$ is made with H_2O so the volume is 4 times the original volume of stripping solution removed (80 ml). Final pH is adjusted to 3.0 with 1% NH_3 or 1% HCl.

Resin-Loaded Paper X-Ray Fluorescence

8. The solution is passed 7 times through each of 6 SB-2 resin-loaded paper disks previously saturated with $SO_4^=$.
9. The papers are air-dried, and wavelength dispersive XRF spectrometry is used to determine the U collected on each disk.

The 6th disk was used for background calculations of net intensity based on 99+% collection on the first 4 or 5 disks. These intensities were compared with net intensities of standards prepared by adding known amounts of uranium to solutions containing 0.5 M $SO_4^=$ at pH 3.0 and collecting uranium on SB-2 papers.

EXPERIMENTAL

Prior to analysis of phosphate materials, the extraction and stripping of uranium using TOPO was studied to determine the minimum level for effective uranium recovery. Prepared solutions of 1N HCl were spiked with 20, 10, and 5 μg of uranium, extracted with TOPO, and stripped with pH 1.0, 2 M $SO_4^=$ solution as outlined in the procedure. The pH was adjusted to 3.0, the $SO_4^=$ concentration adjusted to 0.5 M, and the solution passed through SB-2 pa-pers as previously described. These papers were analyzed by XRF, and net intensities calculated and compared with intensities of a 100-μg uranium standard. This standard was prepared by adding 100 μg uranium to a pH 3.0, 0.5 M $SO_4^=$ solution and collecting the uranium on SB-2 papers following the procedure. Excellent recovery (>98 pct) was obtained at the 5-μg uranium level.

RESULTS AND DISCUSSION

A roundrobin evaluation of methods for determining uranium was conducted on phosphate waste materials provided by the Bureau's

Albany Research Center. Four Bureau of Mines research centers
participated in the roundrobin study, using the method of analysis preferred at each center. The Avondale Research Center
analyzed these materials using the dibenzoylmethane (DBM) method,
as described by Jones (2), and by the new procedure described in
this paper. The results obtained using both methods at this
center are compared in table 1 with the results obtained by the
other three centers using the DBM method or fluorescence methods.

The National Bureau of Standards (NBS) Standard Reference Material
(SRM) Phosphate Rock 120b has a certificate value for uranium of
128.4 ±0.8 µg/g. Analysis of this SRM by the RLP and XRF procedure
and by the DBM method gave good agreement between the certificate
value and values obtained by the two methods as listed in table 2.

TABLE 1. – Uranium in Albany Research Center
roundrobin samples, µg/g

Sample	Avondale (RLP)	Avondale (DBM)	Albany	Reno	Salt Lake City	Avg
Albany-1	70	76	76	75	71	75.5
Albany-2	84	87	83	90	85	86.2
Albany-3	77	82	79	81	75	79.2
Albany-4	70	73	68	67	67	68.8
Albany-5	67	67	66	65	60	64.5

Selected phosphate materials were analyzed by the RLP method and
compared with DBM results obtained on the same samples (1). Table
2 shows good agreement between the two methods over a wide range
of uranium values from 10 to 200 µg/g. Sample weight ranged from
0.1 to 0.5 g with the 0.5-g sample size used most frequently. However, this method should be directly adaptable to large samples (10
and 50 g) as the fusion procedure is the same as the one outlined
by Sill et al. (7).

SUMMARY AND CONCLUSIONS

A method has been developed for determining uranium in phosphate
rock processing and waste materials using resin-loaded paper
disks and X-ray fluorescence analysis. Calcium interference was
successfully eliminated by extracting the uranium from the aqueous solution with TOPO and using an anion collecting resin to
preconcentrate uranium prior to the analysis.

The procedure developed takes the filtrate containing uranium as
UO_2^{2+}, extracts the uranium with 0.1 M TOPO in cyclohexane, and

TABLE 2. - Determination of uranium in phosphate materials by resin-loaded paper X-ray fluorescence and DBM, µg/g

Sample type and ID	RLP	DBM
Matrix, ore grade.........M-1	80	81
Matrix, low grade.........M-4	28	26
Matrix, low grade.........M-6	22	22
Clay, slimes..............C-1	194	202
Gypsum.................. G-3	10	10
NBS-SRM-120b, rock concentrate[1]	130	131

[1]Certified to be 128.4 ±0.8 µg/g.

ous solution with TOPO and using an anion collecting resin to preconcentrate uranium prior to the analysis.

The procedure developed takes the filtrate containing uranium as UO_2^{+2}, extracts the uranium with 0.1 M TOPO in cyclohexane, and strips the TOPO with a 2 M $SO_4^=$ solution at pH 1.0. The efficiency of this extraction and stripping is >98% for concentrations of >5 µg. The strip solutions are converted to 0.5 M $SO_4^=$ and pH 3.0 for collection on $SO_4^=$ saturated SB-2 papers, and the papers are analyzed for uranium using the uranium Lα line in wavelength dispersive XRF spectrometry.

REFERENCES

1. Haynes, B. W., G. W. Kramer, and J. A. Jolly. BuMines RI 8576, 1981.
2. Jones, M. M., J. S. MacDuff, and A. B. Whitehead. BuMines RI 8433, 1980, 12 pp.
3. Woodis, T. C., Jr., J. R. Trimm, J. H. Holmes, Jr., and F. J. Johnson. JAOAC, v. 63, 1980, pp. 208-210.
4. Adams, J. A. S., and W. J. Maeck. Anal. Chem., v. 26, 1954, pp. 1635-1639.
5. Jablonski, B. B., and D. W. Leyden. Anal. Chem., v. 51, 1979, pp. 681-683.
6. Sill, C. W. Anal. Chem., v. 52, 1980, pp. 1452-1459.
7. Sill, C. W. Anal. Chem., v. 33, 1961, pp. 1884-1886.
8. Sill, C. W., and R. L. Williams. Anal. Chem., v. 41, 1969, pp. 1624-1632.
9. Campbell, W. J., E. F. Spano, and T. E. Green. Anal. Chem., v. 38, 1966, pp. 987-996.
10. Law, S. L., and W. J. Campbell. Advances in X-Ray Analysis, v. 17, 1974, pp. 279-292.
11. Horton, C. A., and J. C. White. Anal. Chem., v. 30, 1958, pp. 1779-1784.

THE X-RAY ANALYSIS OF URANIUM ORES FOR IRON SULFIDE MINERALS

A. J. Durbetaki, R. H. Carlson, and T. F. Quail

FMC Corporation, Chemical Research and Development
Center, Princeton, New Jersey 08540

INTRODUCTION

Hydrogen peroxide is used to extract uranium by the in situ
leaching of sandstone ore deposits containing uraninite (UO_2). Since
FeS_2 minerals, marcasite and pyrite, also occur in these deposits and
they consume hydrogen peroxide in their oxidation, it is important to
determine their concentration.

A quantitative X-ray diffraction (XRD) method was therefore
developed in order to monitor the concentration of marcasite and
pyrite in sandstone ores.

EXPERIMENTAL

The XRD measurements were made with an automated powder diffractometer (APD 3500, Philips Electronic Instruments Inc.). Integrated
peak areas were measured using a step-scanning mode. Silver membrane
filters (0.45 μm pore size) were used as sample supports.

All samples and standards, simulating the uranium ore matrix,
were ground to a particle size <10 μm under hexane in an inert
atmosphere. A portion of the ore or standard, 0.0250 g, was then
reacted with 20 mL dilute aqueous acetic acid (10% V/V) using an
ultrasonic vibrator and deposited quantitatively on the silver
membrane.

A long fine focus copper-target X-ray tube operated at 50 kV and
30 mA, a 1° divergence and 0.3° receiving slits were used for this
study.

113

RESULTS AND DISCUSSION

Table 1. Least Squares Curve Fit for Pyrite and Marcasite

X-Actual	Y-Actual	Y-Calc.	Pct. Diff.
0.25	3300	3354	-1.65
0.50	7051	6589	6.55
1.00	13749	12723	7.46
1.25	15450	15634	-1.19
1.50	17475	18447	-5.56
1.75	20400	21168	-3.77
2.00	22214	23802	-7.15
2.50	27007	28821	-6.72

$Y=X/(A+BX)$ where $A=7.8170 \times 10^{-5}$ and $B=1.0 \times 10^{-5}$, Correlation Coefficient = 0.99671, X = pyrite concentration (%), Y = integral counts of analyte (311) reflection

X-Actual	Y-Actual	Y-Calc.	Pct. Diff.
0.50	992	982	1.05
1.00	2025	2215	-9.38
1.25	3050	2958	3.00
1.50	3952	3811	3.55
1.75	4965	4800	3.32
2.00	6150	5959	3.10

$Y=X/(A+BX)$ where $A=5.7231 \times 10^{-4}$ and $B=1.1 \times 10^{-4}$, Correlation Coefficient = 0.99565, X = marcasite concentration (%), Y = integral counts of analyte (211) reflection

The interference free (211) and (311) reflections for marcasite and pyrite, respectively, were selected for their quantification. Following solubilization of calcite by acetic acid to enhance the concentration of the iron sulfides in the filter deposit and reduce background noise level, the intensities of these peaks for a series of standards were integrated. The integral intensities were subjected to regression analysis as a function of concentration of analyte. A quadratic least square curve fit was obtained for both FeS_2 phases (Table 1).

The equation $Y=X/(A+BX)$ was used to calculate the concentrations of the FeS_2 phases in the ores.

Table 2. Determination of Marcasite and Pyrite in
Uranium Ores by XRD: Verification of Accuracy[a]

	Concentration, %	
Sample	Marcasite	Pyrite
A	0.53 ± 0.05 (3)	0.30 ± 0.04 (3)
A + FeS$_2$	3.34 [3.27]	2.79 [2.79]
B	0.38 ± 0.03 (3)	0.27 ± 0.03 (3)
B + FeS$_2$	3.16 [3.12]	2.81 [2.76]
C	1.6 ± 0.06 (4)	1.0 ± 0.04 (4)
C + FeS$_2$	2.78 [2.79]	2.65 [2.60]

[a] A and B were spiked with 2.74% marcasite and 2.49% pyrite, C
with 1.19% marcasite and 1.60% pyrite. Figures in parentheses
represent number of replicate analyses and in brackets
calculated concentrations.

The accuracy of the method was verified by standard addition
of pyrite and marcasite in three ore specimens (Table 2).

Marcasite (sp. gr. 4.887) and pyrite (sp. gr. 5.018) also were
isolated along with uraninite as a heavy mineral fraction using
heavy-liquid separation with tribromoethane (sp. gr. 2.89). This
concentrate gave well defined spectra of these minerals which
further confirmed the presence of only these two "pyritic" phases.

In conclusion, the XRD method that has been developed and
reported in this communication, represents a valid method for the
analyses of marcasite and pyrite in uranium sandstone ores.

ACKNOWLEDGEMENTS

We wish to express our appreciation to the Industrial Chemical
Group for permission to publish and to Drs. C. F. Ferraro and
H. Stange for their valuable review of this work.

A STATISTICAL COMPARISON OF DATA OBTAINED FROM PRESSED DISK AND FUSED BEAD PREPARATION TECHNIQUES FOR GEOLOGICAL SAMPLES

Randall H. Dow

School of Oceanography
Oregon State University
Corvallis, OR 97331

ABSTRACT

The analysis of geological samples by x-ray fluorescence analysis is complicated by matrix effects and sample variability. Diluting the sample in a suitable matrix such as lithium tetraborate and then fusing it helps to reduce problems associated with sample inhomogeneities and reduces matrix effects considerably. Unfortunately, many of the trace elements in the sample can be diluted below their detection limits by this technique. Pressing the ground powder into a hard disk avoids this problem and is less time consuming. Matrix effects can be very serious however, but can be calculated and compensated for by fundamental parameters programs.

Possible errors in a sample analysis can be divided into errors associated with sample inhomogeneity, sample preparation errors, errors introduced by the spectrometer, and those introduced by data reduction. Geological samples were collected in a manner designed to separate and compare these errors; as well as to compare the fused bead and pressed disk preparation techniques. The largest possible source of error may be introduced during reduction of the data.

INTRODUCTION

The School of Oceanography at Oregon State University is using a Philips PW1600 simultaneous wavelength dispersive x-ray spectrometer in the analysis of geological samples. The spectrometer measures 25 to 35 elements using 25 fixed and two scanning channels. It is controlled by a Digital Equipment Corporation (DEC) PDP 11/04 minicomputer running the RT-11 operating system and an in-house

117

control program. A DEC 11/34 is used for data processing using an
empirical alpha coefficient matrix correction program and a subset of
the NRLXRF fundamental parameters program. The full NRLXRF program
is installed on a DEC-10 computer available through the Oregon State
System of Higher Education.

Our samples have been measured in batches of 16 unknowns and
four standards. One of the four standards is permanently left in the
spectrometer to monitor drift, and the other three are intended to be
similar to the unknowns' concentrations.

Errors in samples analysis can be divided into errors associated
with sample inhomogeneity, sample preparation, errors introduced by
the spectrometer, and data reduction errors.

DISCUSSION

Spectrometer errors would be typified by short and long term
drift and by random noise. Spectrometer stability can be monitored
by analyzing the measured counts per second for a single standard
monitor over a long period of time. The monitor sample is measured
at regular intervals while the spectrometer is running. Table I
represents measurements made three times, about every 30 minutes,
eight hours per day, for a period of over four months. Over both the
short term of several hours and long term of four months our
spectrometer has proven very stable. A single channel can be
expected to yield measurements with a relative standard deviation of
less than one half percent. Several channels are sensitive to line
voltage spikes and gave relative standard deviations of up to 3.37%.
Upon examining the data used to develop this table it was observed
that only a very few instantaneous spikes generated the poorer
stability on these channels. A method to detect such spikes will be
implemented which will allow immediate and automatic remeasurements.
We consider this to be excellent stability for our spectrometer.

Our samples vary widely in constituent compositions. The
samples range from ocean floor sediments, to manganese nodules, and
to volcanic rock. We use two primary methods of sample preparation.
A fusion disk is prepared using the Claisse fluxor and a recipe of
one gram sample, six grams lithium tetraborate, 1.5 grams lithium
nitrate, and a small amount of a bromide solution as a non-wetting
agent. A pressed powder disk is prepared by grinding the sample,
using a binding agent and pressing with a backing. For most marine
sediments the grinding step is unnecessary, as is the binding agent,
because the particle size is small and the sample need only be
disaggregated, homogenized, and pressed.

To attempt to separate errors associated with sampling, sample
preparation and data reduction, a set of samples was collected and

Table I -- Spectrometer stability

Elem	Counts/s	%Rel Dev	Elem	Counts/s	%Rel Dev
11 Na	1409 ± 8	0.56	27 Co	867 ± 3	0.32
12 Mg	1533 ± 13	0.85	28 Ni	880 ± 7	0.76
13 Al	37535 ± 189	0.51	29 Cu	876 ± 3	0.37
14 Si	74952 ± 327	0.44	30 Zn	1593 ± 5	0.29
15 P	1848 ± 6	0.32	37 Rb	20446 ± 64	0.31
16 S	407 ± 12	2.89	38 Sr	2889 ± 95	3.27
17 Cl	406 ± 8	1.86	56 Ba	87 ± 1	1.05
19 K	21887 ± 102	0.46	57 La	488 ± 2	0.47
20 Ca	19139 ± 47	0.24	58 Ce	1309 ± 4	0.28
22 Ti	284 ± 2	0.55	82 Pb	3557 ± 12	0.34
24 Cr	111 ± 1	1.03	90 Th	6847 ± 21	0.30
25 Mn	2137 ± 9	0.44	92 U	6956 ± 15	0.22
26 Fe	56529 ± 182	0.32			

Table II -- Silicon concentration (per cent)

A1-Fus	7.42	B1-Fus	5.37	A1-Pres	4.00	B1-Pres	4.25
A2-Fus	7.64	B2-Fus	5.43	A2-Pres	3.98	B2-Pres	4.23
A3-Fus	7.35	B3-Fus	5.49	A3-Pres	4.02	B3-Pres	4.22

Table III -- Results of standards

		Pressed		Fused	
	Std.	Meas.	%Diff.	Meas.	%Diff.
Si	27.9	25.3	−9.2	30.0	7.4
Al	6.19	5.44	−12.1	7.39	19.4
Ca	5.93	5.91	−0.2	5.98	0.9
Fe	4.53	4.29	−5.2	4.45	−1.7
K	3.49	3.13	−9.3	3.04	−13.1
Mg	1.59	1.87	12.6	2.81	77.7
Mn	0.256	0.187	−8.8	2.81	1001
P	0.240	0.226	−5.7	2.81	1072
Ti	0.129	0.121	−6.2	2.81	2094
S	0.050	0.051	1.7	2.81	5538
Zn	0.024	0.020	−15.8	2.81	11579
Cu	0.0019	0.0094	392.2	2.81	148536
Co	0.0015	0.0035	136.7	2.81	192253
Ni	0.0012	0.0082	594.7	2.81	233471

processed. Four samples were taken from a sediment core,
disaggregated, and homogenized. Each of the four major samples was
split into six subsamples. Three of the subsamples in each group
were fused and three pressed. The samples were then measured and the
intensities converted to concentrations. Table II shows silicon
concentration for two of the major samples.

By examining the results the errors can be assigned to sampling
type errors, sample preparation errors, and/or data reduction errors.
The spectrometer has already been proven to introduce minimal errors.

Random sampling errors would produce large variances between
groups. Random sample preparation errors would produce large
variances within groups. Data reduction errors would simply give
wrong answers.

The results in Table II show small variances within groups
demonstrating at least consistant handling of the samples during
sample preparations. The large variances between groups should be
interpreted as measuring different samples. This would be a sampling
error if one considered the four original samples to be equal. There
are large data reduction errors which will be discussed later.

To determine the extent of the data reduction errors being
introduced by the simple model we presented to NRLXRF, three Canadian
standards were prepared as glasses and pressed powders. Each was in
turn treated as an unknown while the other two were presented as
standards. Our simple model resulted in concentrations from NRLXRF
which differed significantly from the published values. Minor and
trace elements failed completely for the glasses and did poorly for
the pressed disks. In this part of our experiment glasses did
slightly worse than pressed disks. Table III presents the results.

Returning to the results from the four samples in Table II two
things need to be noted. First, the pressed disks gave less
intersample variation, which is more correct for these samples.
However, our simple NRLXRF model may have exaggerated what
differences there were. Second, the actual concentrations obtained
from the glass beads are more accurate demonstrating that dilution of
the matrix is useful for samples which are truly "unknown".

CONCLUSION

In conclusion, we have demonstrated that our spectrometer is a
stable analytical instrument, and that random sample preparation
errors are small. We have shown that sampling errors are a very
potential problem. And finally, the correct use of data reduction
software is very important. In particular, fundamental parameters
software can be used to obtain very poor results or very good results
depending upon the completeness of the model presented to it.

TRACE AND MINOR ELEMENT ANALYSIS OF OBSIDIAN FROM THE SAN FRANCISCO VOLCANIC FIELD USING X-RAY FLUORESCENCE

Suzanne C. Sanders,* John D. Zahrt, and Graydon Bell**

Northern Arizona University
Chemistry Department
Flagstaff, Arizona 86001

ABSTRACT

Obsidian from eight localities in the San Francisco volcanic field of Northern Arizona was analyzed for 21 minor and trace elements (Ti, Mn, Fe, Co, Zn, Rb, Sr, Y, Zr, Nb, Mo, Tc, Ru, Rh, Pd, Ag, Cd, W, Au, Hg, Th) using x-ray fluorescence analysis. All samples exhibited irregular analyzing surfaces. This was a more in-depth study of a previous study (1) in which a very small number of standards were used.

The data, in the form of intensity ratios relative to iron, were statistically analyzed and the trace and minor element patterns established. The obsidian clustered into four well-defined groups each consisting of two localities: Government Mountain/Obsidian Tank, Slate Mountain/Kendrick Peak, Robinson Crater/O'Leary Peak, and RS Hill/Spring Valley. The four distinct groups were treated individually to refine the separation between the similar sites. Classification function coefficients were calculated for each locality, then used to identify the source of thirteen obsidian artifacts recovered from a Northern Arizona archaeological site.

INTRODUCTION

X-ray fluorescence (XRF) analysis has many applications, but most require that the material studied has a flat, regular

*Now at Lawrence Livermore National Laboratory, Livermore, CA
**Department of Mathematics, Northern Arizona University
Work performed under the auspices of the U.S. Dept. of Energy
by the Lawrence Livermore National Lab under contract #W-7405-ENG-48.

surface. This study presents an example of the use of XRF to study and classify obsidian samples, materials with irregular surfaces. Since the creation of an artificial smooth surface forfeits the advantage of nondestructibility offered by XRF analysis, a scheme must be devised to study these specimens.

Obsidian, or volcanic glass, was used in prehistoric times for the manufacture of weapons and ornaments, and many obsidian artifacts have been found in the excavation of archaeological sites. If the geologic origin of these artifacts can be established, one can determine important gathering habits and trade information about prehistoric peoples. Determination of the geologic source of obsidian is also important in the relatively new technique of obsidian hydration dating. Since the rate of hydration is different for different obsidian outcrops and hydration rates have only been established for certain localities, it is important before cutting an artifact to verify that it is indeed datable using current methods. The motive for this characterization study is therefore largely archaeological in nature.

By studying the patterns of minor and trace elements in obsidian samples, one can establish a fingerprint for a rock outcrop and trace the origin of an obsidian artifact to its geographic source. This technique assumes that both the obsidian flows and obsidian samples are homogeneous.

Obsidian from eight localities in the San Francisco volcanic field of Northern Arizona were analyzed for 21 minor and trace elements using XRF. The data, in the form of intensity ratios relative to iron, were analyzed using statistical techniques to establish the unique trace and minor element patterns.

A less extensive previous study (1) was done which relied on a visual method of analysis of the data of three elements (Zr, Rb, and Sr). The validity of these conclusions is somewhat questionable since only 18 specimens from nine localities were used, in some cases employing only one sample as the standard for a site. Due to the variability which can exist in the obsidian in each sample site, it was felt that too much importance was placed on too little data. The present work extends the previous study.

PROCEDURES AND ANALYSIS

Little sample preparation was involved. The sample was thoroughly cleansed with distilled water and inserted into the sample holder, exposing the flattest surface to the x-ray beam.

Each element was analyzed on the General Electric XRD-6 wavelength-dispersive spectrometer for a fixed time of 100 seconds, and the analysis was repeated at least once. Experimental conditions were 40 kV and 35 mamp.

The working data were background-corrected intensity ratios relative to iron; this proved to be quite effective for the irregular surfaces exhibited by the obsidian. Iron was chosen as the standard since it was the element with the highest number of counts detectable in the system used.

Eight samples from each of the eight localities were analyzed for iron and the 20 listed elements and the data were entered into a stepwise discriminant analysis program (2). This statistical analysis forms one or more orthogonal linear combinations of the discriminating variables (i.e., the 20 elements) that maximize the separation of the groups. The discriminant analysis derives a set of classification function coefficients (listed in Table 1) that can be used directly on the data from an unknown. Note that Mo and Ru are not significant in this first separation and have not been listed in the table. The classification score for each group is computed by multiplying the intensity ratio for each variable by the coefficient, summing the products, and adding a constant. The case is assigned to the group with the highest classification score, for this is the group for which it has the highest probability of membership (3).

RESULTS AND DISCUSSION

This study indicated that the obsidian outcrops studied are indeed separable using the method employed, at least into four major groups (Government Mountain/Obsidian Tank, Slate Mountain/Kendrick Peak, Robinson Crater/O'Leary Peak, RS Hill/Spring Valley). These four groups can each be further subdivided into two localities which are similar both geographically and chemically. Two separate discriminant functions were derived; the first to give a general geographic area of the obsidian source, and the second (not included) as a more refined function to discriminate between two closely-linked groups. Although the groups are so similar that the classification is not perfect between these closely-linked groups, use of the second set of discriminating functions does introduce more reliability into the analysis.

Some samples of known origin were used to test the validity of the results. In all but one case, the origin was correctly identified, at least into the correct major group. The one error indicated a slight overlap which occurs between the Government Mountain/Obsidian Tank and Slate Mountain/Kendrick

Table 1. Classification Function Coefficients for Overall Sample Discrimination

Element	Government Mountain	Obsidian Tank	Robinson Crater	O'Leary Peak	Spring Valley	RS Hill	Slate Mountain	Kendrick Peak
Ti	35.8306	27.3159	9.4950	10.9007	-8.9989	-10.9044	14.4467	12.4570
Mn	14.6432	14.6799	8.3888	8.7525	12.5528	13.2457	10.7469	10.7797
Co	283.1394	259.6340	188.2349	186.7226	76.2673	88.7811	236.6442	233.1938
Zn	-112.3397	-102.3899	-52.6395	-49.7325	150.2964	153.4175	-98.1310	-103.0769
Rb	8.0620	7.4109	-8.5316	-17.3425	278.3721	288.3374	12.2669	16.5759
Sr	66.3368	58.8925	47.4582	35.9798	54.0490	55.5235	20.5849	14.2521
Y	-259.1375	-221.6270	-115.1793	-117.6852	-62.6950	-71.0491	-158.1539	-150.2392
Zr	-2.6695	10.0550	75.5772	73.9628	75.0969	83.9170	62.5340	70.3092
Nb	275.6536	240.6895	123.0223	133.6535	371.4312	395.6172	118.9657	115.2981
Tc	-247.0171	-222.4306	-137.8018	-141.3871	-201.3693	-203.9501	-128.8960	-144.9788
Rh	69.1081	57.0757	9.7086	9.8695	-243.2314	-242.9341	55.2188	77.8560
Pd	60.7102	48.0139	-0.9605	10.0590	10.0328	14.2418	-18.1342	-18.2048
Ag	145.7926	134.0448	35.9083	39.2622	-31.9494	-50.9485	96.2861	106.5003
Cd	576.1226	555.8491	269.9356	268.4441	450.8201	454.2756	428.3254	408.2263
W	36.5832	31.9459	17.5863	17.9804	19.9850	18.7220	24.8987	24.7483
Au	12.3459	9.7648	4.2982	4.8429	7.9047	8.0158	5.0154	4.0637
Hg	-18.6951	-14.8474	-9.3174	-10.4501	-108.3723	-105.9304	-1.1920	-0.9687
Th	-325.9815	-288.2678	-150.9553	-155.9372	-360.3232	-375.4148	-216.2087	-215.0611
Constant	-541.6379	-466.7952	-185.1625	-179.8427	-1160.6223	-1276.7341	-319.4063	-331.0503

Peak localities, and could have been caused by an unusually high value of titanium in the last Government Mountain sample analyzed.

Finally, thirteen obsidian artifacts were obtained from an archaeological site (NA 10101) near the obsidian sources. The origin of each of these was identified as the Government Mountain/Obsidian Tank locality. Although this locality is not the obsidian source closest to the archaeological site, this chemical analysis partially confirms archaeologists' beliefs that this was the preferred obsidian site.

The previous study, using as few as one sample per locality, cannot be considered reliable although it suggested some trends that were verified in the present work. Even our eight samples per group is well short of the recommendation that there be approximately 20 times more samples than the number of variables. Although our coefficients were somewhat unstable and fluctuated on the introduction of additional samples, good separation was achieved.

REFERENCES

1. Jack, Robert N. "The Source of Obsidian Artifacts in Northern Arizona," Plateau 43 (3): 103-14 (1970).

2. Dixon, W. J. and M. B. Brown Biomedical Computer Programs P-Series. University of California Press: Berkeley (1979).

3. Nie, Norman H., C. Hadlai Hull, Jean G. Jenkins, Karen Steinbrenner, and Dale H. Bent. Statistical Package for the Social Sciences. 2nd ed. McGraw-Hill, Inc: New York (1975).

QUANTITATIVE DETERMINATION OF Ga,Zn,Cu,Ni,Mn,AND Cr BY X-RAY FLUO-

RESCENCE IN LATERITES AND BAUXITES USING TWO EVALUATION METHODS

Hasso Schorin

IVIC,Centro de Ingeniería y Computación

Apartado 1827,Caracas 1010-A,Venezuela

INTRODUCTION

Laterites and bauxites are residual products derived from a
wide variety of rocks by intensive chemical weathering under strong
ly oxidizing and leaching conditions. Generally the main constituents of these residues are Fe,Al and Ti present in form of hydrox-
ides and/or oxides. The content of silica depends on the thorough-
ness of the leaching process. The behaviour of trace and minor elements as Ga,Zn,Cu,Ni,Mn and Cr during the laterite/bauxite formation is not yet well established. Precise and accurate analyses of
these elements are a prerequisite for such investigations. In this
paper the standard addition method[1] is presented and the results
are compared with those obtained by calibration standardization.The
accuracy and the precision of the methods as well as the distinction limit[2] and the limit of detection[2] for each element are given.

SAMPLE PREPARATION AND MEASURING CONDITIONS

As to the standard addition method 10g of dried material(104°C)
were ground for about half an hour in an alumina ceramic vial(<100μm).
After grinding,4g of sample were mixed with 0.4g of "Hoechst Wachs C"
and homogenized in an agate mortar. The homogeneous mixture was then
briqueted in a press at 10t for half a minute. Hoechst wax was used
as a binder since it was found that it gives higher counting rates
and has better preparation properties than other similar materials;
the stability of the briquettes is also supposed to be better[3]. Additionally to another 4g of ground sample standard solutions of the
elements to be determined were added. For the elements Zn,Cu,Ni,Mn,
and Cr commercialy available standard solutions were used. The solution of Ga was prepared by dissoliving Ga_2O_3(1g metal/ℓ).After evap-

orating the solutions at 40°C,0.4g of wax were added to the dry samples,the mixture homogenized in an agate mortar and briqueted as before. The quantity of the elements to be added depends on their concentration in the original material and should not exceed a 3-4 fold
amount[3] .

On the other hand,to set up the calibration curves for the
six elements to be investigated 14 synthetic mixtures of the oxides
Al_2O_3(20-80%),Fe_2O_3(15-80%) and TiO_2(1-10%) were prepared. To these
mixtures solutions of the trace elements were added;their concentrations ranged between 50-1000ppm. After evaporation of the solvent 0.5
to 1.0ml of in acetone diluted glue was added and the homogenized sam
ple briqueted in aluminum capsules of 37mm in diameter at 5t for half
a minute.

All samples were analyzed with a Siemens SRS 100 X-ray spectrometer equipped with a logic controller for automation. The target
excitation for the Cr tube was of 55KV and 40mA. Graphite sample hold
ers of 34mm diameter were used. For the intensity measurements the Kα-
lines of all elements were employed using a scintillation counter;for
Mn and Cr a flow counter was applied. The optimal counting time was
determined for 2% relative standrd deviation according to Neeb[4] : 100
sec. were counted in case of Ga,Cu,Mn,Cr,and 200sec.in case of Ni and
Zn,respectively. For the determination of Zn,Cu,Ni,Mn and Cr an Al-
filter was put between the X-ray tube and the sample.

RESULTS
        The standard addition method is applicable if there is a lin
ear relationship between the incremental change of the concentration
of the element analyzed and the increase of the intensity[3]. To ascertain this condition,Ga,Zn,Cu,Ni and Cr were added to a laterite sample.
The concentration amounted 50ppm,100ppm,250ppm and 1000ppm. For the 5
elements calibration equations of the form:

$$C_i = A_i + mI_i$$

were obtained in which C_i is the concentration of element i,I_i the cor
responding intensity,"m" the slope of the line and A_i the intercept on
the intensity-axis. The values for m and A_i calculated for each element are summarized in Table 1.

Table 1. Coefficients for the calibration straight
         lines obtained by standard addition.

Element	Ga	Zn	Cu	Ni	Cr
m(cps/ppm)	0.67	0.48	0.35	0.29	0.11
A_i	- 9	+ 18	- 10	- 3	- 4

For a second laterite of different composition similar values were obtained.

As to the synthetic mixtures the calibration curves were ascertained by linear regression analysis using the Programm P88[5] and a PDP11/45 computer. According to the criteria established by Plesch[6] the intensities for all 6 elements were corrected for background by Fe. The calibration equation obtained in this way was:

$$C_i = A_i + B_i I_i + B_{ij} I_j$$

where A_i is the background coefficient, B_i the slope, I_i the intensity of the trace element, B_{ij} the coefficient for the background influence and I_j the intensity of the matrix element. The corresponding values A_i, B_i and B_{ij} obtained for each element are presented in Table 2.

Additionally, applying the programm P88 the standard deviation for the calibration is given which comprises all possible errors[2]. Based on these values the real detection and distinction limit[2] for each analyzed element were calculated for a confidence level of 95%[7]. The results are shown in Table 3.

To ascertain the precision 5 natural and, independently, 5 samples with standard additions were prepared from one laterite. On the one hand the contents of the trace elements were determined applying the standard addition method and on the other hand using the calibration standardization. The obtained results are summarized in Table 4. The calculated standard deviation is valid for a 95% confidence level.

Table 2. Calculated coefficients for the calibration equations.

Element	Ga	Zn	Cu	Ni	Mn	Cr
A_i	-159	-272.8	-35	-260	-141	-17
B_i	0.94	3.21	5.05	2.90	3.70	10.05
B_{ij}	$0.22 \cdot 10^{-2}$	$0.34 \cdot 10^{-2}$	$0.14 \cdot 10^{-3}$	$0.44 \cdot 10^{-2}$	$0.26 \cdot 10^{-2}$	$0.11 \cdot 10^{-2}$

Table 3. Statistical parameters for the calibration standardization.

Element	Ga	Zn	Cu	Ni	Mn	Cr
Standard deviation (ppm)	17	19	36	21	20	17
Detection limit (ppm)	40	45	86	50	47	40
Distinction limit (ppm)	57	62	120	69	65	55

Table 4. Data on the precision(n.d.:not determined)

Standard addition method

Element	Ga	Zn	Cu	Ni	Mn	Cr
Mean(ppm)	52	38	103	n.d.	n.d.	300
Stand.deviation(ppm)	+28	+ 6	+39	n.d.	n.d.	61

Calibration standardization

Mean(ppm)	40	46	90	n.d.	258	317
Stand.deviation(ppm)	+11	+ 6	+ 6	n.d.	+11	+22

Table 5. Data on accuracy(n.d.:not determined)

Element	Ga	Zn	Cu	Ni	Mn	Cr
Recommended value(ppm)	n.d.	59	20	200	387	280
Standard addition(ppm)	62	84	56	255	n.d.	281
Calibration standardization(ppm)	(10)	(13)	29	172	326	325

The data show that the precision of the analytical results for the calibration standardization is quite satisfactory. The larger values for the statistical error for the other method may be attributed to the preparation of 5 additional samples with the admixture of standard solutions.

As to the accuracy only the international standard reference material BX-N from France was analyzed. The results obtained are compared with those given by Abbey[8] and are listed in Table 5.

As can be inferred from Table 5 the agreement among the published values and the obtained results is satisfactory as the differences among them never surpass the distinction limit at the 95% confidence level given in Table 3.

CONCLUSIONS

The present investigation shows that Ga,Zn,Cu,Ni,Mn and Cr can be determined precisely and accurately by X-ray fluorescence in laterites and bauxites. The standard addition method may be employed if only a few samples are to be analyzed. Otherwise the calibration standardization is recommended.Lower detection limits will be obtained using a more appropiate X-ray tube.Furthermore it is recommended to apply these methods in the ranges of middle Fe_2O_3 concentrations. At contents larger than 60% effects appear which have to be investigated more in detail.

REFERENCES

1. E.P.Bertin,"Principles and Practice of X-Ray Spectrometric Analysis",Plenum Press,New York (1975).

2. R.Plesch,Praktische Fehlertheorie der Roentgenspektrometrie,X-Ray
   Spectrom,7:156 (1978).

3. R.Lichtfuss und G.Bruemmer,Roentgenfluoreszenzanalyse von Umwelt-
   relevanten Spurenelementen in Sedimenten und Boeden,Chem.Geol.
   21:51 (1978).

4. R.Neeb,Analytical Chemistry of the Platinum Metals I.X-Ray Fluo-
   rimetric Determination of Microgram Amounts of Osmium,Z.Anal.
   Chemie 179:21 (1961).

5. R.Plech,Analytische Grundlagen der Siemens Rechenprogramme fuer
   die Roentgenspektrometrie,Siemens Zeitschrift 48:355 (1974).

6. R.Plech,Empirical Matrix Correction in Practical X-Ray Spectrometry
   5:142 (1976).
7. R.Kaiser und G.Gottschalk,"Elementare Tests zur Beurteilung von
   Messdaten,Bibliographisches Institut,Mannheim (1972).

8. S.Abbey,Calibration Standards,X-Ray Spectrom. 7:99 (1978).

A COMBINED DILUTION AND LINE-OVERLAP COEFFICIENT SOLUTION FOR THE DETERMINATION OF RARE EARTHS IN MONAZITE CONCENTRATES

T. K. Smith

Institute of Geological Sciences
London, United Kingdom

INTRODUCTION

The rare earth elements (REE) together give quite complex X-ray emission spectra with a considerable number of overlaps at analytical energies by lines of other REE with lower atomic numbers. Where the concentration of REE is high, as in lanthanide minerals, this interference is more difficult to rectify. Smith and Gold[1] resolved a similar problem with lower atomic number elements in energy dispersive microprobe analysis by establishing a series of overlap coefficients. They asserted that accurate corrections were necessary because of the relatively poor overall resolution of the instrument and that these should not be limited to the major coincidences. Some of the smaller values had probably been ignored because they were considered statistically insignificant. The mathematical matrix of Smith and Gold covered 22 elements from fluorine to barium, with intensity coefficients (other than intra-element) quoted from 0.01% to 282.1% and with ZAF corrections necessary in cases of K to L conversion. The overlap coefficients were also adjusted for matrix effects.

Fassel[2] illustrated that besides line overlap problems there were others caused by the overlap of the absorption edges and fluorescent spectral lines of the rare earths. Not only were the exciting and fluorescent radiation selectively absorbed but enhancement of one element by another might occur and line overlap factors might not remain constant. Because of these interactions and concomitant mutual interdependence it was necessary to obtain correction factors for the interelement effects if mixtures were to be accurately analysed and a substantial series of binary standards was required. Other, non-REE elements can also cause spectral inter-

ference and absorption-enhancement effects and it is therefore
common to isolate the REE by chemical separation in order to remove
these complications. Some workers have used an internal standard,
but the energy range is too great for adequate coverage. Others
have resorted to second order lines e.g. Dixit and Deshpande[3] or
have chosen which line-overlap corrections to apply on the basis of
composition of the sample.

In this illustration the REE occur as monazite in the form of
ellipsoidal nodules with a maximum dimension of about 2 mm and
small samples were extracted by panning from stream sediments collected as part of a mineral reconnaissance programme.

METHOD

Various steps were taken to minimise the analytical problems
outlined above. The nodules were hand-picked from the pan concent-
rates to maximise the purity of the samples and thereby reduce
variation caused by other elements. The pellet size was increased
by inert dilution to improve the "thickness", to minimise line
overlap corrections and to exert a desirable, approximately constant
dominance on the mass absorption coefficients.

Determinations were carried out using a Philips PW1450/20
automatic software-programmed XRF spectrometer with tungsten anode
tube (Table 1). For each sample approximately 100 mg of nodules was
made up to 12 g with silica, ground (with 3 g of a methacrylate
copolymer binder) in an agate ball mill and pressed into a 40 mm
disc. Appropriate peak ($L\alpha_1$) and background positions were selected
(Table 2) and a blank silica disc used to calculate background
weighting factors to produce zero net peaks.

Table 1.   Instrumental Parameters

Spectrometer	Philips PW1450/20ASP
Anode	W
Voltage	60kV
Current	45mA
Path	Vacuum
Collimator	Fine
Crystal	LiF (220)
Detector	Flow proportional with P10 gas

Table 2.   Background Positions

$^{\circ}2\theta$ relative to $K\alpha_1$ peaks

La	-2.38
Ce	-2.28
Pr	-2.36
Nd	-2.77
Sm	-0.74
Eu	-2.38
Gd	-1.49 a
Tb	+2.53 a
Dy	-0.78
Ho	-0.86 b
Er	+2.32 b
Tm	-1.14 c
Yb	+1.53 c
Lu	-0.89

a, b, c indicate coincidence

Standards, each containing approximately 1000 ppm of a single rare earth element in silica, were used to obtain values of their respective calibration slopes. They were also used to obtain a complete set of values for line overlap for the other REE (Table 3). A self-consistent mathematical solution of a multiple linear regression system (Table 4) was then obtained by microcomputer (Diehl Alphatronic 332; method of least squares) to give corrected rare earth concentrations for the samples containing monazite.

Table 4.   Multiple Linear Regression System

LaLa Σ LaCe Σ LaPr Σ .... LaLu Σ %La Σ

CeLa Σ CeCe Σ CePr Σ .... CeLu Σ %Ce Σ

PrLa Σ PrCe Σ PrPr Σ .... PrLu Σ %Pr Σ

.

.

.

LuLa Σ LuCe Σ LuPr Σ .... LuLu Σ %Lu Σ

Table 3. Multiple Linear Regression Coefficients

	La	Ce	Pr	Nd	Sm	Eu	Gd	Tb	Dy	Ho	Er	Tm	Yb	Lu
La	100	.04	.07	.40	.04	.03	.05	.01	.02	.01	-.01	.04	.02	.01
Ce	-.29	100	-.10	-.17	-.08	-.01	-.06	-.08	-.04	-.09	-.10	-.16	-.20	-.10
Pr	26.33	.10	100	.00	.17	-.01	.02	-.03	.01	-.01	.02	.04	.03	.04
Nd	.08	.90	.00	100	.19	.10	-.06	.05	-.03	-.02	.01	.02	.02	.03
Sm	-.01	.76	-.27	-.19	100	.11	.09	.10	.03	.04	.03	.03	.09	.04
Eu	.22	.06	14.91	2.16	-.01	100	-.12	-.03	.05	.59	.00	.09	-.01	.01
Gd	2.06	8.65	-.29	.59	-.31	.06	100	.04	.09	-.01	-2.16	-.05	.00	.01
Tb	.07	.00	-.25	-.63	.35	-.06	-.37	100	.14	-.03	-2.61	.09	.02	.05
Dy	.01	-.59	-.11	-.44	-1.09	-.12	-.03	.33	100	.19	.04	-.22	-2.45	.10
Ho	-.17	-.22	.06	.11	.22	-.30	66.99	-.10	-.03	100	.09	-.08	.22	.00
Er	-.09	-.25	-.06	-.02	.16	-.25	-.71	12.92	-.02	-1.48	100	.30	.04	-.58
Tm	-.11	-.33	-.07	-.04	9.03	-.16	.57	-.16	.38	.00	.03	100	.04	-.03
Yb	-.90	-.79	-.53	-.49	-.10	-.06	-.60	-.28	-3.91	.09	-.35	-2.27	100	-.39
Lu	.52	-.10	-.07	.31	-.17	-1.80	-2.50	-.08	5.15	8.24	-3.82	.04	-.02	100

CONCLUSION

The advantages of this method are that sample preparation is straightforward, that relatively small amounts of material may be analysed, that the approximate rare earth composition need not be known in advance and that no binary standards are required. The mathematical treatment may have wider application in similar cases.

There are disadvantages in that some common elements (e.g. iron and manganese) cause interference; in the case described they were low. Tungsten carbide cannot be used for grinding if it contains cobalt.

ACKNOWLEDGEMENT

This paper is published with the permission of the Director, Institute of Geological Sciences (NERC).

REFERENCES

1.   D.G.W. Smith and C.M. Gold, A scheme for fully quantitative energy dispersive microprobe analysis, Adv. X-Ray Anal. 19:191 (1976).

2.   V.A. Fassel, Analytical spectroscopy of the rare-earth elements, in: "The Rare Earths," F.H. Spedding and A.H. Daane, eds, John Wiley, New York (1961).

3.   R.M. Dixit and S.S. Deshpande, X-ray fluorescence analysis of high-purity dysprosium oxide for Y, Eu, Gd, Tb, Ho and Er, Z. Anal. Chem. 303:111 (1980).

FEASIBILITY STUDY FOR ON-STREAM X-RAY ANALYSIS OF BARITE

Yury M. Gurvich

Applied Research Laboratories
9545 Wentworth Street
Sunland, California   91040

INTRODUCTION

Modern on-stream X-ray equipment for slurry analysis is now used in hundreds of mines throughout the world. This technique provides valuable information for process control which cannot be achieved by other analytical methods. One of the most recent successful examples is the Brenda Mines, in Canada, where the ARL Process Control X-ray Quantometer (PCXQ) is used for ten different streams in a Cu-Mo concentrator.

The success of slurry analysis is highly dependent upon the analytical method selected, based on the following factors: (i) the features of the analyzed materials, (ii) the concentration ranges, (iii) percent of solids, and (iv) requirements of analytical accuracy and precision. A preliminary investigation of methods and of analytical conditions can be done on a laboratory X-ray spectrometer using solid models of slurry. Such preliminary work significantly simplifies the introduction of the new technology and helps in the preparation of proper specifications.

This article represents an example of such an investigation for on-stream analysis of barite ($BaSO_4$).

SAMPLE PREPARATION

The slurry models were prepared by substituting the liquid part of the slurry with boric acid. The absorption and scattering properties of the models are similar to water slurries of the same

composition.[1] Because BaKα measurements depend on sample
thickness, briquettes were made with a constant weight of powder,
i.e. 6.00 gr. Three different products were analyzed:
concentrate, feed, and tails.

COMPARING BALα VS. BAKα

    Fig. 1 compares the intensity of BaLα line vs. the BaKα line
as functions of barite concentration for 10-30% solids. For high
concentrations of barite, calibration lines become substantially
horizontal due to the self-absorption of Ba radiation in Ba, and
the analysis is not feasible. The conditions are more severe for
a higher concentration of solids, and worse for BaKα than for
BaLα.

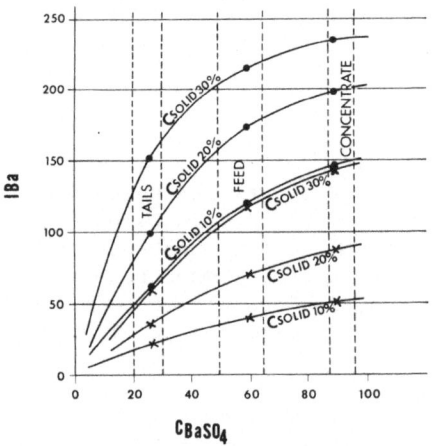

Fig. 1.   Intensity of Ba vs.
          BaSO$_4$ concentration.

Cr TUBE, 60 kV, 10 mA
t = 10 sec.
LiF-CRYSTAL, SCINTILLATION DETECTOR
PHD-WINDOW: 90-95% OF WHOLE

● = BaK$_\alpha$
✕ = BaL$_\alpha$

    Table 1 indicates two other advantages of the BaLα line. Due
to the low level of excitation (6 keV), the BaLα line needs a
lower voltage on the X-ray tube (20 kV), while BaKα requires at
least 60 kV. Requirements for the tube voltage stability for BaKα
are more demanding, as can be seen from the comparison of the
differences between the intensities of both lines at 50 and 60 kV.

Table 1:   Intensity of Ba Lines at Different Excitation Conditions

Sample	BaKα, cps			BaLα, cps		
	40 kV	50 kV	60 kV	40 kV	50 kV	60 kV
C-20	446	7698	19674	5244	6935	8567
F-20	415	6698	17102	4243	5643	6868
T-20	–	–	9870	–	–	3430

However, the BaLα line has three important disadvantages; these are especially significant for slurry analysis:

1. Due to low energy, the infinitive thickness for BaLα is much less than 1 mm. This requires strict stability of the sample surface and geometrical conditions of measurement. Precision of the sample preparation for Ba is 2.3 times worse than for BaKα; daily changing of the film window of the sample cell for on-stream analysis may create additional errors.

2. The intensity of the BaLα line is about three times less than BaKα and will be decreased additionally by the film window.

3. The intensity of the low-energy BaLα line depends strongly on the composition of the sample and on particle size. The analysis of tails may require measurements of other elements which may affect the absorption properties of the sample matrix. This will make both the equipment and the analytical procedure more complicated.

Such disadvantages lead to a preference of the BaKα line for on-stream analysis of barite. The major problem -- lower sensitivity of the BaKα line to a high concentration of barite -- can be eliminated by decreasing the concentration of solids. For example, as shown in Fig. 1, a plot for BaKα at 10% solids is comparable to the plot for BaLα at 30% solids.

SELECTION OF SCATTERED RADIATION

To determine the solids content in X-ray on-stream analysis, scattered radiation is usually used. The intensity of scattered radiation depends on the composition of the sample. Fig. 2 shows three such functions, representing different products. An X-ray tube with Rh target provides strong scattered lines for the correction of the % of solids and the variation of composition.[2]

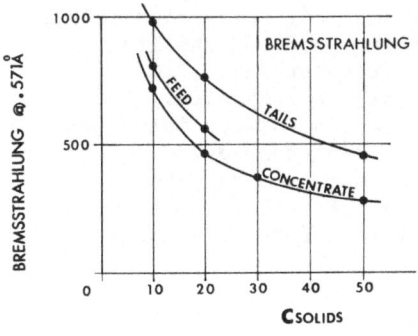

Fig. 2. Intensity of scattered radiation as a function of % solids.

Fig. 3 shows the spectra of two different slurries containing Ba, analyzed with an Rh tube. Concentration of Ba in one of the ores is much higher than in the other (compare intensities of BaKα lines). Incoherent RhKα radiation is highly dependent on sample composition: for high concentrations of Ba the incoherent peak is twice less than for the other slurry. At the same time, the intensity of coherent peaks changed only 7%.

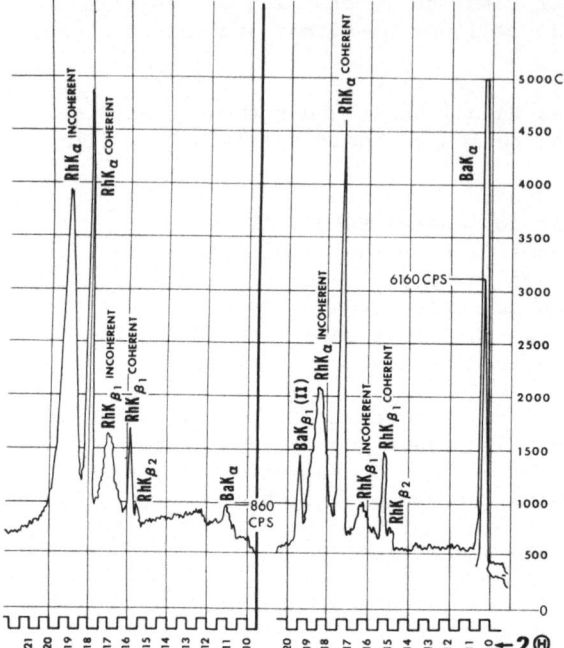

Fig. 3. Spectra of two polymetal ores with Ba.

Rh X-ray Tube
50 kV, 40mA
LiF (200) Crystal
Scan Speed: 2°/min.

Both scattering lines can be used for tail analysis, for simultaneous correction of the variation of solids and slurry composition.

CALCULATION OF ANALYTICAL RESULTS

Although the two most popular mathematical correction models for slurry analysis are the multiple regression and the iteration methods, because of the simplicity of barite products, two more methods of calculating analytical results are suggested. One is shown in Fig. 4, which presents a correlation between intensities of BaKα and scattered radiation for different samples. The percent of solids is also shown. All plots are linear. The equation of straight lines for various percentages of barite can be compared by a computer with actual intensities of Ba line and scattered radiation.

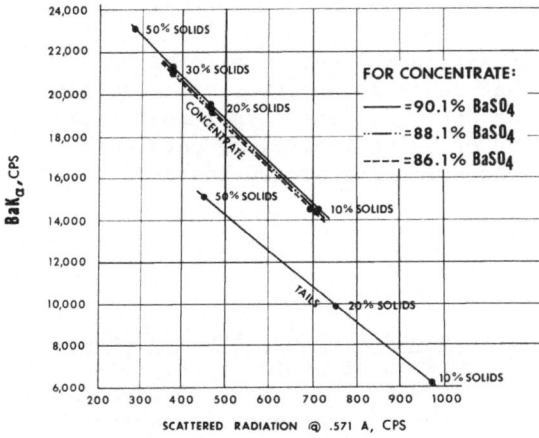

Fig. 4. Ba Kα vs. scattered radiation for concentrate and tails.

Another method is indicated in Fig. 5. Absolute concentration is the barite content in the bulk of a slurry instead of its concentration in solids. Data for both heavy products (concentrate and feed) belong to the same curve, which shows the low matrix effect. The transformation of absolute concentration into barite content in solids can be made by determination of % solids using scattered radiation.

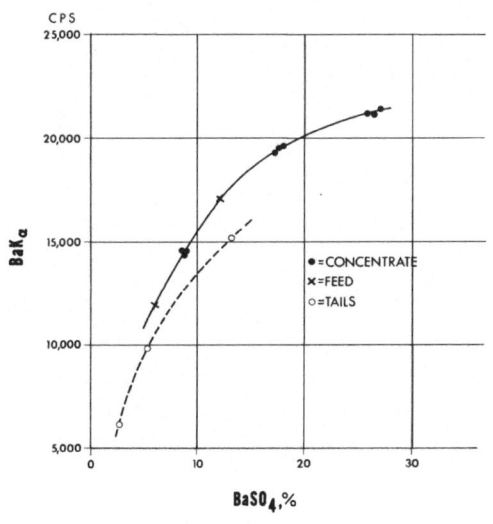

Fig. 5. BaSO₄ absolute concentration vs. Ba Kα intensity.

CONCLUSION

The results of this investigation and the experience of other on-stream applications shows that an X-ray analyzer can give accurate analytical data for process control. The actual analytical characteristics and the strategy of using analytical data for process control can be determined after industrial installation of the X-ray system.

REFERENCES

1. Y.M. Gurvich et al., _Apparatura i Metody Rantgenovskogo Analiza,_ XIV, 53-59, Mashinostroenie, USSR (1974).
2. A.H. Smallbone and Y.M. Gurvich, "The Assay of Gangue for Optimum Control of Flotation Processes," preprint #77-B-351, AIME (1977).

"STANDARD-BACKGROUND" METHOD OF X-RAY SPECTRAL ANALYSIS FOR
QUALITY CONTROL OF NOBLE METALS IN ALUMINA-BASED AUTOMOBILE
EXHAUST CATALYSTS

Yury M. Gurvich

Applied Research Laboratories
9545 Wentworth Street
Sunland, California 91040

INTRODUCTION

Development of matrix correction procedures in the field of
X-ray spectral analysis leads usually to "universal" algorithms
which allow use of a common calibration equation for assay of
various materials with different composition and concentration
ranges. In many applications, such methods can guarantee accuracy
of 3-5% rel.

For X-ray analysis of automotive catalysts, one such
technique[1] is based on the theoretical correction of interelement
effects. The article[1] states that "the overall accuracy of this
approach is about 10% relative error, although this may vary
depending upon the particular specimen being analyzed." However,
narrow specification limits for the quality control of catalysts
require the total error of analysis less than 1-1.5 % rel.

The automotive catalyst is a perfect object for X-ray
quantitative determination. Alumina base is a good pressing media
and does not need an additional binder. Analytical lines of noble
metals (Pt, Pd, Rh) are situated in the middle range of 2θ angles
for an LiF crystal ($16-40°$) whose wavelength resolution allows to
separate analytical peaks of adjacent elements. Concentrations of
noble metals (100 to 1500 ppm) are at least two orders higher than
detection limits of modern X-ray spectrometers. The composition
of catalysts is relatively simple: besides alumina and noble
metals, some catalysts contain first percents of Ce and Fe whose
presence increases absorption properties of the matrix. Since

145

variation of components in individual types or groups of catalysts is limited by specifications, the method of matrix effect correction is supposed to be relatively simple.

Experiments showed that the "standard-background" method developed at ARL by Andermann and Kemp[2] practically eliminates or significantly decreases the most sources of possible analytical errors. The ratio of fluorescent peak (I_F) to scattered radiation (I_{SR}) at a certain wavelength is much less dependent on instrumental drift, sample preparation (thickness, particle size), and matrix effects than fluorescent radiation itself. For instance, if the concentration of Ce and Fe vary in range ±10%, ratios I_F and I_{SR} at a certain wavelength will not change more than ±0.5%.

For analysis of noble metals, the Cr tube is the most suitable, since scattered peaks of other targets can directly interfere with analytical lines. The area 0.57Å at Bremsstrahlung maximum was selected for I_{SR} determination, since the incoherent scattering is the most responsive to changing of specimen composition.

SAMPLE PREPARATION AND PRECISION OF ANALYSIS

The analyzing sample was reduced down to 20 to 30 gr, then ground for 18 minutes in a Shutter Box. Such preparation was necessary because the results of analysis depend on the particle size of the powder. Each briquette was made in Al cup from 4.00 gr of powder. Three parallel briquettes were made for each analyzed sample and the results were averaged. The precision of the sample preparation was determined based on parallel briquettes of 80 samples; the standard deviation for Pt, Pd, and Rh was better than 0.2-0.3% rel.

The total precision of the X-ray method is given in Table 1: on average, the standard deviation is 0.9% rel.

Table 1:   Precision of X-Ray Analysis

Type of Catalyst	Pairs of Analysis	Standard Deviation, % rel.		
		Pt	Pd	Rh
With Pt, Pd, Rh, Ce, Fe	27	0.7	0.8	0.9
With Pt, Pd	12	1.2	0.9	-

INFLUENCE OF CE AND FE

The influence of Ce and Fe on the PdKα line intensity for
seven different types of catalysts is shown in Fig. 1. Data for
the four Pt-Pd catalysts without Ce and Fe belong to the same
straight line. Presence of Ce and Fe decreases intensity of the
Pd line. Fig. 1 also shows the correlation between Pd
concentration and I_{Pd}/I_{SR} ratio for the same catalysts. As shown,
the ratios perfectly fit a linear regression and practically
eliminate matrix effects.

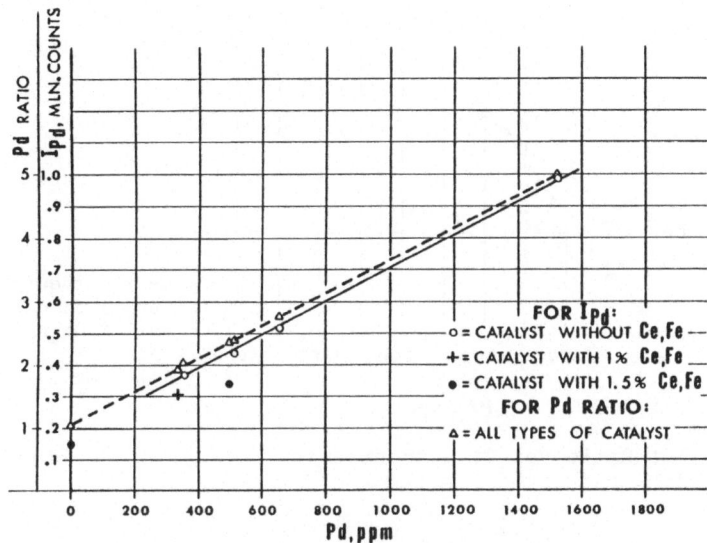

Fig. 1: Analysis of Pd in Catalyst

Similar results were received for the RhKα line. For the PtLα₁
line, however, the ratio did not completely eliminate the
influence of Ce and Fe. Since correlation $C_{Pt} = f (I_{Pt}/I_{SR})$
depends not only on data of X-ray measurements but also on the
results of the reference laboratory, additional experiments were
made with synthetic standards, prepared according to assigned
concentrations of Pt, Ce, and Fe.

The correlation between the Pt ratio and contents of Pt in
synthetic standards with different levels of Ce is shown in
Fig. 2. Deviation of points from the calibration line does not
correlate with the concentrations of Ce and can be explained by
standards preparation. The experiments confirmed, therefore, that
the reference analytical method may be responsible for errors of
Pt in real samples.

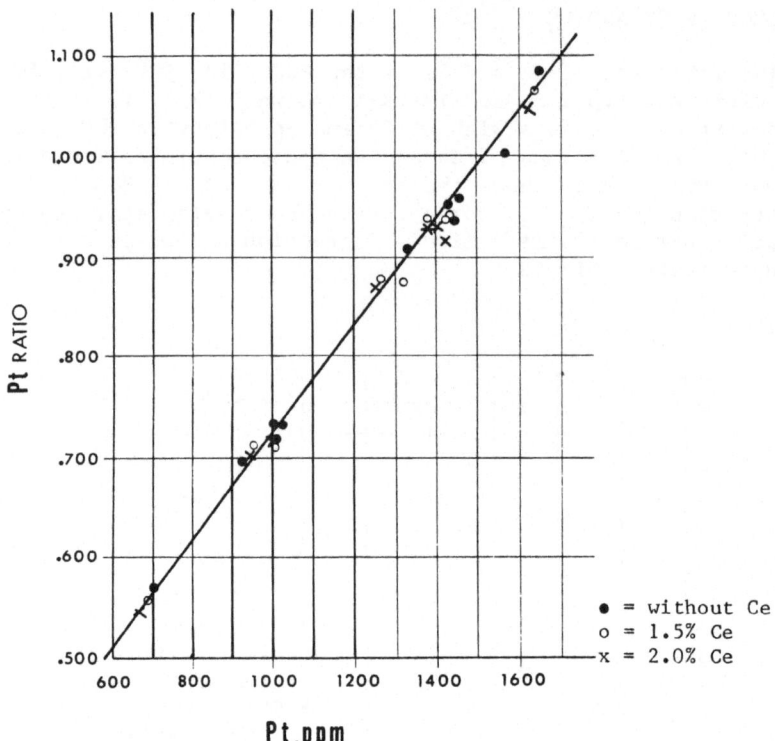

Fig. 2: Pt Ratio vs. Pt Concentration for Synthetic Standards

ACCURACY OF CATALYSTS ANALYSIS

      Results from fifty-four actual samples of various Pt-Pd cata-
lysts (500-1700 ppm Pt, 300-1600 ppm Pd) show that the standard
deviation between X-ray and chemical analysis for Pt is 2.4% rel.
and for Pd is 2.1% rel. The precision of the reference laboratory, determined by the duplicate analyses, was 2.1% rel. for both
Pt and Pd. This means that the errors of X-ray method cannot be
distinguished from the precision level of the reference method.

      Similar data for different types of catalysts with Pt, Pd,
Rh, Ce, and Fe are as follows:

      - for Pt: 3.0% rel. in the range of 500-1400 ppm Pt
      - for Pd: 2.8% rel. in the range of 250-600 ppm Pd
      - for Rh: 4.4% rel. in the range of 70-180 ppm Rh.

A higher level of standard deviation for Pt and Pd can be explained by specific errors of reference analysis[3]. It was found that both precision and accuracy of the reference laboratory depend on Fe content. As a result, in the presence of Fe, concentrations of noble metals were overestimated, sometimes up to 10 to 15% rel. At the same time, the more Fe present in samples, the worse the precision of chemical analysis. Higher standard deviation for Rh can be explained by increasing the deviation of chemical assay for small concentrations of Rh, as well as by a nearness of the detection limit of X-ray analysis (several ppm) to the range of Rh content.

CONCLUSION

The "standard-background" method of X-ray analysis is adequate for production control of noble metals in catalysts.

As an indirect technique, the X-ray method requires good certified standards. Since chemical method cannot always provide an accurate analysis of standards, the alternative is to make synthetic standards with calculated nominals. Preparation of the standards is a complicated process involving an impregnation of alumina base by noble metals solutions at a high temperature. All stages of this process should be planned and performed carefully; their results must be properly corrected, specifically by LOI. Otherwise, losses of material can lead to incorrectness of standard nominals (see deviation of points in Fig. 2).

REFERENCES

1. B.E. Artz, X-Ray Spectrometry, vol. 6, no. 3, 165-170 (1977).
2. G. Andermann, J.W. Kemp, Analytical Chemistry, vol. 30, 1306-1958 (1958).
3. S. Kallman, Talanta, vol. 23, S-79-583 (1976).

SOME ELEMENTAL DETERMINATIONS OF CATALYTIC MATERIALS USING A THIN-FILM INTERNAL STANDARD TECHNIQUE BY RADIOISOTOPE EXCITED X-RAY FLUORESCENCE*

W.C. Parker and J.J. LaBrecque

Instituto Venezolano de Investigaciones
Científicas-IVIC, Apartado 1827
Caracas 1010-A, Venezuela

SUMMARY

Applications of some selected elemental determinations of different catalytic materials by radioisotope excited x-ray fluorescence will be presented. The analytical method, an improved sample preparation and analysis technique were presented at the last Denver Conference (1). Data on the accuracy and precision of each of the following types of catalysts are presented.
1) Platinum catalysts (0.5 - 10% Pt) on alumina and carbon supports used for dehydrogenation, hydrohalogenation, oxidation processes, etc. 2) Palladium catalysts (0.5% - 10% Pd) on alumina and carbon supports employed for alkylation, decarbonization, hydrogenation, etc.

Some of the different types of variations (total, sample preparation, counting, etc) will be shown calculated from experimental data (2).

INTRODUCTION

The determination of noble metals in catalysts requires a higher accuracy and precision than the analysis of other catalytic materials because of the economic

---

* This work was funded in part by CONICIT (Venezuela; S1-1149) which form part of an international project with the NSF (USA).

aspects involved. Agreement between producers and consumers of these types of catalysts (noble metals ones) generally specify the measurements to agree within 1% relative error (3). There are many different methods described in literature (4) for the determination of noble metals in catalysts by atomic absorption spectrometry, differential spectrophotometry, colorimetry, etc. But these techniques all require that the sample be completely dissolved, which is difficult in some cases and very time consuming. Conventional X-ray fluorescence (wavelength dispersive) has been hindered by the high and fluctuating background due to the interaction of the X-ray continuum from the X-ray tube with the various forms of alumina present and because of the lack of meaningful standards. Thus, the possibility of employing radioisotope-excited energy-dispersive X-ray fluorescence to avoid the suggested effects of the various forms of alumina present in the support material was one of the objectives in this work. An analytical method (1) was developed based on a thin-film-internal standard technique in which the measurements are calculated from determined F_{JL}-factor (relative fluorescence ratio factors). Finally the application of this method to Pt and Pd oxidation catalysts is presented.

EXPERIMENTAL

The calculation of the relative fluorescence ratio-factor (F_{JL}) using synthetic standards

    A series of ten synthetic standards each were prepared for Pt and Pd in both an alumina (Al_2O_3) and carbon (graphite) matrix. Since the concentration range was from 0.25% to 10% Pt or Pd in these standards, first a 10% secondary standard was prepared by adding 1 gram of pure Pt or Pd power to 9 grams of either alumina or carbon. The matrix (both alumina and carbon) and the internal standard (MoO_3) were previously ground to less than 90μm and dried at 102°C before mixing thoroughly with the appropriate amounts of the Pt or Pd secondary standard in a SPEX mill/mixer for 30 minutes with 5 steel balls in plastic vials. The relative fluorescence ratio factors (F_{JL}'s) are determined as previously described (1).

Table I. Typical results from four commercial
catalysts for 100 seconds of fluorescent time.

	1% Pd on carbon	1% Pd on alumina	1% Pd on carbon	1% Pt on alumina
Measuring line	K_α	K_α	L_α	L_α
mean	0.93%	0.984%	1.095%	1.042%
total deviation (standard deviation)	0.07	0.05	0.05	0.06
standard counting instrumental and operating errors	0.05	0.04	0.05	0.05
sample preparation error	0.02	0.01		0.01

*Note: the standard error ($\sqrt{N}$) is about 3% for 100
seconds of fluorescent time.

Preparation of catalytic material for photon induced X-ray fluorescence

The catalytic materials are ground to less than 90
μm and oven dried at 102°C before mixing with the internal standard . Again three or more portions of each
sample preparation are determined to ensure a homogeneous
mixture of the sample and internal standard.

Data acquisition and spectrum analysis

The X-ray spectrometer system employed is based on
a PDP-11/05 processor which acts as a multichannel analyzer as well as to manipulate the spectrum and data.
The excitation source is a radioisotope with its appropriate housing; [109]Cd (10mCi) for the Pt determinations
and a [241]Am (100mCi) for the Pd measurements. The detector is a high resolution Si(Li) semiconductor (about
150 eV FWHM resolution at 5.9 KeV) with a thin Be window.
This entire system has been described in detail elsewhere
(5, 6).

Table II. The effect of increased fluorescent time on
the total relative percent error.

Catalyst	Measuring line	Mean ± standard deviation (RSD)		
		100 sec.	500 sec	1000 sec
5% Pd on carbon	K_α	5.181±0.14 (2.70%)	5.050±0.06 (1.19%)	5.072±0.04 (0.79%)
10% Pd on alumina	K_α	9.844±0.26 (2.64%)	9.790±0.09 (0.92%)	9.902±0.08 (0.81%)
10% Pt on alumina	L_α	10.971±0.29 (2.64%)	11.97±0.10 (0.84%)	11.175±0.08 (0.72%)

Finally, the whole system is controlled by a pre-
written computer routine with the following typical commands: 1) An energy calibration (usually 20 ev/channel);
2) data calibration (sets the lower and upper energy
limits as well as the counting time); 3) data smooth
(the spectrum is smooth with a binomial smoothing mod-
ule; 4) a series of background regions (one before and
after each peak of interest); 5) a series of peak regions (peaks of the elements of interest and internal
standard); 6) integrate above background the peaks of
interest; 7) flush (clears the system for the next job).

Applications

A summary of the results of four commercial oxidation
catalysts are presented in Table I with their respective
sample preparation deviation (error). The sample preparation deviation is calculated from the difference of
the total deviation determined from 5 independent portions of the sample preparation and the deviation from
5 independent measurements of one of the sample preparation portion. It can be estimated from Table I that
the largest error arises from the standard counting errors, thus the fluorescent times have been increased
from 100 seconds to 1000 seconds with the result that
both Pd and Pt can be determined by this method with
less than 1% relative error. These results are listed
in Table II.

Finally, a typical spectrum is shown in figure 1 for
a 1% Pd on carbon oxidation catalyst. Dead times were
less than 3% for all samples, while elemental count
rates for 10% noble metal catalysts were about 250
counts/second and 325 counts/second above background
respectively for the PdK_α and PtL_α lines.

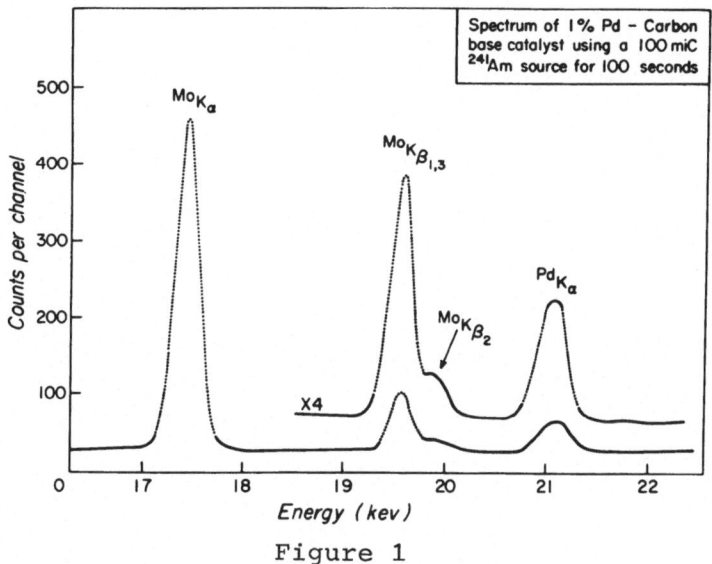

Figure 1

References

1. J.J. LaBrecque, "An improved sample preparation and
   analysis technique for the determination of minor
   elements in catalytic materials by radioisotope
   induced X-ray fluorescence", Adv. X-ray Analysis,
   24 (1980).
2. E.P. Bertin, "Principles and practices of X-ray spectrometric analysis", Plenum Press, New York (1975).
3.   S. Kalman, Talanta 23, 579 (1976), p. 494-500.
4. F.E. Beamish, C.L. Lewis and J.C. Van Loon, Talanta,
   16, 1 (1969).
5. J.J. LaBrecque, Proceedings of the 3rd. International
   Conference on Computers in Chemical Research, Education and Technology, Caracas, Venezuela pp 66-81
   (1976).
6. J.J. LaBrecque, J. Radioanal. Chem., 41, 127 (1977).

ENERGY DISPERSIVE X-RAY MEASUREMENTS FOR CESIUM

AND SILVER IN ZEOLITE ION-EXCHANGE COLUMNS

H. A. Vincent and M. E. Patton

Anaconda Copper Company
P. O. Box 27007
Tucson, Arizona 85726

SUMMARY

Distribution of cesium and silver ion-exchanged onto columns of zeolites may be measured directly and non-destructively on the column with the use of radioisotope-excited energy dispersive X-ray fluorescence. Quantitative accounting of the exchanged metals is obtained without elution.

INTRODUCTION

Several synthetic and natural zeolites have been shown to have high relative affinities for silver and cesium ions.[1,2] Ames has described the absorption of Cs^+, in the presence of several competing cations, on a natural clinoptilolite from Hector, California.[3,4] Similar exchange with synthetic zeolites has been described by Sherry.[5] The absorption of Ag^+ on zeolites was described by Barrer, et al.[6]

In order to assess the usefulness of zeolites for the separation of these cations from solutions of complex composition, it is necessary to have equilibrium isotherms for batch processes or kinetics information for dynamic exchange such as that in column processes.

The usual techniques for acquiring ion exchange data for columns include the measurement of cations in eluted solution fractions, the measurement of cations in column segments after physical separation of those segments, or a non-destructive measurement of cations on column segments by the use of radioactive isotopes of those cations.

A method was desired which would allow precise non-destructive measurements of cation distribution on a column.

In this work energy dispersive X-ray fluorescence was used to describe the distribution of silver and cesium ions exchanged from various solutions which have been passed through columns containing zeolite. The purpose was to assess the utility and quality of the information gained by this technique.

EXPERIMENTAL

X-ray excitation was accomplished using a one curie Americium-241 primary source and a dysprosium secondary target (New England Nuclear Model 496). Glass columns with an inside diameter of 9mm and length of 10cm to 50cm were used for the ion exchange reactions. Solutions for exchange, elution or backwashing were pumped through the column with a peristaltic pump at rates of 2 to 20 column bed volumes per hour. A silver mask .13mm thick with a 5mm wide slit for viewing column segments was used for cesium measurements. A similar lead foil mask was used for the silver measurements. The system is illustrated in Figure 1. Digital count information from a 30mm^2 lithium-drifted silicon detector was routed to a multichannel analyzer for data handling. The portability of the system allowed measurements to be made on columns without disconnecting influent and effluent tubing. The ion-exchange medium was Anaconda natural zeolite 1010A, grain size from -20 to +48 mesh.

RESULTS

When Dysprosium K radiation is used for excitation, both cesium and silver yield strong K X-ray lines which are relatively free from spectral interference. The zeolite matrix before exchange is composed predominately of elements with low atomic numbers. Thus self absorption is the major cause for non-linearity of X-ray intensity versus concentration for both Cs and Ag. An example of this is shown in Figure 2 for Cs with intensity response fit to an expression of the type:

$$I_i = Ao_i + A_i \left(\frac{C_i}{1 + \alpha_i C_i} \right)$$

where: I_i = intensity ratio or number of counts, Ao_i = the intercept, A_i = slope of calibration, α_i = coefficient correcting for effect of the i component. Ao_i is usually zero as the curve is constrained to pass through zero with proper background subtraction.

Calibration of the system can be accomplished by use of reference standards used in columns where the geometries of sample columns and standard columns are equivalent. Some difficulty may be encountered because of differences in packing densities for zeolite particles in

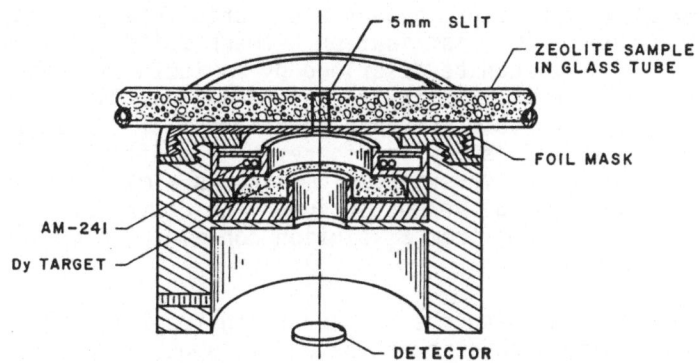

Fig. 1 Excitation system

a column. The density problem can be minimized if the intensity of the scatter of the Dysprosium X-rays is used for normalizing. Calculations of cesium distribution on the column may also be based on the total amount of cesium present if the intensity versus concentration curve has been described for that system.

The values for the α coefficient and the slope, A_i, as shown in Figure 2, allow for the calculation of cesium on the zeolite up to the maximum capacity at 1.55 meq per gram or approximately 20 percent by weight.

Distribution of the cesium ion on the exchanger is governed by several factors including the flow rate of the exchanging solution phase, diffusion, exchange kinetics, solution composition, and the equilibrium of the exchange reaction.

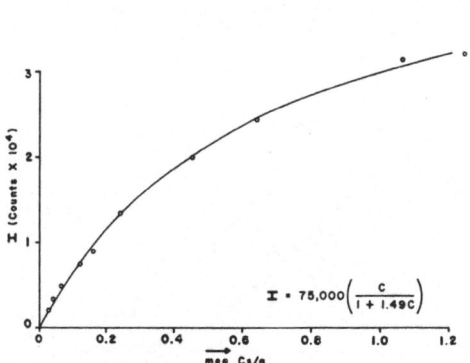

Fig. 2 Cesium determination on zeolite column by EDX

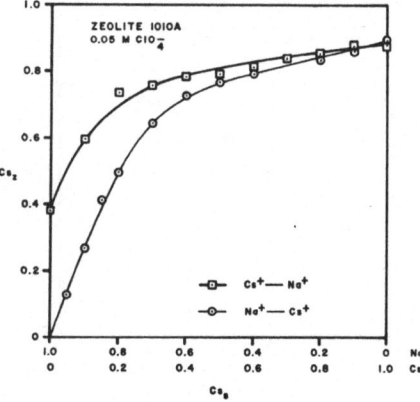

Fig. 3 Exchange of Cs$^+$ vs Na$^+$

At very slow flow rates, a near-equilibrium state exists at all
sites on the column. The distribution between solid and solution
phases at a given site can be described by an isotherm such as that
shown in Figure 3, in which a favorable exchange of Cs^+ versus Na^+
is shown. A less favorable exchange equilibrium of cesium ions
versus potassium ions is shown in Figure 4. The affinity of the
zeolite 1010A for cesium versus calcium ions is similar to that
described for the exchange versus sodium. The distribution of cesium
exchanged onto the column from a solution containing much more calcium
is shown in Figure 5.

It is possible to describe the advancing frontal zone for cesium
on the column and to predict breakthrough volumes. The slopes of the
frontal zones decrease for the exchange from solutions containing
higher concentrations of competing cations or for solutions containing
cations with stronger affinities for the zeolite exchange sites.

An exchange isotherm for silver ions versus sodium ions is shown
in Figure 6. Absorption on a column from 10^{-3} molar solution is
shown in Figure 7 for 24 and 30 liters having passed through the
column. Since the influent solution composition is known, integration
of the segment values must account for all of the silver added. The
change of silver concentration on the column due to backwashing with
a concentrated sodium nitrate solution is illustrated by the lower
curve in Figure 7.

Cesium values obtained from column measurements are compared in
Table I with values obtained after chemical dissolution of the
segments. For a 100 second count time, variation of values due to
random count errors range from 0.8% relative for the higher concentrations to 2.2% relative for the segments of lower concentration.
A significant contribution to variance for this technique may be the

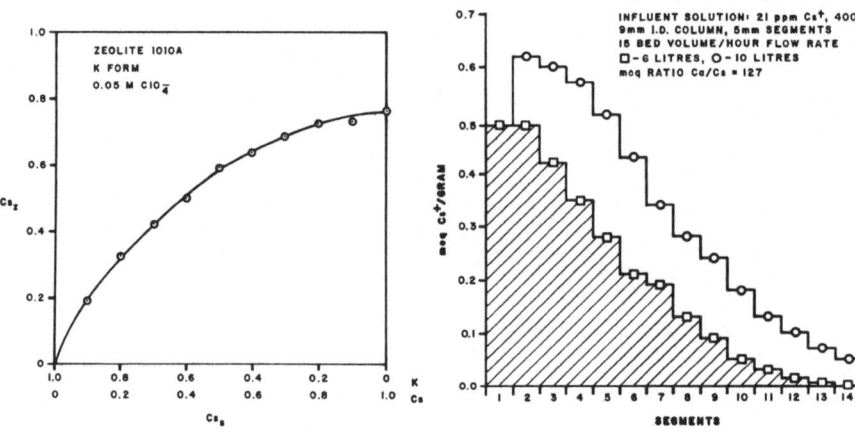

Fig. 4 Exchange for Cs^+ vs K^+ Fig. 5 Cs^+ on zeolite 1010A

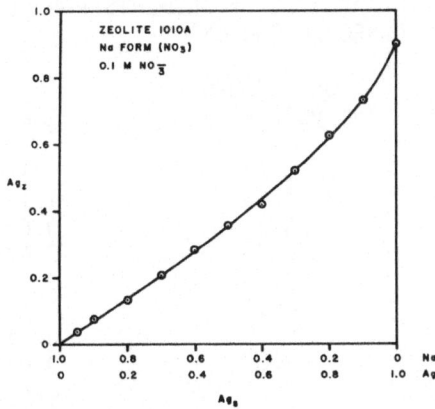

Fig. 6 Exchange isotherm for silver versus sodium

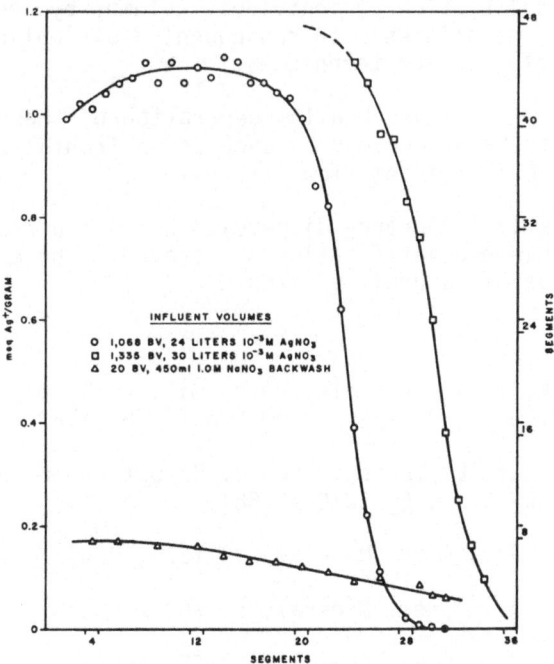

Fig. 7 Silver exchange on zeolite

TABLE I - COMPARISON OF EDX On-COLUMN MEASUREMENTS
WITH ANALYSIS OF SEGMENTS FOR CESIUM
(100 SECOND COUNTS)

Segment No.	OFF-COLUMN Analysis (meq Cs/g)	ON-COLUMN Analysis (meq Cs/g)
1	1.24	1.19, 1.26, 1.26
2	1.06	1.16, 1.15, 1.18
3	0.64	0.65, 0.64, 0.65
4	0.45	0.47, 0.43, 0.44
5	0.24	0.26, 0.24, 0.24
6	0.16	0.18, 0.12, 0.15
7	0.12	0.11, 0.13, 0.12
8	0.07	0.07, 0.07, 0.08
9	0.04	0.05, 0.06, 0.04
10	0.025	0.03, 0.03, 0.03

geometric positioning of the sample material. It should be noted
that a cesium concentration gradient exists across the 5mm width of
each segment. The greatest change occurs for those segments defining
the frontal zone. Because of the variation between columns, it was
found that the calibration of individual columns by ion-exchange
loading of known quantities with subsequent distribution calculation
was the most useful of the techniques.

Maximum flow rates which allow separation of these metal ions
from solution may be determined by successive frontal zone determinations made with increasing flow rate values.

The application of energy dispersive X-ray fluorescence for
in situ ion exchange quantification has proved to be extremely useful
in the evaluation of natural zeolites.

REFERENCES

1. D. W. Breck, W. G. Eversole, R. M. Milton, T. B. Reed,
    and T. L. Thomas, J. Amer. Chem. Soc., 78 5963 (1956).

2. B. K. G. Theng, E. Vansant, and J. B. Uytterhoeven,
    Trans Faraday Soc., 64 3370 (1968).

3. L. L. Ames, Jr., Amer. Mineral, 49 127 (1964).

4. L. L. Ames, Jr., Amer. Mineral, 51 903 (1966).

5. H. S. Sherry, J. Phys. Chem., 70 1158 (1966).

6. R. M. Barrer and W. M. Meier, Trans. Faraday Soc.
    55 130 (1959).

DIRECT ANALYSIS OF PLUTONIUM METAL FOR GALLIUM, IRON AND NICKEL

BY ENERGY DISPERSIVE X-RAY SPECTROMETRY*

H. L. Bramlet and J. H. Doyle

Rockwell International
P.O. Box 464
Golden, CO   80401

ABSTRACT

An x-ray secondary target method for routine determination
of gallium, iron, and nickel in plutonium metal is described that
has significant advantages over wet chemical analysis. Coupons
requiring minimal preparation for analysis are produced as a
breakaway tab on the plutonium ingot. All three elements are
determined on the same coupon. Gallium is determined using an
arsenic secondary target followed by iron and nickel using a
zinc target. The analysis times are 5 minutes for gallium and
15 minutes for the combined iron and nickel. The method of
analysis was evaluated in the range of from 0.5 to 1.5% gallium.
Iron was investigated over the range of 67 to 3000 ppm and nickel
from 64 to 110 ppm.

INTRODUCTION

The room temperature crystal structure of pure plutonium (Pu)
metal is monoclinic. Since monoclinic metals are difficult to
machine, one alternative used is to stabilize the high temperature
face-centered cubic phase of plutonium by addition of gallium (Ga).

Historically, analysis of Ga at the Rocky Flats Plant has
been done by two wet chemistry methods. The first analytical
method, and also the one being used currently, is based upon a
spectrophotometric determination developed by Los Alamos National
Laboratory. For a time in the mid-1960's a volumetric determination based upon first separating the Pu from the Ga by ion

*Work performed under Department of Energy Contract DE-AC04-76DP03533 exchange was used. The iron (Fe) and nickel (Ni) analyses have been difficult over the years. At first wet chemical analysis and emission spectroscopy were used. Atomic absorption is being used at the present time.

Since the weighing of samples and subsequent wet chemistry must be done in gloveboxes using at least 25 mil thick lead gloves, the methods noted above are slow and require especially skilled personnel. Therefore, in recent times a considerable effort has been made to establish reliable XRF methods for determining these elements at the Rocky Flats Plant.

TARGET SELECTION

The importance of choosing the proper secondary target cannot be over emphasized. The energy of the $K\alpha$ lines from the secondary target must just barely exceed the K_{ab} edge of the element being analyzed. After discussion with Kevex Corporation personnel, a zinc (Zn) target was selected for Fe and Ni. An arsenic (As) target was selected for Ga. A comparison of counting efficiency showed that one hundred thousand counts could be accumulated in the Ga peak in five minutes using the As target while the previously used Y target took twenty minutes for the same number of counts.

EQUIPMENT USED

A Kevex 810A secondary target system with computer control of the target and sample was used. A 3.5 Kw molybdenum target was chosen. A new type amplifier was selected for the lithium drifted silicon detector with the capability of discriminating against all radiation above and below the region of interest. All of the data in this paper was taken with a Tracor 880 Pulse Height Analyzer System using direct memory access for counting and containing a Digital Equipment Corporation 11/04 computer with 28 K words core. All spectral data was stored on floppy disks. Appropriate software for sample changing and quantitative analysis was generated in-house by the authors.

GALLIUM

For production support, the samples for analysis are disks 1-1/4" in diameter and 1/8" thick which will be generated as a break away tab on the Pu ingot. For samples requiring only Ga analyses, minimal polishing is necessary before wrapping in mylar film to contain spread of Pu contamination. When Fe and Ni results are desired, polishing to about 600 grit is necessary. Due to the very high mass absorption coefficient of Pu it is necessary to achieve as high a through put of counts as possible since system dead time from radioactive decay alone may run 20%. This is accomplished by operating the tube at 30 Kv and 70 mA, using a large diameter collimator, inserting a 4 mil polyethylene post filter, and adjusting the amplifier window to the edges of the Ga Kα and Pu M lines. A count time of five minutes was used for all Ga data collection (See Figures 1 and 2).

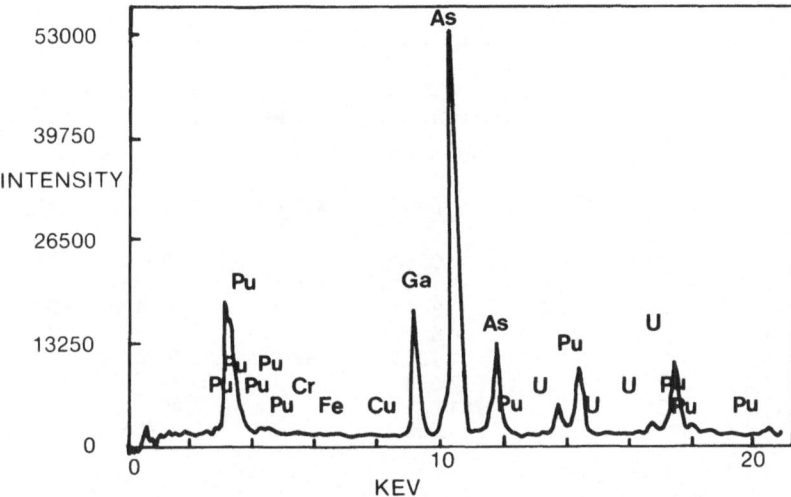

Figure 1. Before insertion of the post filter and amplifier window. (1% Ga in Pu using an As secondary target.)

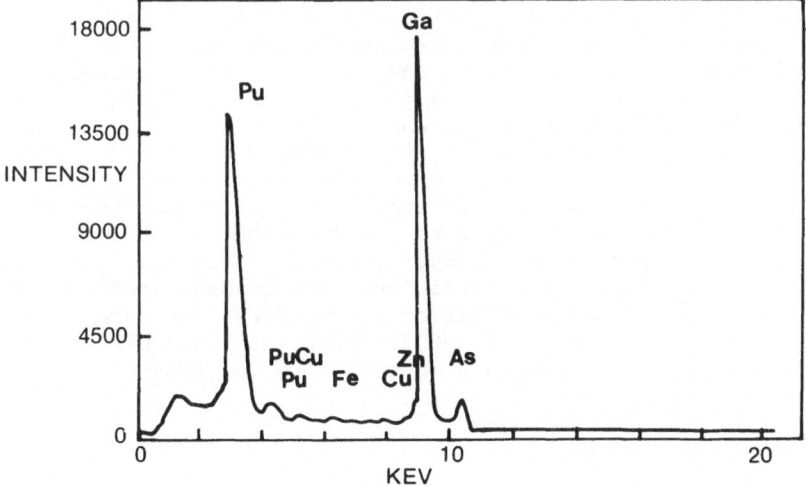

Figure 2. After insertion of the post filter and amplifier window.

A ratio of the counts in the Ga Kα lines to the counts in the Pu Mαβγ complex is plotted against known values of standards determined by wet analysis. The computer accomplishes this calibration using regression analysis. Results of typical calibration calculations are shown in Table 1. The high correlation coefficient proves that the calibration line is linear.

TABLE 1
Regression Analysis Data

	Ga		Fe		Ni	
	Ratio	Wt. %	Ratio	PPM	Ratio	PPM
1.	.27672	.544	.06629	67	.059709	64
2.	.53942	.998	.18983	267	.07308	81
3.	.82279	1.507	.36902	540	.092426	87
4.			.62983	1070	.098982	99
5.			1.7261	3000	.11987	110
(a)	.052676		-77.583		22.878	
(b)	1.7639		1781.8		735.50	
(c)	.99994		.99955		.97785	

(a) Y intercept
(b) Slope
(c) Corr. Coef.

The standard deviation was determined to be 0.006 at one sigma which is comparable to wet analysis at the 1% Ga level. Since plutonium oxidizes rapidly a sample was exposed to air for one week prior to measuring its composition. It's average composition (10 measurements) differed from a freshly polished sample by less than 0.002 which is one-third the standard deviation. Clearly, oxide formation on a standard will not be a significant error if calibration coupons are polished on a weekly basis. Since a high and a low standard are counted with each tray of samples, daily polishing would result in a shorter useable lifetime for the expensive standards. To further test the sensitivity of the analysis to surface conditions, two coupons from the same ingot were polished with different size grits. The spread of the means between a 240 grit and a 600 grit polish was only one-half the standard deviation which demonstrates a low sensitivity to surface roughness.

IRON AND NICKEL METHOD

Samples which also need Fe and Ni must be polished with the utmost care using a final step of 600 grit or finer. Polishing materials should be analyzed to find sources that are as low as possible in Fe. Excessive pressure on the sample during polishing can cause embedding of Fe containing materials into the soft Pu surface.

A ratio of the counts in the Pu M-N doublet just above the Pu Maβγ complex is plotted against known values of standards determined by atomic absorption analysis. Fifteen minutes counting time was used (Figure 3). The regression line (Table 1) is linear as shown by the correlation coefficient. Typical precision results show a standard deviation of 15 ppm at the 500 ppm level and are comparable to atomic absorption standard deviations for the range of interest.

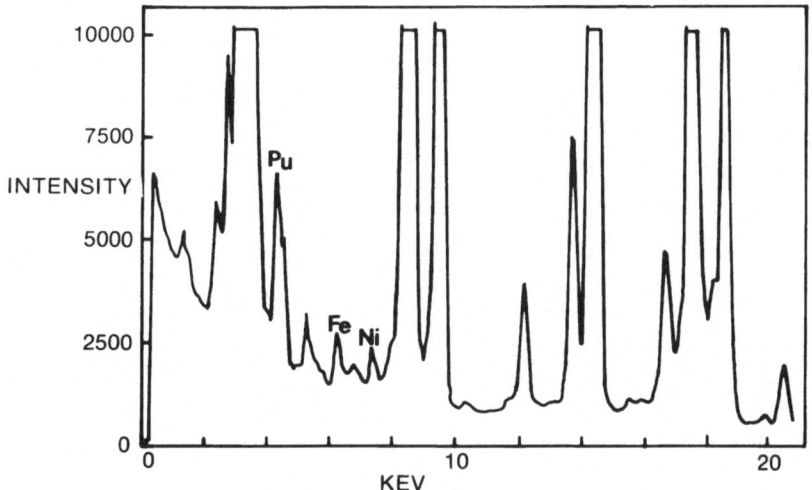

Figure 3. 110 ppm Ni and 205 ppm Fe in Pu metal.

Nickel standards were very difficult to find with a satisfactory range of values. However, a limited range curve which covers our current levels in samples was constructed (Table 1). The correlation coefficient indicates a satisfactory degree of linearity.

CONCLUSION

An x-ray spectrometry method for analysis of Ga, Fe and Ni has been developed that has significant advantages over current practice. No wet chemistry steps are involved which means that

the analysis can be done in a plutonium foundry area if desired.
Weighing is not required and less skilled personnel are needed.
Also less elapsed time per sample along with less actual work
time per analysis is taken than with current wet chemistry methods.

REFERENCES

1. P. L. Wallace, P. K. Hosmer, J. C. Walden, and
    W. L. Haugen, "The Direct Measurement of Ga in
    Binary Pu-Ga Alloys Using X-ray Spectrometric
    Techniques", X-Ray Spectrometry, Vol. 7, No. 4,
    pp. 212, 1978.

THE ANALYSIS OF COPPER ALLOYS BY CHEM-X, LOW POWER WDX MULTICHANNEL SPECTROMETER

J. Lucas-Tooth
B.W. Adamson
Y.M. Gurvich

Applied Research Laboratories
9545 Wentworth Street
Sunland, California 91040

The analysis of copper alloys is a classical case for X-ray fluorescence as an analytical technique. For many years, large wavelength dispersive X-ray spectrometers have been used in process control associated with the manufacture of brasses, bronzes and coinage alloys. For the last two to three years, a new, benchtop WDX multichannel spectrometer, the Chem-X, has been successfully introduced for the same purpose.

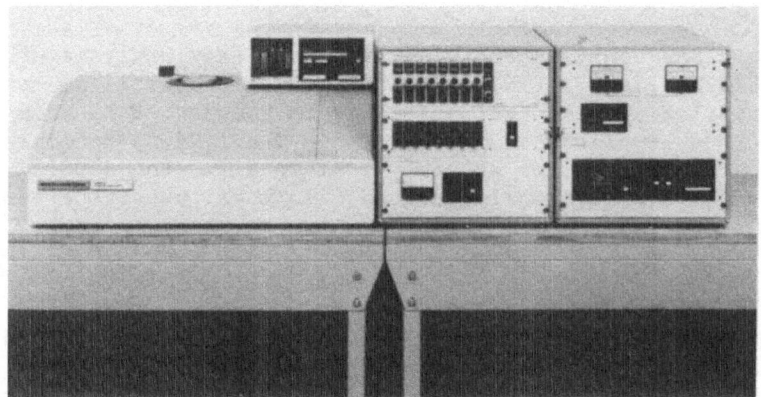

Fig. 1: The Chem-X Multichannel WDX Spectrometer

Use of a low power X-ray tube (up to 200W) eliminates the
necessity for a cooling water circuit. Utilizing close coupled
optics and a He atmosphere, the Chem-X is capable of analyzing
simultaneously up to eight elements with $Z \geqslant 13$ (Al). Simple
design permits the interchange of monochromators; therefore, the
user may have the capability to analyze more than eight elements
by keeping extra monochromators "on the shelf."

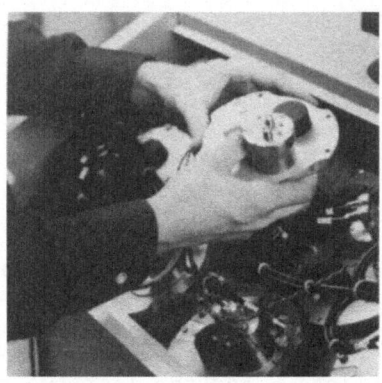

Fig. 2: Plug-In Monochromators for Chem-X

The Chem-X has been successfully used for X-ray analysis of
copper-based alloys at IMI Ltd., SMM Propellors, William Morris &
Sons, A. Fryer & Sons, Ltd. - Springvale Foundry, McKechnie Metal
Powders Ltd., England; Kamani Metals in India, and Trefimetaux in
France. For calculation of analytical results, the SAS-11
software package with SB-11 is used. In most cases, the simple
calibration equation:

$$C_i = a + bI_i$$

gives quite adequate results. For samples with well defined
chemical composition, accuracy of X-ray analysis is usually at the
level of 0.2% rel. for concentrations 50% and higher, 0.5% rel. at
10%, 1-2% rel. at 1% and higher, and 2-5% rel. for 0.002-0.8% (see
Table 1). Precision of analysis is less than 1% rel.

Table 1: Typical Results of Copper-Based Alloys for Chem-X

Application	Element	Concentration Range, %	Time, Sec.	Standard Deviation	
				Precision	Accuracy
Binary Brass	Cu	54.7–66.0	50	0.05	0.10
Phosphor Bronzes	P	0.10–1.0	200	0.004	0.025
	Zn	0.09–0.74	200	0.002	0.015
	Sn	8.9–13.2	200	0.024	0.05
	Pb	0.16–0.80	200	0.008	0.008
Special Leaded Brasses	Al	0–7.3	200	0.03	0.09
	Si	0–1.0	200	0.005	0.03
	Mn	0–3.7	200	0.004	0.04
	Fe	0.03–2.7	200	0.003	0.04
	Sn	0.01–0.59	200	0.004	0.015
	Pb	0.01–2.5	200	0.009	0.035
Impurity Brasses	Fe	0.002–0.084	200	0.001	0.0015

The computer then checks for line overlaps and these are corrected at an intensity level as the next operation. Matrix effects are corrected by the equation:

$$C_{i,corr} = (a_o + b_o C_i)(1 + \overset{j}{\underset{}{\Sigma}} a_j C_j)$$

a_i terms allow the operator to obtain correction for the change of mass absorption coefficient of the matrix; they usually can be supplied for each particular application. Then coefficients a_o and b_o can be determined with a few standards by simple linear regression.

The analysis of copper-based alloys containing lead gives a particularly good examples of the advantages of the Chem-X. When one analyzes binary (Cu/Zn) alloys on a high-power XRF spectrometer, it is not uncommon to achieve a precision of 0.02%. Statistically, the same precision level can be obtained on the Chem-X as well: it requires an analysis time of about 200 sec. However, this high precision can rarely be practically important when matrix effects are considered. For determination of copper in the range 51–78%, variation of Al, Si, Mn, Fe, Sn and Pb should be corrected; the value of coefficients are:

$$a_{Al} = -0.0045 \qquad\qquad a_{Fe} = 0.0242$$

$$a_{Si} = -0.0060 \qquad\qquad a_{Sn} = 0.0147$$

$$a_{Mn} = 0.0248 \qquad\qquad a_{Pb} = 0.0125$$

From the size of the Pb coefficient, it can be seen that any
inaccuracy in determining Pb will be carried through into the
copper equation. In fact, preparation of leaded samples is
notoriously difficult due to the lead being unevenly distributed
in solid solution, and even a clean cut of the surface area gives
an error in lead of about 0.06%, without any consideration of
statistics. Thus, the corrected Cu concentration shows an error
due to this cause alone of approximately 0.045%. Similar effects
occur on the other elements mentioned. Thus, the overall accuracy
of analysis is limited more by the methodology than by the fact
that a low-power tube is used.

Two standards are programmed for long-term drift correction of the
Chem-X. In actual fact, drift is usually less than 0.1% rel. per
4 hours of operation. The DEC SAS-11 computer is sufficient in
the case of a variety of applications and many different alloy
types. However, for many routine QC laboratory applications, a
smaller computer, the HP97S, is more than adequate.

Results of copper-based alloys analyzed by Chem-X are quite
similar to the performance of large WDX systems. The most
startling difference between the Chem-X and these instruments is
the price, typically one-third of multichannel, and one-half of
sequential X-ray spectrometers.

ENERGY DISPERSIVE XRF ANALYSIS OF LUBRICATING OIL ADDITIVES WITH SECONDARY TARGET EXCITATION AND THE EXACT[1] FUNDAMENTAL PARAMETERS PROGRAM

Kenneth C. Stehr

Applications Laboratory, KEVEX Corporation

Foster City, CA 94404

INTRODUCTION

The need for rapid and accurate analysis of lubricating oil additives is firmly established by the petroleum industry. The application of energy dispersive X-ray fluorescence spectrometry techniques to this analysis is a valuable aid in achieving reliable and fast quality control of these product types.

Due to the large concentration variations possible for the elements of interest within these materials (principally Zn, Ca, P, S at 0-20%), interelement effects can become significant. In order to correct for interelement effects, historical XRF methods have relied on a large number of standards for each element to be determined or matrix matching with standards closely bracketing the unknown sample concentrations.

The EXACT fundamental parameters program was developed to account for interelement effects when using monochromatic radio-isotope excitation.[2] This program (EXACT) requires that only one well characterized multielement standard, or one standard for each element of interest, be used for calibration purposes. The matrix effects of carbon and hydrogen are accounted for by specifying their molar ratio within the sample material (normally 1:2, $C:H_2$) and are analyzed by difference. Other elements, such as N and O, which are present and not detectable via XRF must be entered as elements of known concentration.

This paper reports methods of analysis, precisions and accuracies obtainable in the determination of Zn, Ca, P and S in lubricating oil utilizing two successive secondary targets. A germanium secondary target is employed to fluoresce Zn and a titanium secondary target for the analysis of Ca, P and S. The samples analyzed were submitted by two major petroleum companies.

EXPERIMENTAL

The analyses were performed on a Kevex 0700/7077 system with a Si(Li) detector having a resolution of 165 eV at 5.9 keV. This system is equipped with a 60 kV, 2.0 mA Rh-anode X-ray tube. The secondary targets used were high purity Ge and Ti without filters. Data acquisition was accomplished at 15 kV tube potential for 100 seconds live time for each secondary target used. All analyses were performed in a helium atmosphere.

In order to realize increased accuracy, it is necessary to calibrate the software at two levels of concentration. Calibration coefficients derived from 1% solutions of $CaCO_3$, $(NH_4)_2SO_4$, $(NH_4)_2 HPO_4$ and Zn metal were used to analyze samples with additive levels of less than 5%. Briquettes of the same materials or 10-15% solutions may be used to calibrate for samples with additive levels greater than 5%.

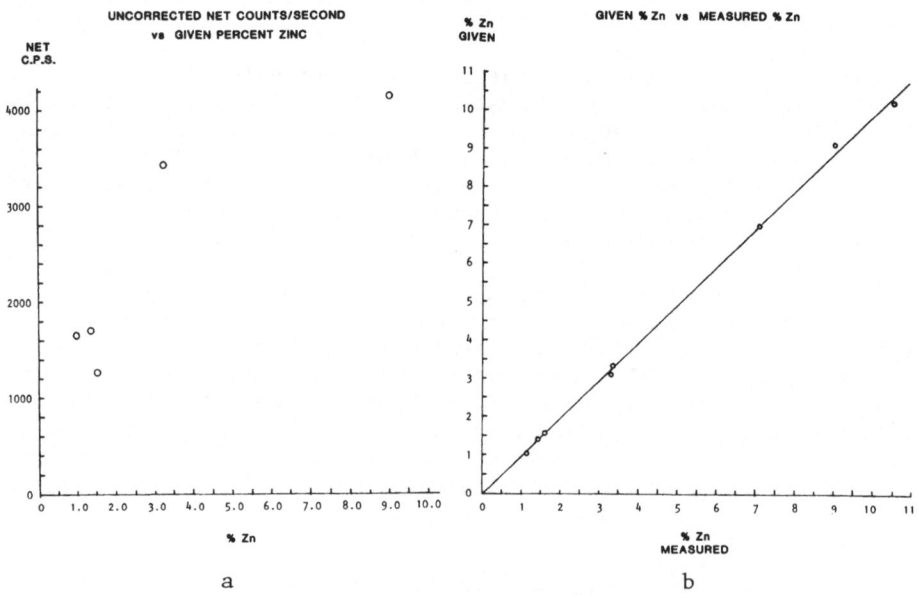

Figure 1. Uncorrected intensity data (a). EXACT results (b).

Sample preparation consisted of pouring the additive materials into disposable sample cups which were equipped with 0.25 mil polypropylene windows. Each cup was filled to the same level using approximately 5 ml. All samples and standards were analyzed with the same operating parameters.

RESULTS AND DISCUSSION

The need for matrix correction is clearly demonstrated by figure 1a. The uncorrected net counts per second plotted against the list concentration of Zn shows considerable scatter of the data points. Figure 1b represents the correspondence between the list and measured concentrations after application of the EXACT program.

Three replicate analyses were performed on two additive samples which are representative of the high and low concentrations encountered in this work (see Table 1). The standard deviations expressed represent the total instrumental and sampling error.

Table 1.   Replicate Analyses

| | Sample I | | | Sample L | | |
	$\bar{X}$	SD	RSD	$\bar{X}$	SD	RSD
Zn	10.7%	0.27	2.5%	1.11%	0.01	0.9%
Ca	ND*	–	–	1.50%	0.006	0.38%
S	18.7%	0.42	2.2%	2.65%	0.03	1.1%
P	8.8%	0.23	2.6%	1.66%	0.042	2.5%

* Not Detected

A summary of the results obtained from the single sample analysis of twelve oil additive samples is provided in Table 2.

CONCLUSIONS

Reliable analyses of lubricating oil additives can be achieved in less than ten minutes by this method. The calibration procedure is facilitated by the utilization of fundamental parameters software. The degree of inter-matrix flexibility afforded by fundamental correction schemes allows the analyst to choose between a wide variety of readily available substances for use as calibration

standards.   Sample preparation is minimal since it is not necessary
to dilute the additives prior to analysis.

Table 2

Summary of Oil Additive Analysis Results

	A List	A Meas.	B List	B Meas.	C List	C Meas.	D List	D Meas.
Zn	1.58	1.59	1.40	1.40	N.G.	N.D.	N.G.	N.D.
Ca	4.40	4.45	3.03	3.08	1.99	1.98	4.95	5.19
S	6.72	6.67	2.87	2.80	11.2	11.6	4.40	4.19
P	1.17	1.09	1.12	1.16	N.G.	N.D.	N.G.	N.D.

	E List	E Meas.	F List	F Meas.	G List	G Meas.	H List	H Meas.
Zn	9.1	9.4	3.32	3.32	N.G.	N.D.	N.G.	N.D.
Ca	15.6	15.2	5.7	5.7	6.7	6.7	15.70	16.78
S	N.G.	N.D.	N.G.	N.D.	2.39	2.38	1.24	1.23
P	7.4	7.4	2.88	2.93	N.G.	N.D.	N.G.	N.D.

	I List	I Meas.	J List	J Meas.	K List	K Meas.	L List	L Meas.
Zn	10.2	10.5	7.0	7.1	3.1	3.5	1.03	1.11
Ca	N.G.	N.D.	N.G.	N.D.	N.G.	N.D.	1.48	1.49
S	19.0	18.8	13.0	12.6	5.8	5.6	2.68	2.64
P	8.9	8.9	6.0	6.0	3.5	2.9	1.69	1.65

N.G. = Not Given
N.D. = Not Detected

REFERENCES

1.   J. W. Otvos, G.E.A. Wyld, and T.C. Yao, "Fundamental Parameter
     Method for Quantitative Elemental Analysis With Monochromatic
     X-ray Sources," paper presented at 25th Annual Denver X-ray
     Conference (1976).

2.   J.C. Harmon, G.E.A. Wyld, T.C. Yao, and J.W. Otvos, "X-ray
     Fluorescence Analysis of Stainless Steels and Low Alloy Steels
     Using Secondary Targets and the EXACT Program," paper presented
     at 26th Annual Denver X-ray Conference (1977).

THE ANALYSIS OF OIL ADDITIVES USING FUNDAMENTAL

INFLUENCE COEFFICIENTS

Mrs H M West

Philips Analytical Dept. Pye Unicam Limited

York Street, Cambridge, England

One of the most important applications of XRF in the oil industry has been in the analysis of oil additives which range from Mg to Ba. In such a low atomic number, hydrocarbon matrix, small varia- tions in the additives cause pronounced changes in the average atomic number of the matrix and hence give rise to strong interelement effects. In recent years a number of mathematical methods (ref 1-5) have been developed to correct the measured X-ray intensity of the analyte line for these considerable effects. This paper describes a method for the determination of Zn,Ba,Ca,Cl,S,P and Mg in oil additives using a concentration based correction model in which influence coefficients (alphas) are applied.

The ALPHAS (6,7) computer programme was designed by de Jongh to derive theoretical influence coefficients which describe the magnitude of the interelement effect from a knowledge of the spectral distribution emitted from the X-ray tube, the spectrometer geometry and the matrix of the material of interest. The influence coefficients referred to as 'alphas' allow interelement effects to be corrected for when entered in the general algorithm.

$$C_i = (D_i + E_i R_i) (1 + \Sigma^n_{j=1} \alpha_{ij} C_j)$$

where C_i = concentration of element i
E_i = slope of the calibration line in concentration units/Kc/s
D_i = intercept of the concentration axis (background)
R_i = measured nett count rate in Kc/s for element i
α_{ij} = influence coefficients for the effect of element j on element i
C_j = concentration of element j.

The correction is therefore a concentration correction and the magnitude of the interelement effect is a product of the 'alphas' and the concentration of the correcting elements. The algorithm separates the instrumental dependent parameters (D+ER) from the matrix dependent parameters $(1+\Sigma\,\alpha_{ij}C_j)$ thus allowing a linear relationship to be derived between the nett intensity and a function of the concentration referred to as the apparent concentration.

$$C/(1+\Sigma\alpha_{ij}C_j) \quad = \text{apparent concentration} = (D+ER).$$

Once the 'alpha' coefficients have been derived, a mini-computer may be used for calibration of the spectrometer. A rapid 2 point recalibration procedure is all that is needed to establish the slope of the calibration curve for each element as the apparent concentration is linearly related to the nett intensity. In order to test the ALPHAS programme a table of fundamental coefficients was calculated for the oil additives based on the following average composition:-

TABLE 1. AVERAGE COMPOSITION OF OIL ADDITIVES

Z		wt %	Z		wt %
1	Oil	87.50	17	Cl	1.50
12	Mg	1.00	20	Ca	2.00
15	P	1.50	30	Zn	1.50
16	S	3.50	56	Ba	1.50

The 'alphas' obtained demonstrate the extent of the effects with values such as 83 for Ba on ZnKα; 32.7 for Ca on BaLα; and 8.4 for Zn on MgKα. The spectrometer used was a Philips PW1400 sequential system fitted with a rhodium target tube. Kα analyte lines were measured for all elements except Ba where the Lα line was used. A suite of synthetic samples was prepared as shown in Table 2. The concentrations were chosen on a high/low basis using the following reagents:

Mg — Magnesium sulphonate (overbased Mg 9%, S 1.6%)
P — Triethyl phosphate
Cl — Cerechlor (contains ≈50% Cl)
Ca — Calcium sulphonate (overbased Ca 11.5%, S 1.5%)
Ba — Barium octoate
Zn — Zinc octoate
S — ditertiary butyl disulphide (plus the Mg & Ca components)
with sulphur free alkyl phenol as the diluent.

Table 2. CONCENTRATIONS OF THE ADDITIVE ELEMENTS IN THE
          SYNTHETIC BLENDS ANALYSED

SAMPLE	Mg	P	S	Cl	Ca	Ba	Zn
1	0.800	1.500	3.629	1.022	2.000	1.000	1.601
2	0.200	1.500	3.533	1.022	2.000	0	0.800
3	0.800	0.500	3.629	0	2.000	1.000	0.800
4	0.200	0.500	3.533	0	2.000	0	1.601
5	0.800	1.523	1.128	0	2.000	0	1.642
6	0.200	1.500	1.032	0	2.000	1.028	0.800
7	0.800	0.500	1.128	1.022	2.000	0	0.800
8	0.200	0.500	1.035	1.022	2.016	1.000	1.780
9	0.800	1.500	3.630	0	1.004	0	0.800
10	0.200	1.500	3.535	0	1.008	1.000	1.601
11	0.800	0.500	3.630	1.039	1.000	0	1.601
12	0.200	0.500	3.533	1.022	1.000	1.000	0.806
13	0.800	1.500	1.130	1.030	1.006	1.000	0.800
14	0.200	1.500	1.033	1.022	1.002	0	1.601
15	0.800	0.500	1.129	0	1.003	1.003	1.601
16	0.200	0.500	1.038	0	1.038	0	0.800

Table 3. EXAMPLES FROM REGRESSIONS

	Sample Number	X-ray int. Kc/s	Concentration		Apparent Conc.	
			Chem.	Calc.	Chem.	Calc.
Sulphur	1	65.655	3.63	3.623	2.062	2.059
	2	73.417	3.53	3.533	2.299	2.301
	6	22.039	1.03	1.036	0.693	0.697
	16	26.727	1.04	1.039	0.844	0.843
	SIGMA =	0.018				
Calcium	1	40.200	2.00	1.994	0.773	0.771
	6	57.048	2.00	1.973	1.120	1.105
	13	26.612	1.006	0.996	0.506	0.501
	15	31.050	1.003	1.002	0.590	0.589
	SIGMA = 0.016					
Barium	1	8.526	1.00	0.996	0.312	0.310
	5	0.135	0.00	0.003	0.00	0.001
	6	11.816	1.028	1.029	0.431	0.432
	13	11.654	1.00	0.996	0.427	0.426
	SIGMA = 0.004					
Zinc	1	138.568	1.601	1.572	0.358	0.351
	2	93.719	0.80	0.788	0.242	0.238
	14	239.951	1.601	1.606	0.606	0.607
	16	161.709	0.80	0.809	0.405	0.410
	SIGMA = 0.016					

Regression analyses were performed on the intensities obtained from the suite of synthetic blends using the 'alphas' theoretically pre-determined. Table 3 shows typical examples of the regression results for the elements S, Ca, Ba and Zn for four pertinent samples per element. The effectiveness of this fundamental influence coefficient approach is clearly demonstrated. Taking the results for Zn as an example, the fact that sample No. 16 gives a higher measured intensity than No. 1 (which contains twice its concentration) is not detrimental to the final or calculated concentration.

The slopes and background intercepts from these regressions were stored in the computer and selected high/low samples used for recalibration of the spectrometer. The final proof of the technique is the analysis of real samples as shown in Table 4. Many of the additives are extremely viscous and dilution with alkyl phenol is necessary.

Table 4.   ANALYSIS OF UNKNOWNS

	Sample 1 (as 1+8 dilution)			Sample 2 (as 1+5 dilution)	
	Chem. %	X-ray %		Chem. %	X-ray %
P	8.26	8.27	Ca	9.22	9.19
S	15.9	15.94	S	3.69	3.67
Zn	9.33	9.34			

CONCLUSION

This approach to the analysis of oil additives provides the analyst with a calibrated spectrometer using a minimum of reference materials.

The author wishes to thank Mr P Bullock, his staff and the Directors of BP Chemicals Limited, Hull, for their assistance in preparing the samples and for permission to present this paper.

REFERENCES

1. Birks et al., Anal. Chem. 22, 1258 (1950)
2. Davis E.N. and van Nordstrand R.A., Anal.Chem.26, 973 (1954)
3. Louis R.Z., Anal. Chem. 201, (5) 336 (1963)
4. Haycock R.F. J Inst. Pet 50 123 (1964)
5. Bird R.J., Toft R.W. J.Inst. Pet 56 550 (1970)
6. de Jongh W.K., X-ray Spectrom 2, 151 (1973)
7. Jenkins et al; Advances in X-ray Ananlysis Vol. 18

THE MEASUREMENT OF LOW CONCENTRATIONS OF ORGANIC AND INORGANIC GASEOUS CONTAMINANTS IN OCCUPATIONAL ENVIRONMENTS BY X-RAY SPECTROMETRY (XRS)

N.G. West, C.J. Purnell, R.H. Brown, and E. Withers

Health and Safety Executive
Occupational Hygiene Laboratories
403 Edgware Road, London NW2 6LN., England

INTRODUCTION

To ensure that workers in certain industries are not exposed to harmful levels of toxic chemicals, it is necessary to provide regular monitoring of the concentrations of chemical contaminants in the workplace air. In the United Kingdom, monitoring is normally carried out on a routine basis by the factory occupier backed up by periodic visits from the Factory Inspectorate acting on behalf of the Government. The main source of guidance for occupational hygienists in assessing conditions in a factory is the list of threshold limit values (TLVs) published annually by the American Conference of Governmental Industrial Hygienists.[1] Threshold limit values refer to airborne concentrations of substances and represent conditions under which it is believed that nearly all workers may be repeatedly exposed day after day without adverse effect.

Many of the hazardous substances are in vapour or gaseous form and the established method of sampling is to draw a known volume of air through either a solid adsorbent or an absorbing solution. Instrumental analysis techniques currently used for pumped samples include chromatography, colorimetry and infra-red spectroscopy. In recent years an alternative to the conventional pumped sampling systems has been developed in the form of the diffusive sampler.[2] This device samples by gaseous diffusion whereby a gaseous contaminant is transferred at a controlled rate across a diffusion barrier onto an adsorbent which acts as a zero sink. The increasing use of diffusive samplers[4] is due to the fact that they have a number of advantages over conventional pumped sampling systems for monitoring the personal exposure of

workers. In particular they are very simple to use, do not
require a heavy and expensive sampling pump and have greater
wearer acceptability. The combination of diffusive sampling with
XRS analysis offers a very simple sampling and analysis procedure.
This paper describes recent work which has demonstrated the
potential of the system for a range of analytes.

DIFFUSIVE SAMPLERS

     Two basic designs of diffusive sampler have been developed,
the badge type and the tube type. For the work described here, a
typical badge type (Porton Down sampler) and a typical tube type
(HSE sampler) were used. (Fig. 1). The Porton Down sampler[3]
developed at the Chemical Defence Establishment, Porton Down, UK
consists of a plastic holder containing the adsorbent and fitted
with a micro-porous polypropylene membrane which acts as the
diffusion barrier. The HSE sampler, designed at the Occupational
Medicine and Hygiene Laboratories, London, is essentially a glass
tube with a draught shield across one end and an adsorbent
compartment at the other, the air gap between the two ends forming
the diffusive barrier. The diffusive uptake of a sampler with an
air gap (or an effective air gap e.g. Porton Down sampler) can be
calculated from a simple application of FICK's first law:

$$\text{Uptake} = \frac{DCAt}{Z}$$

D is the diffusion coefficient of the contaminant in air
C is the ambient concentration of the contaminant
A is the effective cross sectional area of the sampler
t is the time of exposure
Z is the diffusion path length

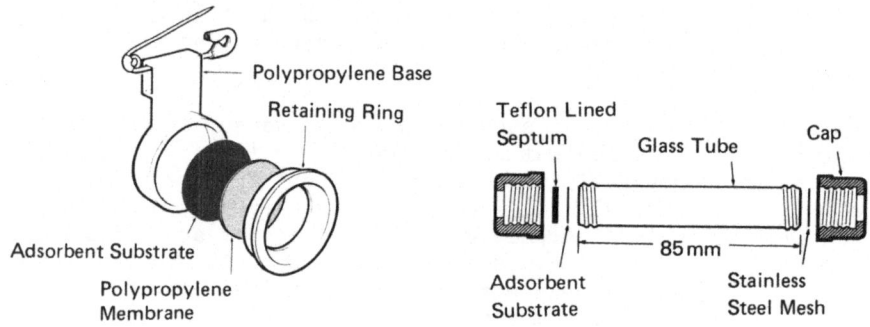

(a) PORTON DOWN BADGE SAMPLER   (b) HSE TUBE SAMPLER

Fig. 1. Two basic designs of diffusive sampler.

The physical characteristics of the sampler are defined by the $A/_z$ ratio which controls the uptake rate for a given contaminant. It is found that the $A/_z$ ratio for the Porton Down sampler is 128 times higher than for the HSE sampler.

ANALYSIS OF DIFFUSIVE SAMPLERS BY X-RAY SPECTROMETRY

In principle, any gaseous species containing an element with atomic number >8 may be monitored by XRS providing it can be sampled onto an adsorbent substrate suitable for presentation to the spectrometer. Charcoal cloth is used as the adsorbent for organic vapours and for inorganic gases reagent-impregnated papers may be employed. The impregnated papers have the advantage that, if the correct reagent is chosen, a coloured stain can be produced by reaction with a particular contaminant. This enables an approximate on-site figure for the concentration of the contaminant to be obtained in the factory prior to accurate XRS analysis in the laboratory.

For practical purposes, the sensitivity of the XRS technique must be compatible with the uptake rate of the diffusive device and the threshold limit value of the analyte. The suitability of the XRS technique for a particular contaminant is assessed on the basis of three factors (1) theoretical sampler uptake rates (2) estimated XRS detection limit and (3) minimum analytical requirement. The minimum analytical requirement is defined as the ability to detect the contaminant when the loading (i.e. concentration (TLV) x exposure time (hours)) is 0.5. Data on uptake rates for the two types of sampler are presented in Table 1 together with the minimum uptake rates required for XRS analysis. For mercury vapour, arsine and tetra-ethyl lead only the Porton Down sampler met the minimum uptake requirements, while for the remaining gases either sampler was suitable. The HSE sampler was chosen for hydrogen sulphide and sulphur dioxide to avoid possible saturation of the adsorbent and the Porton Down sampler was selected for halothane to give maximum sensitivity.

X-ray spectrometric analysis was performed on two wavelength dispersive spectrometers, a Philips PW1410 fitted with a chromium target tube and a Philips PW1450 with a molybdenum tube. The PW1410 spectrometer was used for sulphur analysis and the PW1450 for the other elements. Standards for calibration of the instruments were prepared by generating standard atmospheres for each of the analytes and then exposing samplers for measured periods of .time at known concentration levels. It is important for accurate analysis that the standards are prepared in the same manner as samples are taken in the factory since simpler methods of calibration such as spotting aliquots of analyte onto the adsorbent cannot reproduce the exact distribution of the analyte within the substrate.

Table 1   Theoretical diffusive sampler uptake rates (ng/ppm/min)

Analyte	Adsorbent	Sampler		Minimum uptake rate for XRS Analysis @ 0.5 TLV hours	Emission line
		Porton Down	HSE		
Hydrogen sulphide	silver nitrate paper	117	0.86	0.08	SKα
Sulphur dioxide	Nickel Hydroxide paper	270	2.11	0.16	SKα
Mercury vapour	Cuprous iodide paper	983	7.67	980	HgLα
Arsine (AsH_3)	Silver nitrate paper	341	2.66	115	AsKα
Halothane $(CF_3CH\ Cl\ Br)$	Charcoal cloth	480	3.76	2	BrKα
Tetra-ethyl lead $(C_2H_5)_4\ Pb$	Charcoal cloth	619	4.84	447	PbLα

RESULTS AND DISCUSSION

Calibration curves

    Calibration curves were constructed for each of the analytes
over the loading range from 0 to 10 TLV hours. A typical calibration is shown in Fig. 2 for arsine using standards exposed at
a number of different concentration levels from 0.5 to 6 TLV
(i.e. 0.03 to 0.36 ppm AsH_3). The points lie on a linear plot
which was found by further experiment to extend to at least 40
TLV hours' loading. The range of the calibration was more than
sufficient to measure the time weighted average concentration of
arsine for an 8-hour shift under normal working conditions.
Sulphur dioxide, halothane and tetra-ethyl lead also gave linear
calibrations up to high loadings (30 TLV hours), but hydrogen
sulphide and mercury vapour showed curvature with reduced
sensitivity above 5 TLV hours and 10 hours' loading respectively.

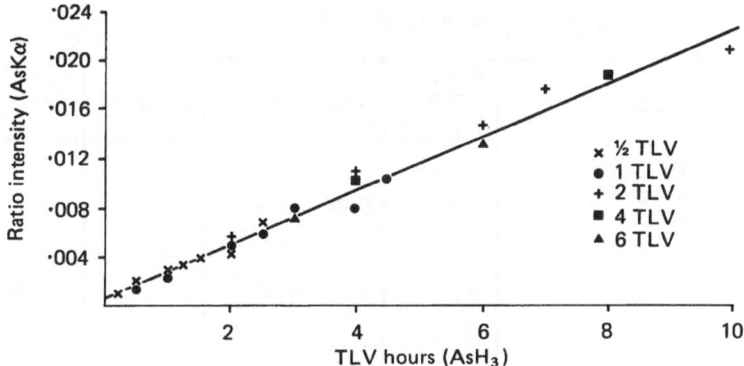

Fig. 2. Calibration graph for Arsine.

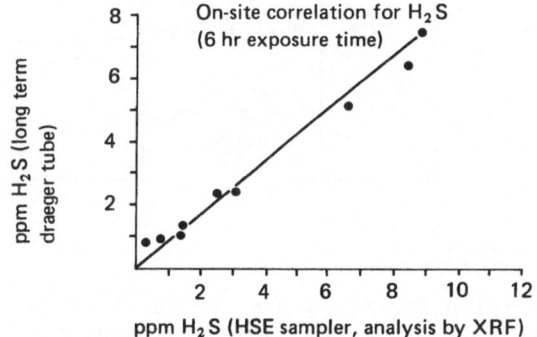

Fig. 3. Comparison of long term detector tubes and HSE tube samplers for the measurement of hydrogen sulphide in air

There are two separate but inter-related factors which can affect the sampler sensitivity, (1) a reduction in the uptake rate due to adsorbent saturation and (2) attenuation of the X-ray intensity caused by matrix effects in the substrate. Adsorbent saturation is related to both the capacity of the adsorbent and the kinetics of the reaction/adsorption process taking place on the substrate while X-ray attenuation is governed by the mass absorption coefficient of the matrix and the depth of deposition of the analyte in the adsorbent. For mercury, the loss of sensitivity was due mainly to adsorbent saturation effects resulting in a reduction in the concentration gradient controlling the uptake rate. In the case of hydrogen sulphide, X-ray absorption is the major factor involved. The fact that sulphur dioxide did not show a similar absorption effect is probably due to the fact that the H_2S penetrated deeper into the adsorbent prior to reaction at an active site.

Table 2   Detection limits (D.L.)

Analyte	TLV	D.L. (TLV hours)	Sampler	Crystal	Path
Hydrogen sulphide	10 ppm	0.1	HSE tube	PE	Vacuum
Sulphur Dioxide	5 ppm	0.4	HSE tube	PE	Vacuum/ He path
Mercury vapour	0.0006 ppm	0.5	Porton Down badge	LiF_{200}	air
Arsine	0.05 ppm	0.06	Porton Down badge	LiF_{200}	air
Halothane	1 ppm (proposed)	0.004	Porton Down	LiF_{200}	air
Tetra-ethyl lead	0.10 mg m^{-3}		Porton Down badge	LiF_{200}	air

Detection limits and precision

The detection limits for the six analytes given in Table 2 are based on the analysis of twelve unexposed samplers. Following Currie[5], the limit is defined as 1.64σ where σ is the standard deviation of the results on the unexposed samplers. The relatively poor detection limit for SO_2 compared with H_2S is due to pick-up of sulphur from the atmosphere and possibly also from the vacuum oil in the spectrometer. A helium path is therefore preferred for this analysis. The sensitivity for halothane determination is very good with a detection limit of 0.004 TLV hours which is well below the gas chromatography figure of 0.1 TLV hours.

Measurements of the precision of the diffusive sampler method were made for both the Porton Down sampler and the HSE sampler by exposing ten of each type of sampler simultaneously in a standard atmosphere. The analytes chosen were arsine and sulphur dioxide. The overall precision (expressed as a coefficient of variation) for the Porton Down sampler at a loading of 8 TLV hours was found to be 6.4% and for the HSE sampler 9.8%. The equivalent figures for the sampling precision alone were 6.3% and 8.6% respectively which compares well with pumped sampling methods.

Comparison of factory results with conventional methods

Comparisons have been made with conventional sampling and analysis methods for hydrogen sulphide, mercury vapour and halothane, and good agreement has been found. Results for H_2S are shown in Fig. 3.

CONCLUSIONS

The combination of diffusive sampling with XRS analysis provides a simple, elegant sampling and analysis system with the potential to cover most inorganic gases and a large number of organic vapours. The analysis, which requires no sample preparation, is fast, precise and non-destructive and compares favourably with more conventional analysis techniques for these analytes.

ACKNOWLEDGEMENTS

The authors wish to thank Mr M.D. Wright and Mr C.J. Wells for valuable assistance in obtaining experimental data.

REFERENCES

1.   American Conference of Governmental Industrial Hygienists TLVs, Threshold Limit Values for Chemical Substances and Physical Agents in the Workroom Environments. ACGIH, Cincinnati (1980).

2.   E.D. Palmes and A.F. Gunnison, Personal Monitoring Device for Gaseous Contaminants, Am. Ind. Hyg. Assoc. J., 34 : 78 (1973).

3.   A. Bailey and P.A. Hollingdale-Smith, A Personal Diffusion Sampler for evaluating time weighted exposure to organic gases and vapours, Ann. Occup. Hyg., 20 : 345 (1977).

4.   C.J. Purnell, M.D. Wright and R.H. Brown, Performance of the Porton Down Charcoal Cloth Diffusive Sampler, Analyst, 106 : 590 (1981).

5.   L.A. Currie, Detection and Quantitation in X-ray Fluorescence Spectrometry in "X-ray Fluorescence Analysis of Environmental Samples" Thomas G. Dzubay ed., Ann Arbor (1977).

THE APPLICATION OF X-RAY FLUORESCENCE AND DIFFRACTION TO THE CHARACTERIZATION OF ENVIRONMENTAL ASSESSMENT SAMPLES

Albert C. Censullo and Frank E. Briden

U.S. Environmental Protection Agency
Industrial Environmental Research Laboratory
Technical Support Staff
Research Triangle Park, NC   27711

INTRODUCTION

The Technical Support Staff is called upon for analysis of a wide variety of sample types many of which have little sample history. However, it is usually necessary to account for all elements present. For these reasons, x-ray fluorescence spectrometry (XRF) has been a useful tool. Unfortunately, XRF requires the use of a range of standards for each element, the preparation of which could become so time consuming that the advantages of XRF would soon be diluted. Consequently, the utility of the J.W. Criss fundamental parameters computer program[1] was evaluated for samples in which only one standard per element was used and where the standard matrix did not strictly resemble the unknown matrix. Some of the results of these tests on environmental assessment samples are reported here.

The environmental significance of a sample is dependent not only on its elemental composition, but also on the species or phases which the elements comprise. X-ray powder diffraction (XRD) may be used to advantage for speciation. It will be seen that multi-phase environmental assessment samples are amenable to XRD interpretation. This may be performed using the Joint Committee on Powder Diffraction Standards (JCPDS) Hanawalt or Fink manual search methods[2] (tedious for multi phase samples) or the computer search program[2]. Some results of the application of the JCPDS computer interpretation of typical environmental samples will be discussed.

EXPERIMENTAL

All x-ray fluorescence measurements were made using a Siemens SRS-1 wave length-dispersive spectrometer. In all cases, excitation was provided by a chromium target x-ray tube operated at 50 kV In most cases, the sample was mixed 3:1 with a binder mix. Reduction of x-ray fluorescence data was effected by the XRF-11 fundamental parameters program (Criss Software) executed on a PDP-11 computer.

The powder diffraction patterns were obtained using a Siemens type F diffractometer with a copper x-ray tube operated at 35 kV. Samples were pulverized until able to pass a 400 mesh sieve. Samples were either backloaded onto a glass slide or applied to a glass slide as a slurry in amyl acetate. Interpretation and search of powder diffraction patterns was performed by the JCPDS search program, run on a UNIVAC 1100.

RESULTS AND DISCUSSION

For the evaluation of x-ray fluorescence data processed by the fundamental parameters (FP) program it was necessary to compare results with those for a sample which was rigorously characterized. The NBS 1633a fly ash satisfied this requirement. Only five single analyte standards and one double analyte standard were used and these differed in elemental composition from the unknowns by factors of two to seventy. The results for the fly ash analysis are shown in Table 1. The relative error ranged from 2% for calcium to 22% for sulfur.

Table 1. Elemental Composition in NBS SRM 1633a Fly Ash

Element	Accepted Comp. (%)	XRF-FP Comp. (%)	Standard Comp. (%)	Standard Phase	Error (%)
O	41.9	50.0			19
Al	14[a], 14.0[3]	12.1	52.9	Al_2O_3	13
Si	22.8[b]	26.0	46.8	SiO_2	14
S	0.27[a]	0.21	23.6	$CaSO_4$	22
K	1.88[b]	1.68	34.7	K_2CO_3	10
Ca	1.11[b]	1.08	29.4	$CaSO_4$	2
Ti	0.80[a], 0.84[3]	0.83	60.0	TiO_2	3
Fe	9.40[b]	7.91	69.9	Fe_2O_3	15

[a]NBS information value.   [b]NBS certified value.

The oxygen accepted value for the fly ash was calculated assuming all the major elements were present as oxides. The XRF-FP program generated the oxygen value by difference.

In the analysis of fuels by XRF, there is a complication in that the major elements carbon, hydrogen and nitrogen cannot be determined by XRF. Consequently, logical assumptions were sought which could simplify the determination yet produce reasonable values for those elements for which XRF data were available. For the determination of NBS 1635 coal, the standard was NBS 1632 coal. In the first set of analyses, hydrogen was set at 3 percent, nitrogen at 1 percent and the following carbon variations in SRM 1632 were specified: 0, 50, 64% (NBS accepted value), and 70%. Table 2 shows the results of this study. For comparison of the results, attention will be focused on phosphorus, whose behavior was quite typical. The NBS value for phosphorus was 0.012%. The calculated values for P were 0.013, 0.014, and 0.015% respectively for 50, 64 and 70% carbon – all slightly high but reasonable. However, if no carbon is assumed in the standard only 0.006% P was obtained. Another analysis was performed, specifying the correct carbon composition in the standard (64%) and the unknown (58.9%). The resulting calculation gave 0.027% P. This study concludes that it is necessary to make a rough assumption of carbon composition in the standard but that constraining the carbon composition in the unknown leads to significant errors in elemental composition.

In another study, results from XRF were compared with those from atomic absorption spectroscopy (AAS), differential pulse polar-

Table 2. Elemental Composition (%) in NBS SRM 1635 Coal
          Using SRM 1632 Coal as Standard

Element	Specified % C in SRM 1632				Specified Correct % C in 1632 and 1635	True 1635 Composition
	0%	50%	64%	70%		
C	90.5	92.4	92.3	92.4	58.9	58.9
Mg	0.041	0.106	0.110	0.108	0.119	0.098
Al	0.269	0.290	0.300	0.300	0.350	0.320
Si	3.370	0.619	0.420	0.652	1.03	0.900
P	0.006	0.013	0.014	0.015	0.027	0.012
S	0.416	0.394	0.421	0.431	0.838	0.330
Cl	0.025	0.008	0.009	0.009	0.021	0.010
K	0.117	0.014	0.015	0.016	0.039	0.009
Ca	0.128	0.487	0.531	0.549	1.49	0.560
Ti	0.103	0.017	0.019	0.019	0.069	0.020
Fe	0.970	0.132	0.146	0.152	0.625	0.239

ography (DPP), ion chromatography (IC), neutron activation analysis
(NAA) and spark source mass spectrometry (SSMS) for a particulate
sample from a copper smelter (Table 3). For this study, only one
standard per element was used in the XRF analysis. If the true
value for each element is taken as the mean of the values from all
of the methods used for that element then the errors for the XRF
analysis are 2, 4 and 5% respectively for cadmium, lead and anti-
mony. The worst case was iron with an error of 55% average. For
this set of elements the average error was 22%.

The JCPDS computer search program was used to identify phases
containing the major elements in the sample. In all XRD analyses,
elemental composition from XRF was specified to the search program.
The program searches for line matches, then performs a sequential
subtraction of each selected phase from the total unknown pattern.
Phases that do not significantly alter the unknown pattern after
this subtraction are "rejected." Other "accepted" phases are ranked
by means of a "reliability factor". An "average" match has a reliability factor of 40-50, while a perfect fit would have a reliability factor of 100. Output for the copper smelter sample is shown
in Table 4. The arsenic phase was identified as arsenic trioxide.
Lead sulfate and cadmium sulfate were found, but later rejected by
the multiple subtraction scheme. These and other rejections of
likely phases could be due to a variety of factors, including pro-

Table 3.   Elemental Composition (%) in Copper Smelter
Electrostatic Precipitator Particulate Catch

Element	AAS	DPP	IC	NAA	SSMS	XRF	Standard Comp. (%)	Standard Phase
Al						11.03	52.9	Al_2O_3
Si						1.15	46.8	SiO_2
S						5.73	23.6	$CaSO_4$
K					0.23	0.16	34.7	K_2CO_3
Ca					0.16	0.09	29.4	$CaSO_4$
Fe	2.19				3.12	4.31	69.9	Fe_2O_3
Cu	5.88	6.23		2.90	7.92	2.25	59.3	Brass
Zn	2.19	1.85		2.90	2.40	2.01	35.7	Brass
As			31.2		13.10	41.00	75.7	As_2O_3
Pb	5.25	5.71			18.20	10.22	86.6	PbO_2
Sb				1.56	2.68	2.28	83.5	Sb_2O_3
Cd	0.23				0.39	0.32	87.5	CdO

Table 4. Phases in Copper Smelter ESP Particulate from XRD

Phases Identified	Reliability of Accepted Phases	% Lines Matched Rejected Phases
As_2O_3	57	–
$ZnSb_2SO_4$	19	–
$Sb_2(SO_4)_3$	15	–
$CuFeS_2$	14	–
K_2O	10	–
$PbSO_4$	–	83
$CdSO_4$	–	71

gram parameter selection, preferred orientation of the sample, low concentration of rejected phase in unknown, and differences between the unknown phase and the JCPDS standard phase patterns. Tables 4, 5 and 6 also show the percentage of lines present for rejected but likely phases.

The XRF and XRD procedures were also applied to several other samples of environmental interest including a dual alkali scrubber sample. The only elements detected in scrubber solids by XRF were Na, S and Ca. The major by-product anticipated in the process is the hemihydrate of calcium sulfite. This species was readily identified by the JCPDS search. The other species shown in Table 5 are consistent with the chemistry of the scrubbing process.

Table 6 lists the results of XRF and XRD applied to an iron

Table 5. Phases and Elemental Composition of Dual Alkali Scrubber Solids

XRD Phases Identified	Reliability of Accepted Phases	% Lines Matched Rejected Phases	Element	XRF (%)
$CaSO_3 \cdot 1/2\ H_2O$	60	–	Na	9.2
$Na_2Mg(CO_3)_2$	21	–	S	22.4
$Ca(OH)_2$	20	–	Ca	24.1
$CaCO_3 \cdot H_2O$	10	–		
MgO	3	–		
Na_2SO_4	–	75		
$NaHCO_3$	–	67		

Table 6.   Phases and Elemental Composition of
Iron Smelter ESP Particulate

XRD Phases Identified	Reliability of Accepted Phases	% Lines Matched Rejected Phases	Element	XRF (%)
Fe_2O_3	63	–	O	34.1
K_2SO_4	39	–	Mg	1.29
			Al	6.93
$Pb_2(SO_4)O$	24	–	Si	2.81
			P	0.14
$PbGa_2O_4$	18	–	S	1.65
$Fe_4(As_2O_{11})$	10	–	K	0.58
$MgSiO_3$	–	88	Ca	7.95
			Ti	0.09
$MgSO_4$	–	80	Fe	44.4
$CaAlSiO_6$	–	75		

smelter electrostatic precipitator (ESP) sample. The phase of the
major element, iron, was identified as ferric oxide by the JCPDS
search of the powder diffraction data. The combination of these
techniques again yields results that are consistent with the known
processes involved.

CONCLUSIONS

     In general, the fundamental parameters program has proven to
be quite capable of yielding meaningful results when applied to
the analysis of environmental samples. Relative errors ranged from
2 to 20% for a variety of sample types. This accuracy was obtained
using only one readily available pure compound (usually a simple
inorganic oxide) as a standard per element determined. The advantage of this approach to a laboratory that receives a variety of
"unique" samples is apparent. Additionally, the JCPDS computer
search of powder diffraction data of environmental samples was
shown to contribute to the speciation of the complex samples that
are encountered in environmental assessments.

1.   J. W. Criss, Fundamental Parameters Calculations on a Labora-
     tory Microcomputer, Advances in X-ray Analysis, 23:93, Plenum
     Publishing Corp., 1980.

2.   G. G. Johnson, Jr., "User Guide", Publ. by JCPDS, 1975.

3.   E. S. Gladney, Analytica Chimica Acta, 118:385 (1980).

ACCURATE PIXE ANALYSIS OF THIN SAMPLES, AEROSOL LOADED FILTERS AND SURFACE LAYERS OF THICK SAMPLES*

U. Wätjen and F.-W. Richter

Fachbereich Physik, Universität Marburg

D-3550 Marburg, Fed. Rep. of Germany

INTRODUCTION

Particle induced X-ray emission (PIXE) analysis is well established as a multi-element, non-destructive technique to measure the trace element content of thin samples ($\lesssim 1$ mg/cm^2). The wide range of its analytical applications was shown in numerous reports at the 2nd International PIXE Conference in Lund (1). For thin sample analysis routinely done at the Marburg PIXE facility (2), we calibrated our spectrometer empirically at proton energies of 2 and 4 MeV using two separate sets of calibration standards, one purchased from MicroMatter Co., the other prepared by precipitate exchange (3). Both sets of standards were checked by AAS and CMP. K-shell ionization cross sections, calculated from our measured x-ray yields, agree very well with recent literature values (4), and will be reported elsewhere.

In general, biomedical and aerosol samples meet the constraints of small aereal density, which allow quantitative analysis with an accuracy of 5 to 10 % in the ppm range, using absolute calibration coefficients only, without the need of internal standards or even matrix correction calculations. Heavily loaded aerosol filters - routinely analyzed in our laboratory now - cannot be considered as thin samples, though. Correction procedures for this case will be discussed in the last section after the derivation of matrix effect calculations with model samples in the following.

---

*Supported by Bundesminister für Forschung und Technologie, Bonn

SURFACE LAYERS OF THICK SAMPLES

In trace, minor and major element analysis of surface layers of thick samples (projectile range up to 60 um in steel with E_p=4 MeV) the raw analytical results, derived from thin sample calibration parameters, are corrected for matrix effects employing a fundamental parameters approach. Our correction code is accounting for the decrease of x-ray yields due to slowing down of projectiles in the target, the absorption of characteristic x-rays produced in deeper layers of the target, and the enhancement of x-rays by secondary x-ray fluorescence.

   The basic formulae governing thin sample analysis and their modifications, due to the penetration of protons into a thick sample, are published elsewhere (5) together with a brief outline and graphic representation of the correction algorithm using primary slices of equal proton energy loss, superimposed by secondary slices and spherical layers of equal photon attenuation. The basic volume element in the calculation of the enhancement effect is shown in more detail in fig.1. As part of a spherical layer it is characterized by a certain incident intensity depending on the radius of the sphere, and since it is part of a secondary slice, all x-rays produced in it are attenuated by the same factor according to the constant distance from the sample surface. For the same secondary slice all volume elements – different by the incident intensity $(\Delta I_x)^{n+.5}$ – are calculated, and the sum of all secondary slices gives the amount of enhancement due to x-rays produced in one primary slice.

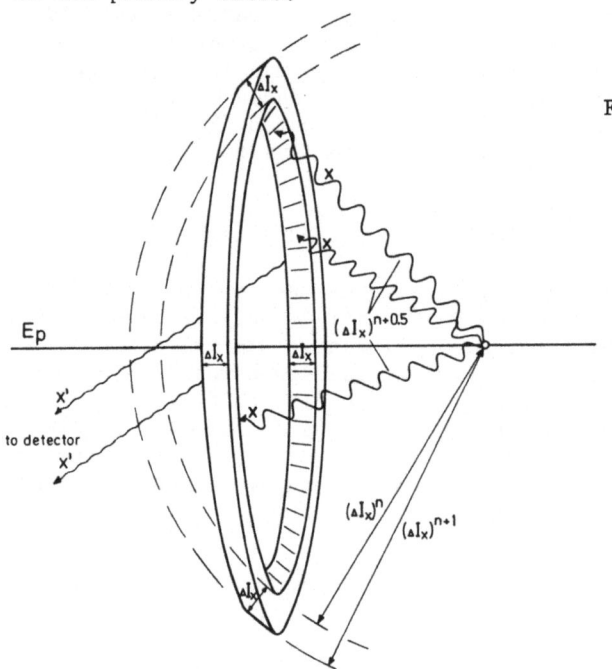

Fig. 1. Basic volume element for calculation of enhancement effect, showing primary x-rays x from location of primary production o incident on the slice of hollow sphere with constant x-ray attenuation factor ΔI_x to produce secondary photons x'.

Table 1. Certified mass content, PIXE analysis and PIXE simulation
results of British Chemical Standard No. 310/1
Nimonic '90' alloy. A: without secondary excitation,
B: taking account of second. excit.(by K_α photons only).

Element	Mass Content	Aereal Density PIXE ANALYSIS ($\mu g/cm^2$)	Aereal Density SIMULATION A ($\mu g/cm^2$)	Aereal Density SIMULATION B ($\mu g/cm^2$)
Al	.0106	–	0.36	0.36
Si	.0046	–	0.53	0.53
Ti	.0243	75	61.6	65.4
Cr	.194	720	623	693
Mn	.0035	26*	12.5	14.8
Fe	.0025	8.5	7.0	8.3
Co	.170	515	536	536
Ni	.586	1980	2026	2026

*value too high due to x-ray overlap with Cr

Tests of this simulation code were made with pellets of NBS organic material and homogeneous steel samples of known composition,
yielding results in the ppm to % range within 10% of PIXE analysis.
A sample rendering enhancement as high as 20% of the total count rate
of an element (Fe) is given in table 1. It has to be noted that our
calculations are absolute ones, there is no fit to a reference sample
of similar composition. In this context deviations of 10% can well
be accepted.

To apply the code to unknowns there must be a rough idea of the
low Z matrix, and the PIXE results of the other elements are entered
as first estimate in an iterative procedure.

HEAVILY LOADED AEROSOL FILTERS

The correction code applied above to thick homogeneous samples
is suited very well for targets not infinitely thick with respect to
the proton range, and having a non-uniform elemental distribution.
With heavily loaded aerosol filters (one week samples of $\sim$.5 mg/cm²
dust on 5 mg/cm² cellulosic acetate) this code is used to derive
self absorption correction factors for the analysis of light elements.

The usual assumption for filter penetration is an exponential
distribution in the filter with exponents dependent on the particulate
sizes, verified only recently by direct electron microprobe measurements (6). Using such elemental profiles for filter sample slices,
and simulating irradiations from the front and the back side, we can
derive those distribution profiles within the filter, which correspond to the measured front/back ratios (fig. 2). The range of ratios

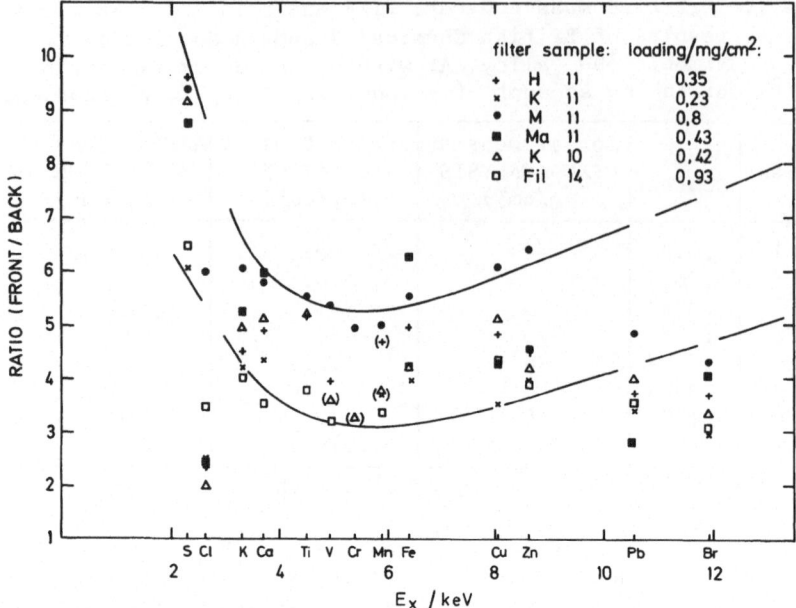

Fig. 2. Measured ratios for irradiations from front and back side of
different filters (all cellulosic acetate). Curves calculated
for two extreme distribution profiles spanning all the range
of observed ratios.

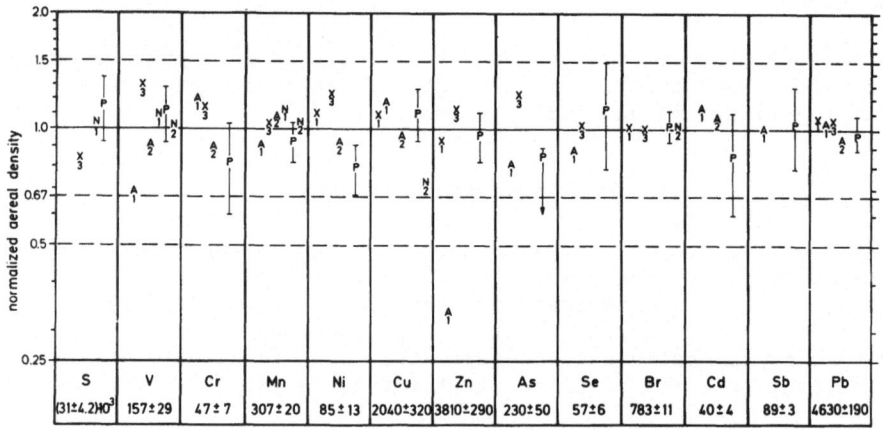

Fig. 3. Preliminary interlaboratory comparison of 1 week aerosol
sample analyzed by atomic absorption (A), neutron activation
(N), XRF (X), and PIXE (P) (taken from a meeting of partici-
pants in the aerosol project of the German Association for
the Promotion of Radio Nuclide Techniques).
Bottom line gives average of reported values in ng/cm².

seems to be very large, but it can easily be explained by a variation
of only 20% in blank filter thickness (which in fact is found) or a
change in deposition profiles. The profiles yielding the extreme
curves of fig. 2 are used to calculate the attenuation factors due to
these distributions in the filter. We derive an average correction
factor of 1.59 for S with a deviation of only ±11% within the full
range of observed front/back ratios.

Fig. 3 compares the analytical results of an aerosol filter,
where we applied the sulfur self absorption factor derived above.
The results are normalized to the average of reported values, most of
them agree very well. The sulfur comparison is not conclusive, yet,
since XRF faces the same attenuation problem as PIXE, and NAA is close
to its detection limits with this S mass. A comparison with other
independent sulfur analysis techniques is being prepared. Since we
put special emphasis on relatively good detection limits of $Z \approx 50$
by K radiation (using 4 MeV projectiles and thick mylar absorbers in
a second analysis run), we are able to analyze for Cd and Sb –
further values on these elements are reported by AAS only.

CONCLUSIONS

The matrix correction code, derived for homogeneous samples infinitely thick with respect to proton range, can well be adapted for
more complex cases such as heavily loaded aerosol filters of intermediate thickness with elemental distributions varying with filter
depth. Simple correction factors can be derived, improving the accuracy of PIXE analysis of samples not meeting the constraints of thin
layers.

REFERENCES

1. S.A.E. Johansson, ed., "Particle induced x-ray emission and its
   analytical applications", Proc. 2nd Int. Conf., Lund, Sweden, 1980,
   Nucl. Instr. and Meth. 181, 1-546 (1981).
2. I. Hasselmann, W. Koenig, F.W. Richter, U. Steiner, U. Wätjen, J.C.
   Bode, and W. Ohta, Nucl. Instr. and Meth. 142, 163-170 (1977);
   W. Koenig, F.W. Richter, U. Steiner, R. Stock, R. Thielmann and
   U. Wätjen, Nucl. Instr. and Meth. 142, 225-229 (1977).
3. A. Disam, P. Tschöpel and G. Tölg, Fresenius Z. Anal. Chem. 295,
   97-109 (1979).
4. Proc. Workshop on theories of inner shell ionization by heavy
   particles, Nucl. Instr. and Meth. 169, 249-317 (1980).
5. F.-W. Richter and U. Wätjen, Nucl. Instr. and Meth. 181, 189-194
   (1981).
6. H. Seiler, U. Haas, I. Rentschler, H. Schreiber, P. Wieser and
   R. Wurster, Optik 58, 145-157 (1981).

ENERGY DISPERSIVE ANALYSIS OF ACTINIDES, LANTHANIDES, AND OTHER ELEMENTS IN SOIL AND SEDIMENT SAMPLES

G.R. Laurer, J. Furfaro, M. Carlos*,
W. Lei, R. Ballad, and T.J. Kneip

Institute of Environmental Medicine
New York University Medical Center
New York, New York   10016

*Pontificia Universidade Catolica do Rio de Janeiro
Rio de Janeiro, Brasil

INTRODUCTION

A model is under development at this laboratory to predict the mobilization of Th from an estimated 20,000 metric ton deposit, in place and weathering as long as 80 million years on the Morro do Ferro, a hill on the Pocos de Caldas plateau in Minas Gerais, Brazil. The significance of this deposit lies in the marked similarities, chemically, between Th and Pu with reference to the problem of Pu waste management from the nuclear fuel cycle. One of the objectives of this program is to determine the elemental composition of the Morro do Ferro soil and its relationship to the mobilization of Th from the deposit. X-ray fluorescence is used for the analysis of Th, U, Ba, La, Ce, Pr, Nd, Sm, Dy, Eu, Gd, Yb, Hf, Fe, Zr, Nb, Mo and Pb. This paper describes the XRF system and methodology used for the determinations and presents the results of 72 core, soil and sediment sample analyses for the above 18 elements. Analyses of three United States Geological Survey (USGS) rock standards are also presented.

MATERIALS AND METHODS

K shell X-rays are used because they are more energetic, minimizing absorption effects due to differences in matrix density or heterogeneity.[1] To minimize calibration problems, [57]Co (122 keV/ 85.6%; 136 keV/11%) is used as the excitation source for the lanthanides and neighboring elements of lower Z as well. The activity is 40 millicuries, fabricated in annular form. The detector is 16 mm diameter, 10 mm deep, hyperpure germanium (HPGE) having a resolution of 484 eV, FWHM at 122 keV.

To reduce scatter interference with the Th K_α X-ray region (U K = 98.4 keV), the scatter angle is maintained as closely as possible to 180°. The collimator is 1 3/8 in. high and composed of five sections, 1/5 in. thick, with a 3/8 in. neck on the fifth. As the source decays, sections of the collimator can be removed allowing the geometry to be reoptimized. A Tracor Northern 4000 computer-assisted (22K, LSI 11/23) analyzer is used with a printer/keyboard and plotting software. This system was acquired with two data reduction programs: Super ML[2] for least-squares fitting and a matrix correction program which uses the Rasberry-Heinrich method.[3] A 4096 channel data memory is used for each ADC, eight of which can be used with this system. Currently, two ADC's are in place, one for XRF, the other for alpha spectrometry.

The Super ML program requires single-element reference samples with concentrations in the mid-range of the values of the unknown. The mid-range values for this study were estimated from earlier semi-quantitative, spectrographic analyses of Morro do Ferro soils[4] and corrected as preliminary data was generated. The average major concentrations of compounds from 36 sampling stations on the Morro do Ferro were SiO_2-13.4%, Al_2O_3-23.6%, Fe_2O_3-24.3%, TiO_2-0.84%, and P_2O_5-0.38%.[5] From this, a powder consisting of 2 parts aluminum oxide and 1 part silicon dioxide was defined as the matrix for the references. The weighing form for all compounds was as the oxide. About one kilogram of matrix was prepared initially by passing aluminum and silicon through a 200 mesh (75 μm) sieve and shaking for 30 minutes. For each additional use, it was reshaken. Master mixes were then prepared by adding at least 50 milligrams of the compound, sieved at 200 mesh, to the standard matrix and mixing for 30 minutes in a mixer/mill. One or more aliquots are taken and mixed with more standard matrix for 30 minutes until the desired concentration is reached. Seven grams are then taken, mixed with 1.5 grams of methocel for 20 minutes, and pelletized in a 30 mm x 9 mm spec-cap at 25 tons.

The Rasberry-Heinrich method for correcting interelement and matrix effects is an empirical procedure that requires multi-element reference samples which mimic the unknown. Twenty-three, with varying elemental concentrations which bracketed the ranges of the unknowns, were prepared from the single-element master mixes discussed in the preceding section. Estimates of the concentrations expected in Morro do Ferro samples were obtained from results of the earlier analyses cited above and by preliminary analyses performed by this laboratory. Each element was prepared based on a 10 gram total sample and poured into a polystyrene jar. When all compounds were added, it was mixed for 30 minutes, a 7 gram aliquot taken, mixed with 1.5 grams of methocel for 20 minutes, and pelletized. The difference between the actual weights and the calculated weight of each was, on the average, 1.5%.

The Morro do Ferro solid samples, after drying at 110°C overnight, are passed through a 35 mesh sieve to remove twigs and grass and crushed for 5 minutes in a tungsten-carbide mixer/mill. It is then sieved at 200 mesh, shaken for 15 minutes, and a 7 gram aliquot mixed with 1.5 grams of binder is pelletized as before.

Tests for sample homogeneity were performed on eight aliquots of a master mix of U and six pellets pressed from a Morro do Ferro core sample. Using the net counts under the $U_{K\alpha 1}$ peak for the U aliquots, a Chi-squared test[6] showed the eight values for the $U_{K\alpha 1}$ peak to be normally distributed at an α of .05. The Q test,[7,8] which assumes a normally distributed population, tests for outliers when N, the sampling population, is less than or equal to 10. The lowest and highest counts (255, 337) were found not to be outliers with a 90% confidence. A Student t test[9] showed no significant difference between the two lowest and two highest counts at an α error of .02. Since the master mix is homogeneously distributed, it may be assumed that the aliquot is homogeneous as well. The same tests, performed on the six pellets prepared from the core sample, showed all 11 quantitated elements to be homogeneously distributed.

To assure reproducible peak positions when fitting unknown sample spectra to standard spectra, a statistically well-established spectrum is used to which all spectra - unknown and standards - are compared and, if necessary, shifted in gain and/or zero until the reference peaks overlap. $K_{\alpha 1}$ and $K_{\alpha 2}$ peaks from the tungsten collimator and Zn K_α X-rays from an annular piece of zinc attached to the sample holder are the reference peaks. In use, once the energy calibration is fixed, a 50,000 sec. background normalized to a 1,000 seconds, became the spectrum to which all other spectra were compared. To account for the decay of the ^{57}Co source, each spectrum was corrected to time zero, i.e., the day of the 50,000 sec. background, before gain and zero correction.

Figure 1 is a spectrum of a Morro do Ferro core sample, C-5-5. Before fitting, background was stripped from all sample and standard spectra. Although not necessary in all situations utilizing the Super ML program, for Th analyses using ^{57}Co, in which the $K_{\alpha 1}$ analyte peak appears on the side of a relatively steep backscatter peak, we have obtained more consistent results by reducing the size of the scatter peaks with a background strip.

The Rasberry-Heinrich method for correcting interelement effects is well known. The programmed version used limited us to 11-element

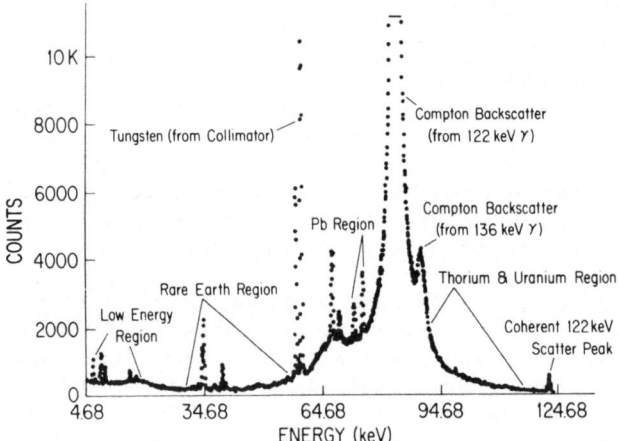

Fig. 1. Spectrum of Morro do Ferro core sample,
         C-5-5 (1000 seconds).

standards, of which 23 were prepared with concentrations which bracketed the range encountered in the samples. Matrix-corrected concentrations for the seven additional elements, Nb, Mo, Eu, Gd, Dy, Yb, and Hf, seen during analyses of Morro do Ferro samples are calculated based on the assumption that Nb and Mo have the same matrix effects as that of Zr, and similarly Gd, Dy, Eu, Yb, and Hf; of Sm. Thus, the uncorrected concentration, as obtained from the K ratio of the unknown and single-element reference is multiplied by the ratio of the corrected to uncorrected concentration of Zr or Sm, depending on which element is being determined.

RESULTS

USGS rock standards were analyzed to confirm the calibration and sample preparation techniques and verify the software utilized in data reduction and quantitation of the elemental concentrations. Table 1 compares the XRF results with the USGS values[10] for andesite, granite, and granodiorite. The quantitated and most of the estimated XRF results are in good agreement with the USGS values.

A total of 52 core samples, 14 soil samples, and 2 sediment samples from locations near the ore body and in the ambient Morro do Ferro basin, and 4 soil samples from off the plateau were analyzed. Table 2 lists the results of three of them and displays the relatively large ranges of elemental concentrations encountered from the Morro do Ferro. Also shown is the 3 σ lower limit of detection for the system with a 1000-second count time.

Thorium concentrations obtained by XRF for several soil and core samples were verified by two other analytical techniques, viz., alpha and gamma spectrometry (Table 3). For alpha spectrometry, thorium was extracted by radiochemical procedures,[11] precipitated on a membrane filter, then placed on a surface-barrier, silicon detector for alpha radioactivity determinations. For gamma spectrometry, the sample, after sieving to 200 mesh, was put in a 20 mℓ plastic vial and gamma counted in a NaI(Tl) well-crystal.

CONCLUSIONS

An energy dispersive X-ray fluorescence system has been constructed and calibrated to analyze 18 elements in 72 core, soil, and sediment samples taken from the Morro do Ferro in Brazil. Quantitative results have been obtained for 11 elements and semi-quantitative results for an additional 7 elements. Much time and effort was spent in preparing single and multi-element references properly. Gain and zero correction of each spectrum to a reference spectrum was also a vital part of the calibration procedure.

Table 1. Comparison of USGS Standard Values with XRF Results and the Lower Limit of Detection

Element	Andesite: AGV-1			Granite: G2			Granodiorite: GSP-1			LLD 2.93 σ**
	AGV-1 (μg/g)		XRF (μg/g±1σ)*	G2 (μg/g)		XRF (μg/g±1σ)*	GSP-1 (μg/g)		XRF (μg/g±1σ)*	(μg/g)
Fe	47300	r	40910±1558	18500	r	14896±559	30300	r	29424±1228	10770
Zr	225	r	266±18	300	r	268±5	500	r	479±10	98
Ba	1208	r	1262±2	1870	r	2072±1	1300	r	1359±2	11
La	35	m	32±1	96	a	72±1	191	a	191±1	11
Ce	63	a	72±1	150	m	176±1	394	a	426±1	10
Pr	7	m	25±1	19	m	17±1	50	m	67±1	10
Nd	39	a	45±1	60	a	81±1	188	a	231±1	10
Sm	5.9	a	7±1	7.3	a	9±.4	27.1	a	29±1	8
Pb	35.1	a	44±1	31.2	a	37±1	51.3	a	67±1	23
Th	6.41	r	5±2	24.2	r	26±1	104	r	111±1	21
U	1.88	a	3±1	2.0	a	.1±.5	1.96	a	4±1	12
Nb	15	r	9±19	13.5	r	2±6	29	r	< 90	90
Mo	2.3	m	123±52	.36	m	25±13	.90	m	5±30	78
Gd	5.5	m	2±1	5	m	3±.2	15	m	5±4	5
Dy	3.5	r	11±1	2.6	r	9±1	5.4	r	14±1	16
Eu	1.7	a	8±4	1.5	r	5±1	2.4	a	10±2	9
Yb	1.7	a	2±1	.88	a	1±.3	1.8	a	1±1	8
Hf	5.2	a	1±1	7.35	a	6±.3	15.9	a	16±1	13

* Counting error only.

r = recommended value; m = magnitude; and a = average.

Counting time: Andesite, 23511 seconds; Granite, 79534 seconds; and Granodiorite, 46581 seconds.

** LLD for 1000 seconds: α error = .10 and β error = .05.

Table 2. Solid Samples

Element	So-2 (μg/g±1σ)[a]	So-3 (μg/g)	C-69-1 (μg/g)
Fe	121230±7956	152140	47950
Zr	1309±189	2208	2988
Ba	1773±24	946	105
La	1935±38	8937	239
Ce	70758±73	22629	452
Pr	429±28	1518	23
Nd	1110±22	5897	89
Sm	215±22	495	16
Pb	253±11	154	42
Th	6976±24	12786	74
U	151±10	96	5
Nb	873±189	518	907
Mo	< 78	< 78	< 78
Gd	38±6	127	5
Dy	< 16	135	23
Eu	138±15	212	12
Yb	84±17	71	12
Hf	41±17	98	52

[a]Counting error only.

Table 3. Comparison of Thorium Concentrations by Three Different Analytical Techniques

Sample Number	Alpha Spectrometry (conc. μg/g)[a]	Gamma Spectrometry (conc. μg/g)	X-ray Fluorescence (conc. μg/g)
C-25-0	50±4	57±8	42±7[a]
C-45-1	258±42	248±11	279±8
C-64-2.5	115±23	131±10	123±1
C-71-0	10±1	11±7	10±1

[a]Counting errors only: ±1 σ.

Elemental concentrations obtained with the XRF system have been verified by two separate analytical techniques, alpha and gamma spectrometry, and by comparison to known concentrations of these elements in three USGS rock standards. These results verify the calibration procedures and methods used with the XRF system for the analysis of elements in a soil matrix over a wide range of atomic number and concentration, utilizing K X-ray excitation and a radioisotopic source.

ACKNOWLEDGMENTS

This research was undertaken with the support of the Department of Energy (Contract # DE-AC97-79ET46606) and Comissao Nacional de Energia Nuclear and was assisted by core grants No. ES 00260 of the National Institute of Environmental Health Sciences and No. CA 13343 by the National Cancer Institute, DHHS.

REFERENCES

1. J. Kuusi, M. Virtanen, and P. Jauho, Heavy Element Analysis by Isotope-Excited X-ray Fluorescence, Nucl. Technol. 13:216 (1972).
2. F. Schamber, N. Wodke, and J. McCarthy, Least-Squares Fit with Digital Filter: The Method and its Application to EDS Spectra, in: "Eighth International Conference on X-ray Optics and Microanalysis," Boston (1977).
3. S. Rasberry and K. Heinrich, Calibration for Interelement Effects in X-ray Fluorescence Analysis, Analyt. Chem. 46:81 (1974).
4. H. Wedow, The Morro do Ferro Thorium and Rare Earth Deposit, Pocos de Caldas District, Brazil, U.S. Geological Survey Bull. 1185-D (1967).
5. F. Roser, H. Gomes, N. Leal da Costa, L. Hainberger, and T. Cullen, A Study of Natural and Artificial Radioactive Contamination in Brazil, NYO-2577 Report No. 2, Instituto de Fisica, Pontificia Universidade Catolica, Rio de Janeiro (1965).
6. R.D. Remington and M.A. Schork, "Statistics with Applications to the Biological and Health Sciences," Prentice-Hall, Inc., Englewood Cliffs, New Jersey (1970).
7. R.B. Dean and W.J. Dixon, Simplified Statistics for Small Numbers of Observations, Analyt. Chem. 23:636 (1951).
8. W.J. Dixon and F.J. Massey, Jr., "Introduction to Statistical Analysis," 2nd Ed., McGraw-Hill Book Co., Inc., New York (1957).
9. B.W. Brown, Jr. and M. Hollander, "Statistics: A Biomedical Introduction," John Wiley & Sons, Inc., New York (1977).
10. F.J. Flanagan, 1972 Values for International Geochemical Reference Samples, Geochim. Cosmochim. Acta. 37:1189 (1973).
11. M. Eisenbud, W. Lei, R. Ballad, K. Krauskopf, E. Penna Franca, T.L. Cullen, and P. Freeborn, Mobility of Thorium from the Morro do Ferro, in: "International Symposium on Migration in the Terrestrial Environment of Long-Lived Radionuclides from the Nuclear Fuel Cycle," IAEA-SM-257/49 (1981).

X-RAY FLUORESCENCE ANALYSIS OF WELDING FUME PARTICLES

Thomas P. Carsey

National Institute for Occupational Safety and Health

Cincinnati, Ohio 45227

The technique of x-ray fluorescence is ideal for the analysis of metal fumes such as are produced in various welding and brazing operations. Although the constitution of welding and brazing fumes is dependent on the nature of the base metal, flux, and welding material used, they typically contain the oxides of iron, nickel, chromium, and manganese (stainless steel welding)[1,2]. The fumes are in the form of very small particles, usually 1.0 um,[2] and are often agglomerated into long chains. Filter samples of these fumes, as are used in industrial hygiene monitoring, may contain 30-300 ug/filter of particles on one surface of a low-mass, 37-mm membrane filter[3].

The NIOSH laboratories have been involved with the development of an analytical technique for the rapid analysis of filter samples of welding and brazing fumes by wavelength dispersive x-ray fluorescence spectroscopy (WDXRF). These fumes contain a number of elements known to be toxic[1], and are of a particle size that can be absorbed by the lung[2].

In this study, a commercial standard filter set and two laboratory-made standard filter sets were compared via the analysis of generated welding fume samples by XRF. The latter standards were made by (1) hydrophobic-edge membrane filters spiked with prepared metal ion solutions, and (2) filters through which a dispersion of metal oxide powder (welding shop dust) in isopropanol has been drawn. They are discussed individually below.

(1) Spiked filters: Hydrophobic-edge filters are ordinary cellulose ester membrane filters with a wax impregnation throughout a 6-mm annular area around the edge. An aqueous solution

will not flow into the annular area of the filter, but will
remain in the inner circular region   35 mm in diameter. Spiking
solutions were made containing appropriate metals (iron, nickel,
chromium, manganese) at low, medium, and high concentrations, by
combining stock solutions of metal ions, and diluting to 50 mL
final volume with 5% nitric acid. The spike solutions were designed so that 50 uL aliquots would contain the proper amounts of
the desired metal ions. Each filter was pre-wetted by placing on
the surface of water in beaker, removed, placed between sheets of
clean absorbent paper, gently blotted to remove excess water,
mounted in the holder unit, and covered. After a half-hour, 50
uL of the premixed spike solution was applied with a micropipette
onto the center of the filter, the cover was put on, and the assembly set aside to allow the filter to dry. Triplicate filters
at each level were made and examined; they were visually quite
homogeneous.

    (2) Dust Dispersion filters[4] Several grams of welding shop
dust (settled welding fume particles) were freezer-milled, sieved
(10-um sieve), and dispersed into about one liter of isopropanol. Before use, the dispersion was ultrasonicated for about
one-half hour. During use, the dispersion was magnetically
stirred. A Gelman GA-6 (0.45-um pore size) membrane filter was
placed into a Millipore Buchner funnel assembly and approximately
50 mL of isopropanol placed above the filter. Two milliliters of
the dispersion was withdrawn with a pipette, added to the ispro-
panol, and stirred with the pipette tip until the liquid was
homogeneous. A low vacuum was then applied, drawing the liquid
through the filter. The filter was then removed and placed in an
XRF filter holder to allow drying without distortion. Four
filters at different deposition level were made (i.e., using 2,
3, and 4 mL of dispersion). Two at each level would be wet-ashed
(P&CAM 173)[5] for elemental analysis (ICP-AES and atomic absorption), the other two would be used in the XRF spectrometer for
standardization. The ICP-AES analysis indicated that the filters
contained at least minimal amounts of Fe, Mn, Ni, Cr, Ti, Zn, and
Mo. The overall process could be repeated with excellent repro-
ducability; the pooled coefficient of variation between filters,
determined by XRF analysis, was about 5%. Also, plots of XRF
intensity vs. volume of dispersion data were very linear,
indicating that overloading of the filters was not occurring.

    Welding fumes samples were obtained in a generation system
built over a welding station. This generation system extracts a
stream of ventilation air above the welding area, diffuses the
stream, and directs the stream upwards in a laminar flow past a
collector manifold containing six field monitor cassettes. The
air flow through these cassettes is maintained at 1 L/m via
critical orifices. Six filters can be sampled with practically
uniform (6%RSD) depositions. To obtain replicate samples at

four filter loadings, six filters at four different deposition
loadings were obtained; these levels are denoted low, medium,
high, and very high. The degree of loading was evaluated
visually. Standard three-piece closed-faced cassette monitors
containing Millipore AAWP 37-mm membrane filters with backup pads
were used. The welding fume samples were obtained using Type
310-15 Chromend rods on Type 304 3/8" stainless steel. Analyte
breakthrough, and interelement, filter absorption, and particle
size effects were carefully investigated and found to be not
significant.

The XRF analysis employed the internal standard technique; a
cobalt-impregnated filter was placed below all standard and fume
samples and the analyte-line intensities of all analytes was divided by that of the cobalt. The cobalt-impregnated Gelman
Acropor filters were produced according to the method described
in reference 6. The twenty-four welding fume samples were carefully removed from their cassettes and inspected. All appeared
to be quite homogeneously deposited, with very few black ash
particles. The fume samples, the four sets of standard filters,
and the blanks were analyzed by XRF. Following the analysis, the
blank XRF intensities were subtracted from the analyte intensities in all samples and standards, and calibration parameters
were developed using the element depositions and XRF intensities
of the standards. Using the calibration parameters from each
standard set, the elemental depositions (in ug/filter) were calculated. Then, the fume samples were wet-ashed and analyzed by
ICP-AES[6]. These results are used in the table below. The
amount is the XRF result; bias is [(ICP-XRF)/ICP]x100; precision
(Pre) is the relative standard deviation of the six samples.

Dust Dispersion Standards

	LOW Amnt (ug)	LOW Bias (%)	LOW Pre RSD	MEDIUM Amnt (ug)	MEDIUM Bias (%)	MEDIUM Pre RSD	HIGH Amnt (ug)	HIGH Bias (%)	HIGH Pre RSD	VERY HIGH Amnt (ug)	VERY HIGH Bias (%)	VERY HIGH Pre RSD
Fe	6.26	+1.76	7.8	29.25	-3.51	3.8	40.30	-10.1	3.1	65.30	-8.3	3.6
Mn	*4.39	+6.44	5.6	18.90	+4.41	5.0	21.9*	5.14	3.3	34.36	-2.2	3.7
Cr	4.75	-5.93	5.6	22.90	-3.89	5.2	30.78	-20.9	3.2	50.27	-17.6	4.2
Ni	1.83	+2.22	7.2	8.52	+2.75	4.4	10.58	-12.9	3.4	16.75	-9.23	4.5

Spike Filter Standards

	LOW Amnt (ug)	LOW Bias (%)	LOW Pre RSD	MEDIUM Amnt (ug)	MEDIUM Bias (%)	MEDIUM Pre RSD	HIGH Amnt (ug)	HIGH Bias (%)	HIGH Pre RSD	VERY HIGH Amnt (ug)	VERY HIGH Bias (%)	VERY HIGH Pre RSD
Fe	6.45	-1.14	7.8	30.19	-6.85	3.9	41.63	-13.6	3.1	67.37	-11.7	3.7
Mn	*5.38	-14.6	5.6	23.04	-16.5	5.0	26.7*	-28.1	3.3	41.85	-24.5	3.7
Cr	6.21	-39.1	5.3	30.02	-36.2	5.2	40.36	-58.5	3.2	65.90	-54.1	4.2
Ni	2.15	-14.6	7.4	10.01	-14.3	4.4	12.43	-32.6	3.4	19.69	-28.4	4.5

Columbia Scientific Standard Filters

	LOW			MEDIUM			HIGH			VERY HIGH		
	Amnt (ug)	Bias (%)	Pre RSD	Amnt (ug)	Bias (%)	Pre RSD	Amnt (ug)	Bias (%)	Pre RSD	Amnt (ug)	Bias (%)	Pre RSD
Fe	6.34	+0.52	7.8	29.73	−5.20	3.9	40.99	−11.8	3.1	66.33	−9.98	3.7
Mn	−	−	−	−	−	−	−	−	−	−	−	−
Cr	5.68	−27.2	5.3	27.27	−23.7	5.2	36.66	−44.0	3.2	59.87	−40.0	4.2
Ni	1.98	−6.04	7.4	9.26	+5.88	4.4	11.50	−22.6	3.4	18.25	−22.3	4.5

asterick refers to set in which one outlier was deleted.

The main conclusions of the study are as follows:
(1) Corrections for interelement effects, filter absorption, and particle-size effects, were investigated in detail. They were found to be not significant due to the very light depositions and the small particle size[3]. Linear XRF intensity vs. deposition relationships were obtained with all metals.
(2) Although quite precise (RSD usually 7%), WDXRF results differed somewhat with those obtained with ICP-AES and AAS analysis of the filters following dissolution. The variation in the calibration curves derived from the various standard filters is reduced by the use of the internal standard method, with cobalt as the internal standard element.
(3) The commercial standard filters did not produce significantly better analytical results than the laboratory-made standard filters (compared to AAS and ICP-AES). Although all three standardization systems showed biases, the dust dispersion standard filters performed better than the others.
(4) Measured breakthrough was zero for iron, manganese, nickel, and chromium, and is not considered to be problematic for fume samples on membrane filters.

REFERENCES

1. Criteria for a Recommended Standard...Occupational Exposure to Welding, Brazing, and Thermal Cutting," (Draft), DHHS-NIOSH (1979).
2. "Fumes and Gases in the Welding Environment," American Welding Society, Miami, Florida, 1979.
3. T. P. Carsey, "Feasibility Study for X-ray Fluorescence Analysis of Welding and Brazing Fumes," submitted to National Technical Information Service (NTIS).
4. R. A. Semmler, R. D. Draftz, and I. Puretz, "Thin-Layer Standards for Calibration of X-ray Spectrometers," in "X-ray Fluorescence of Environmental Samples," T. Dzubay, Ed., Ann Arbor, 1977, p. 181.
5. D. G. Taylor, Coord., "NIOSH Manual of Analytical Methods," 2nd Ed., Vol. 5, 1979, p. 173.
6. H. Kingston and P. A. Pella, Anal. Chem., 53, 223 (1981)

A NEW COMPUTER ALGORITHM

FOR QUALITATIVE X-RAY POWDER DIFFRACTION ANALYSIS

T. C. Huang and W. Parrish

IBM Research Laboratory
San Jose, California 95193

ABSTRACT

An effective and practical computer algorithm has been developed for rapid and precise phase identification of polycrystalline materials by X-ray diffraction methods using the JCPDS database and/or user created standard files. The entire JCPDS file was reorganized for efficient search. Identifications are facilitated by a number of options: automatic correction of systematic errors using internal standard reflections, selectable window widths for file searching, elemental restrictions (chemical prescreening), handling preferred orientation, match without using intensities, match with 3 reflections, and others. A comprehensive algorithm for calculating a figure-of-merit (FOM) is used so that the "correct" phases can easily be identified with highest FOMs. This method has been tested extensively on a wide variety of analyses and is applicable to either a host or a minicomputer.

INTRODUCTION

A number of programs have been developed for computer search/match powder pattern identification.[1-3] However, recent evaluation indicates the computer methods have had mixed success.[4] Their overall performance was stated to be no better and in some cases worse than manual searching according to the results of the JCPDS second round robin test. Among the drawbacks were the inability to resolve the correct phases from a large list of candidates, the need for more precise ds, and the slow speed in searching the entire JCPDS database.

Our experience with the new search/match program described in this paper and used in a wide variety of analyses, indicates that the criticisms of previous computer searching methods have been avoided and that this method far exceeds manual methods in all significant aspects. A number of special features have been incorporated to make this possible: reformatting of the JCPDS file, use of internal

213

standard reflections for automatic correction of systematic errors, proper use of the Bragg law to set up search windows and error corrections, a comprehensive figure-of-merit calculation for each potential match, and interactive options to aid the user. This method is applicable to either a host or minicomputer. The use of this method with the IBM Series/1 minicomputer is described in the accompanying paper.[5,6]

REFERENCE STANDARDS

The reference file used in Search/Match may either be the JCPDS standard file or user created files. The well-known JCPDS database of Sets 1-30 containing approximately 34,000 phases was reformatted, ordered and packed for the efficient use in our program. To minimize the search time, the database was rearranged into four subfiles: inorganics (IN), organics and organometallics (OR), minerals (MI), and metals and alloys (ME). Each subfile was reordered into three sections: the frequently encountered phases (FEP) were placed first, the remaining quality and starred cards second, followed by the remainder of the subfile. The numbers of patterns in each section and the total for each subfile are listed in Table 1. Some phases may have more than one pattern in the same or different subfiles and hence the sum of the total standards (44,835) exceeds the number of phases in the file. This arrangement allows the search process to be terminated at the end of any section when identification has been made. In many cases, the search of the FEPs is sufficient, thus the time for search is greatly reduced.

The quality of the JCPDS standard data which were compiled from many sources varies over a wide range. For greater certainty and speed, the user may search his own standard files created with patterns analyzed by the IBM Peak Search or Profile Fit Programs.[7] He may also include in the user's file published data, patterns calculated from crystal structure, or other sources.

Table 1. The Rearrangements of JCPDS Database

Subfile	FEP	Remaining Starred	Remainder	Total
IN	2332	3870	19193	25395
OR	1346	878	8547	10771
MI	608	313	2099	3020
ME	363	471	4815	5649

SEARCH/MATCH PARAMETERS

There are a number of options which may be selected for the Search/Match procedure. Their proper use may be important in obtaining correct identifications.

Systematic Errors

These may be corrected by using internal standards. When the correct ds of several observed reflections are known, their values can be entered and the computer uses them as a standard scale to adjust the ds of the unknown specimen, prior to the Search/Match. Table 2 is an example showing a column of experimental d values and the entered internal standard values d_{IS}; the change Δd and the corrected value d_c are listed in the last two columns. The expressions in terms of θ used for calculated Δd and d_c are shown at the bottom.

The internal standard reflections can be from any number of phases as long as their d's are accurately known. This option will reduce the deviations of the unknown d's from their "true" values and better matches between the "true" standards and the unknown can thus be obtained.

Table 2. Systematic Error Corrections with Internal Standards

#	$d(\text{Å})$	d_{IS}	Δd	d_C
1	5.000		0.028	4.972
2	4.280	4.260	20	4.260
3	3.750		15	3.735
4	3.355	3.343	12	3.343
5	3.000		10	2.990
6	2.500		7	2.493
7	2.090	2.085	5	2.085
8	1.000		1	0.999

$$\theta_C = \theta + \Delta\theta$$

$$\Delta\theta = [\theta_{IS}(1)-\theta(1)] + \Delta\theta_{IS} \times [\theta-\theta(1)]$$

$$\Delta\theta_{IS} = \{[\theta_{IS}(h)-\theta(h)] - [\theta_{IS}(1)-\theta(1)]\}/[\theta(h)-\theta(1)]$$

Random and Other Errors

These are accounted for by selecting an error limit either in $\Delta°2\theta$ or Δd (Å) at a certain d to define the windows for matching reflections of a standard to the unknown. The Bragg law is used to generate error windows for all observed reflections. This program allows the windows to overlap and thereby increases the precision of the match process which is especially important for multiple phase mixtures. The selected error limit should be large enough to account for the discrepancies (including those from the systematic errors if internal standards are not used) between data of the unknown and the standards.

Chemical Pre-Screen (Elemental Restrictions)

The value of chemical information as an aid in identification has been well-known since the inception of the method. The computer Search/Match makes it easy to add chemical restrictions which can greatly reduce the number of standards matched and thereby reduce the analysis time and avoid listing extraneous standards. The menu provides for listing up to eight elements present and eight absent; three groups of absent elements may be entered to avoid long typing by listing the lowest and highest in a series of consecutive elements. The elements H to F are unrestricted because they are not generally detectable by routine X-ray fluorescence analysis; their presence may be eliminated by the element absent option. The search process skips those standards that do not have at least one of the elements listed as present and those containing one or more of the absent elements. The chemical screening must be used with care to avoid errors.

Preferred Orientation

This option allows matching phases which have a significant degree of preferred orientation and causes one or more of the three strongest reflections within the experimental range of the pattern to be absent. The option directs the program to bypass the general requirement that all three of the strongest reflections be present.

Match with Less than Three Reflections

This option makes it possible to match standards which have only one or two reflections in the experimental range.

Match Without Intensity

This option is used for cases where the intensities of the unknown or standard patterns are unreliable as in electron diffraction patterns. Under this option, the intensity data will not be used in the match process to evaluate the goodness of a match.

SEARCH/MATCH PROCESS

The observed ds which are automatically transferred into the system from the output of our Peak Search/Profile Fit program,[7] or manually entered by the user, may be first corrected for systematic errors using internal standard reflections as described above. The error windows for all observed reflections are then calculated from the user selected window and the Bragg law.

The selected standard file is searched in sequence. Each standard pattern is examined by the user-selected prescreens to be qualified for the match calculation. The correct chemistry is checked by elemental restrictions. For a specimen without strong preferred orientation, the three strongest standard reflections located within the experimental range must be present in the unknown. The proper use of these options in the search process greatly reduces the number of standards for the match calculation and provides a concise list of most likely phases.

In the match process, each reflection of the standard is matched to those of the unknown. A standard reflection is considered to match a reflection in the unknown if the former is the only one within the corresponding error window, or if multiple reflections are within this error window and it has the highest intensity. Since overlapping windows are used in our program, the probability and precision of a match are greatly increased. Generally, the program will only further consider in the match calculation those standards which have a minimum of three reflections present in the unknown. However, the user may eliminate this requirement to allow for matching or phases with less than three reflections in the case that these reflections are all the standard has inside the experimental range.

The figure of merit (FOM) is then calculated to determine the goodness of a match between a potential standard and the unknown. The FOM calculation takes into account all the necessary factors by which one is able to judge how close the match is. These factors are: fraction of matched reflections in the unknown pattern, fraction of matched reflections in the standard, weighted match in ds, match in Is (optional) and ratio of the sum of matched to total intensities of standard reflections in the experimental range. The relative weight of the above factors has been carefully determined and set so that the goodness of a match can be truely represented by the FOM. The higher the FOM, the higher the probability of a match. The FOM of a perfect match is 100 and it would be rarely attained in practice.

When the search of a section of a selected subfile is completed, the highest FOM standards are listed in descending order for identification. The user may decide from this list either to continue or to end the Search/Match process.

RESULTS AND DISCUSSION

This program has been tested extensively on a wide variety of analyses including published data in the literature and those from the round robin test of JCPDS. Two of the latter examples are given to demonstrate the method.

Table 3. Search/Match Results of an Inorganic Mixture of
K_2SiF_6, LiCl and NaBr

#	Standard Formula/Name	Phase	PDF#	FOM	NU	NS
1	K_2SiF_6 / HIERATITE SYN	*4	7-217	80	11	12
2	LiCl	*0	4-664	72	5	5
3	NaBr	*0	5-591	56	6	8
4	Na(Cl,CN)	0	2-778	47	4	6

The first example is a multi-phase inorganic mixture prepared by JCPDS.[3] No internal standards were used to correct for systematic errors. Element restrictions require the Search/Match be limited to those standards containing at least one of the five elements (K, Na, Si, Cl and Br) entered as present, and rejecting all standards containing one of the elements inside the element absent group from Mg to U except those five elements entered as present. This is an example of using the kind of element information which can be obtained from the X-ray fluorescence analysis. All other Search/Match parameters were also similar to those used in the JCPDS Manual.[3]

It took less than 4 seconds of CPU time of an IBM 370/168 computer to analyze this 20-reflection mixture by searching the FEP section of the JCPDS in subfile. In these 2332 FEP standards, only four survived the Search/Match analysis and the results are given in Table 3 ranked according to their FOMs. The phase column lists the JCPDS designation for alloy phase, data mark and subfile notation.[3] The set and card numbers are printed for reference. The number of matched reflections (NU) and the number of possible standard reflections inside the experimental range (NS) are also listed to assist identification. Note that all three correct phases of this mixture were ranked with highest FOMs.

Another example is an organic mixture of m-dinitrobenzene, 2,4- and 2,6-dinitrophenol prepared by JCPDS for the second round robin test (Problem 6). The computer Search/Match had problems analyzing this.[4] Recently, the CIS system was also used to analyze this mixture with Search/Match limited to phases containing only H, O, C and N.

Without using element restrictions, our program successfully identified these three phases with highest FOMs. Table 4 lists the highest five FOM phases. The FOMs of standards other than the three correct phases are noticeably lower. This break in FOMs allows for the separation of the true components from the rest of the list. Again, the total CPU time was only four seconds for an IBM 370/168 computer to analyze this 53-reflection mixture by searching the FEP section of the OR subfile.

For both of these examples, there was no need to search beyond the FEP section of their respective subfiles since no new phase was added to the highest FOM listings. In fact, this is generally the case for many of the analyses done in our laboratory.

Table 4. Search/Match Results of an Organic Mixture

#	Standard Formula/Name	Phase	PDF#	FOM	NU	NS
1	2,4-Dinitrophenol	*2	23-1670	72	22	23
2	2,6-Dinitrophenol	I2	26-1826	70	21	21
3	m-Dinitrobenzene	*2	11-855	69	21	21
4	p-Hydroxybenzonic Acid Hydrate	'2	22-1756	53	8	8
5	Poly n-Butyl Isocyanate	'2	22-1591	53	13	14

This method has been incorporated into the IBM Series/1 Minicomputer Search/Match System (see the following paper[5]). A number of useful features have also been added to this system including the interactive facility to assist the user in making an intelligent identification, and a utility program for rapid plotting of selected JCPDS standard patterns at the graphic terminal.

Acknowledgement. We are indebted to Monte C. Nichols of Sandia Laboratories, Livermore, CA, who discussed the initial phases of this project with us.

REFERENCES

1. L. K. Frevel, A Fast Search-Match Program for Powder Diffraction Analysis, J. Appl. Cryst. 9:199-204 (1976)

2. M. C. Nichols, "A Fortran II Program for the Identification of X-Ray Powder Diffraction Patterns," UCRL-70078, Lawrence Livermore Laboratory, CA (1966).

3. G. G. Johnson, Jr., "User Guide: Data Base and Search Program," JCPDS, Swarthmore, PA (1975).

4. R. Jenkins and C. R. Hubbard, A Preliminary Report on the Design and Results of the Second Round Robin to Evaluate Search/Match Methods for Qualitative Powder Diffractometer, Adv. in X-ray Anal. 22:133-142 (1979).

5. W. Parrish, G. L. Ayers and T. C. Huang, A Versatile Minicomputer X-Ray Search/Match System, Adv. in X-ray Anal. 25: (following paper) (1982).

6. W. Parrish, G. L. Ayers and T. C. Huang, A Minicomputer and Methodology for X-ray Analysis, Adv. in X-Ray Anal. 23:313-316 (1980).

7. W. Parrish and T. C. Huang, "Accuracy of the Profile Fitting Method for X-ray Polycrystalline Diffraction," Proc. Symp. on Accuracy in Powder Diffraction, NBS Special Publ. 567, 95-110, Washington D.C. (1980).

8. R. G. Marquart, I. Katsnelson, G. W. A. Milne, S. R. Heller, G. G. Johnson and R. Jenkins, A Search-Match System for X-ray Diffraction Data, J. Appl. Cryst. 12:629-634 (1979).

A VERSATILE MINICOMPUTER X-RAY SEARCH/MATCH SYSTEM

W. Parrish, G. L. Ayers, and T. C. Huang

IBM Research Laboratory
San Jose, California 95193

ABSTRACT

The operation of a new polycrystalline phase identification method using the IBM Series/1 minicomputer is described. Data of the unknown can be entered by automatic transfer of previous runs, stored data sets and manually. Full screen menu selections are provided to facilitate operations and correct entries. Typical S/M time for a multi-phase inorganic mixture containing 43 reflections using a 0.3° window averaged 11 sec per 100 standards and with simple chemical prescreening less than 4 sec including program initialization and calculations of comprehensive figures of merit. Interactive options provide graphics terminal comparison of the unknown pattern with selected standards which appear as diffractometer patterns, subtraction of identified standards from the unknown and others. Utility programs permit storing data sets for later analysis, user created files and a program to display any file standard as a diffractometer pattern.

INTRODUCTION

This computer Search/Match (S/M) method has been successfully used in our laboratory in a large number of analyses.[1] The programs are a portion of a system of automation, data reduction and analysis for wavelength dispersive X-ray powder diffraction and X-ray fluorescence analysis developed to operate on the IBM Series/1 minicomputer.[2,3] One of the most important advantages over manual S/M methods is that it is feasible to make comprehensive searches of large files with various search parameters in a short time. For example, using the modern high speed computer methods described here, it is possible to collect and reduce the data of an "average" powder pattern and run it through S/M in 10 to 15 minutes.

The certainty of the identifications can be evaluated from the calculated figures-of-merit which quantitatively determine the goodness-of-match of each

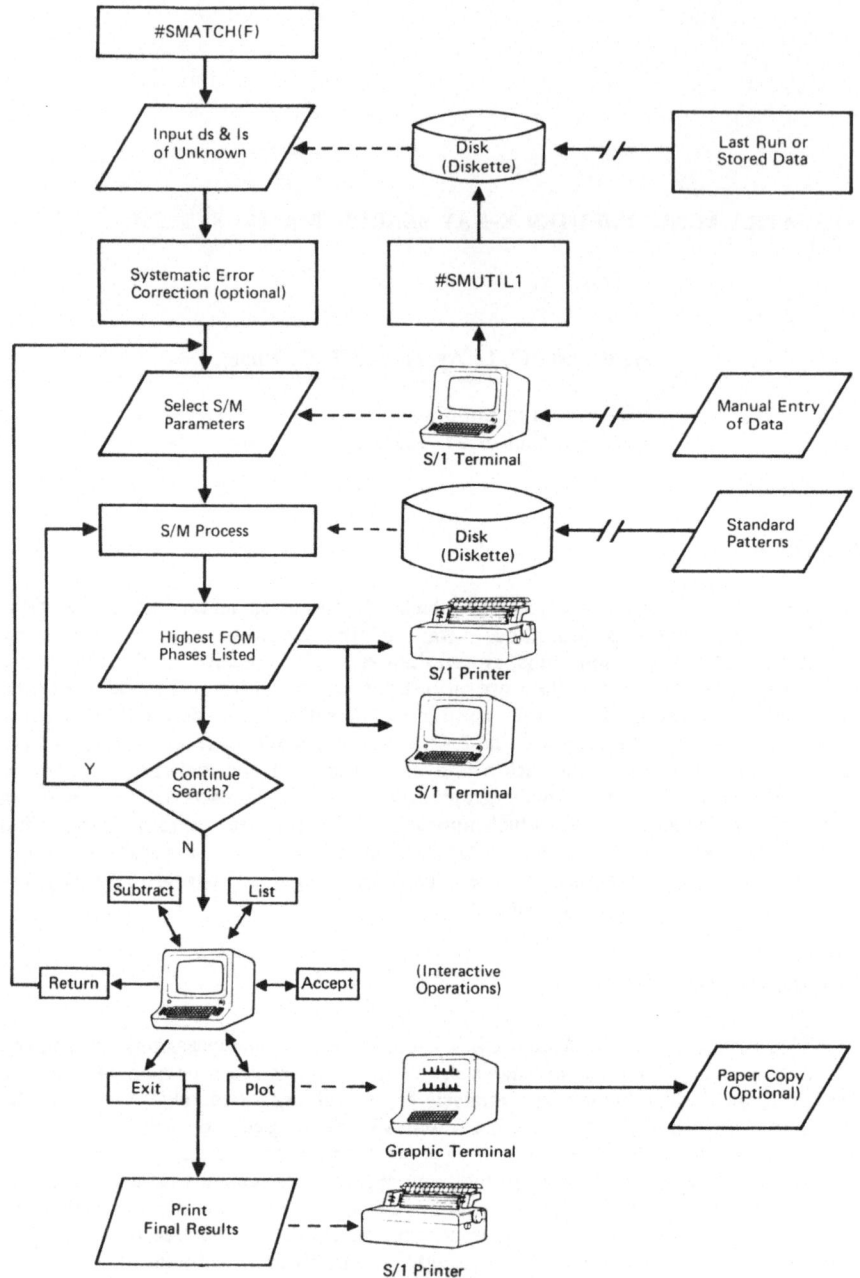

Figure 1. The IBM Series/1 X-Ray Search/Match Method.

```
*****************IBM SEARCH/MATCH PROGRAM****************
NOTE: 1. HIT "TAB" KEY TO MOVE CURSOR TO NEXT QUESTION.
      2. LEAVE BLANK QUESTIONS THAT DO NOT APPLY.
      3. HIT "ENTER" FOR FULL SCREEN INPUT.

ORIGIN OF THE UNKNOWN DATA:

    1. LAST PS OR PFM RUN?  (Y)     = N

    2. #SMDATA?  (Y)                = Y      UNIT #    =    1
                                            LOG #     = 1105


    3. MANUAL INPUT?  (Y)           = N NO. OF REFL=
                     ENTRY?  (D=D, T=2 THETA)        =
                     MINIMUM 2-THETA (DEG):          =
                     MAXIMUM 2-THETA (DEG):          =
                     WAVELENGTH (A):                 =

    USE INTERNAL STANDARDS?  (Y)= N

    LIST OF PRINTER?  (Y)           = Y
```

Figure 2. First full screen menu for data entry into Search/Match program. Entries are shown in bold face.

potential standard.   These data are supplemented by interactive graphics which provide a direct comparison of the unknown pattern with the selected standards. The general scheme of the method is shown in Fig. 1.

The Series/1 is a stand-alone computer whose configuration for this application contains at least 96 kB internal memory and 13.9 mB fixed disk storage to accommodate the programs and the entire JCPDS file of standard patterns. The programs are written in EDX (Event Driven Executive) and Fortran.

FULL SCREEN ENTRIES

Full screen menu entries were developed to guide the user, and facilitate data entries and parameter selections.

There are three ways to enter the data of the unknown pattern into the S/M program (Fig. 2): (1) Automatic transfer of data from the last diffractometer run. These are the set of ds and Is computed by Peak Search or Profile Fitting methods with the $K\alpha_2$ peaks stripped out.   (2) Automatic transfer of the results of runs stored in the #SMDATA file (see below). These runs are identified by the X-ray unit and computer log numbers. (3) Manual entry of ds and Is.

The corrections for systematic errors, described in the accompanying paper,[1] can be made for any type of data input.

The menu for selecting the S/M parameters is shown in Fig. 3. At this point, the user selects the subfile to be searched. The search window may be entered in either $°2\theta$ or Δd at a selected d value; the default value is $0.2°(2\theta)$. The selection of the options on intensities, preferred orientation, identification with less than three reflections and chemical pre-screening are described in the accompanying paper.[1] The option FWHM (full width at one-half peak height above background) may be used if the peaks are broadened. A maximum of 20 standards can be listed. If the chemistry option is selected, another menu appears for those data entries.

The printer lists the input ds, the maximum and minimum ds corresponding to the search window selected, and the Is of all reflections that will be used in the S/M program, Fig. 4.

COMMENT: = $NH_4BF_4+PBCL_2+CDCO_3$,#25,6/3/81

STANDARD SUB FILE TO SEARCH? (IN OR MI ME U1 U2)   = IN

SEARCH WINDOW:	DELTA 2-THETA (DEG)	= 0.10
	DELTA D (A)	=
	AT D(A)	=

USE INTENSITIES FOR MATCH? (Y)   = Y

USE PREFERRED ORIENTATION? (Y)   = N

PHASE MATCH WITH LESS THAN 3 REFL? (Y)   = N

FWHM (DEG 2-THETA) AT 40 (DEG)   =

W*G DATA MEMBER: = 1

NO. OF STANDARDS TO BE LISTED: = 20

USE CHEMISTRY FOR MATCH? (Y)   = N

Figure 3. Menu for selecting Search/Match parameters. Entries in bold face

SEARCH/MATCH PROCESS

The computer time for S/M increases with increasing number of reflections in the unknown pattern, the search window width, and the number of standards in the subfile. The use of even the simplest chemical pre-screen greatly reduces the time. For example, a three phase mixture pattern containing 43 reflections required 257 seconds in the search of the 2332 frequently encountered phases of the inorganic subfile with a 0.3° window, and the time was reduced to 90 seconds with element restrictions. These are typical search times and include the program initialization, calculation of the FOMs, etc.

In the arrangement of the JCPDS file used in this method, the frequently encountered phases are searched first.[1] When the search is completed the names of the potential standards are listed in descending order of the FOMs together with the number of peaks in the unknown pattern NU which match those of the standard pattern NS in the angular range used, Fig. 5. In all cases, NU is no greater than NS.

If the unknown is a single phase the FOM may approach (but rarely reach) 100 which would be a perfect match of all the ds and Is. In mixtures the FOMs will be lower because not all the reflections in the unknown will be in one standard. In practice, the FOM values are dependent on the quality of both the input data and the standards.

INTERACTIVE OPERATIONS

After the FEP search has been completed the interactive options shown in the lower portion of Fig. 1 may be used to verify the identifications. It is generally good practice to start with the PLOT option which displays the unknown pattern on the upper portion of the graphics terminal screen and the standard selected from the S/M list on the lower portion, Fig. 6. The tick marks corresponding to the positions of the peaks in the unknown appear in a row below that pattern. Each matched reflection in the selected standard pattern has a tick just above that peak and another tick below the ticks of the unknown to facilitate comparison. The printer also lists the ds and Is of the selected standard and an M is printed on those reflections matching the unknown, Fig. 7. The weak reflections omitted from Peak Search due to the standard deviation and minimum peak height selections in the diffractometer run[3,4] may appear in the standards list and of course they will not be marked as a match. This procedure may be repeated for all the likely standards until there is a reasonable certainty that all the phases have been identified.

There are two important features of the plots of the standards: (1) the pattern appears as if run on the user's diffractometer, and (2) the intensities are scaled to their peak heights in the unknown pattern.

Another PLOT option is the display of small bar pattern of the unknown and all the standards found in the S/M. The height of the matched reflections of each standard are plotted twice the height of the unmatched reflections; these bar heights

Refl	2-theta	D(input)	D(max)	D(min)	Intensity
1	15.009	5.8981	5.9763	5.8200	2.
2	15.100	5.8626	5.9398	5.7854	1.
3	15.190	5.8282	5.9045	5.7519	2.
4	15.575	5.6850	5.7575	5.6124	13.
5	19.085	4.6466	4.6948	4.5984	2.
6	19.244	4.6086	4.6560	4.5612	2.
7	19.573	4.5318	4.5776	4.4859	21.
8	19.792	4.4821	4.5269	4.4372	24.
9	21.887	4.0577	4.0943	4.0211	13.
10	22.167	4.0070	4.0428	3.9713	1.

Figure 4. Partial printer listing of input ds and their error limits (based on $0.1°$ 2θ) for Search/Match. First ten reflections of inorganic mixture.

Number	Standard Name	Phase	PDF No.	FOM	NU	NS
1	NH4BF4	*0	15- 745	62.	27	30
2	PBCL2/COTUNNITE	*4	26-1150	59.	23	29
3	CDCO3/OTAVITE	*4	8- 456	51.	8	9
4	(AL4C3)	7R*0	11- 629	47.	5	9
5	NA2BEF4	*0	6- 637	47.	9	11
6	(MOB)	8Q*0	6- 644	44.	4	5
7	CO2GEO4	*0	10- 464	42.	3	4
8	CR2(SO4)3	0	18- 393	41.	12	17
9	ZNCR2O4	*0	22-1107	41.	3	6
10	PBCLF/MATLOCKITE	*4	26- 311	41.	9	13

Figure 5. Top ten standards found in S/M of IN FEP file. Top three were the correct phases. Chemical prescreening would have reduced the list.

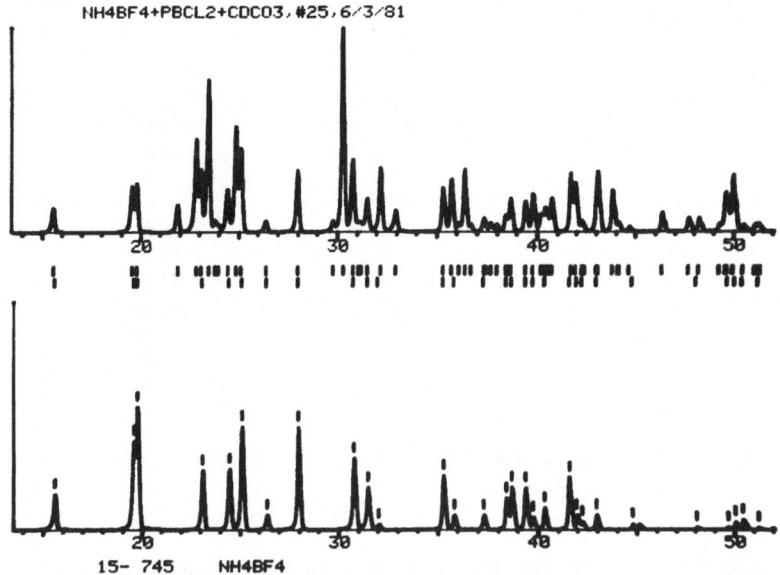

Figure 6. Graphics display of unknown pattern (above) and first FOM standard displayed as a diffractomer pattern (below).

thus have no correlation with the intensities. This display is useful for making a quick visual survey of the likely standards listed.

The IDENTIFY option may then be used to enter each standard selected as a constituent of the unknown.

The SUBTRACT option may be used to remove all identified phases from the unknown pattern. The remaining intensities are the differences between the sum of the identified standard patterns (whose intensities have already been scaled by the computer to the unknown) and the pattern of the unknown. This remainder, listed by the printer and displayed on the graphics screen, may arise from relative intensity differences between the standard and that phase in the unknown, or the presence of additional phases. The intensity differences may be due to incorrect specimen preparation, different instrument geometries, poor standard data and similar factors.

After using the S/M and the interactive options the specimen was unequivocally identified as a mixture of the three top standards listed in Fig. 5.

Refl No.	... STD: 15- 745 ... D(A)	I
M 1	5.668	30.0
M 2	4.528	60.0
M 3	4.482	100.0
M 4	3.844	50.0
M 5	3.635	50.0
M 6	3.542	85.0
M 7	3.376	12.0
M 8	3.186	85.0
M 9	2.902	60.0
M 10	2.839	35.0
M 11	2.792	4.0
M 12	2.541	45.0

Figure 7. Partial listing of matched reflections in first standard, NH_4BF_4.

The RETURN option can be used to start at the beginning of the S/M for another round using the remainder data or all or a portion of the original data. If complete identification has not been made the remaining portions of the subfile or another subfile may be searched and the procedure described above will be repeated.

UTILITY PROGRAMS

Three utility programs greatly facilitate the computer S/M method. The #SMUTIL1 program can save up to approximately 380 data sets of unknown patterns in #SMDATA for later analysis. The sets contain the ds and Is determined by Peak Search, Profile Fit or manual entry. They can be transferred to the S/M by entering the instrument and run log numbers which identify the run.

Utility program #SMUTIL2 is used for the purpose of managing the user created standards files. These files, which can be virtually any size, may consist of runs made on the user's diffractometer with his own standard materials, manually entered JCPDS cards or data from the literature. Special standards frequently encountered in the user's analytical work such as mixtures, materials not in the JCPDS file, thin films with preferred orientation, solid solution series and other special specimens may be incorporated. A user's file may also be created to include only those standards likely to occur in the user's analyses.

The user's files have many advantages. The materials used for standards can be properly characterized for the particular conditions of preparation. Because the runs are made on the user's diffractometers he has complete control of the quality of the data and can duplicate the experimental conditions when running the unknowns. Differences in the instrument geometry, calibration, preferred orientation and similar factors which may cause uncertainties when using the JCPDS file are thereby avoided. Identifications can be made much faster because the file is smaller, narrower search windows can be used, high figures-of-merit will be obtained, and subtraction of identified phases can be done with greater precision. Chemical information is also included.

The #SMLOOK utility program provides a unique method of viewing any standard in the files as a diffraction pattern run on the user's diffractometer. The standard is selected by entering the subfile, set and card numbers. Also required is the WG file which contains a set of standard profiles that define the instrument function for profile fitting.[5] The parameters give the profile shapes as a function of 2θ and are stored in the computer. The d and I of each reflection in the standard pattern is converted to the profile for that particular 2θ. The entire pattern or any portion of it expanded to full screen width is displayed on the graphics terminal and a paper copy can be made. The ds and Is and the calculated 2θs can be listed by the printer. The patterns of the standards generated in the interactive PLOT option described above are made in the same manner.

REFERENCES

1. T. C. Huang and W. Parrish, A New Computer Algorithm for X-Ray Powder Diffraction Analysis, Adv. in X-Ray Anal. 25 (previous paper) (1982).

2. W. Parrish, G. L. Ayers and T. C. Huang, A Minicomputer and Methodology for X-Ray Analysis, Adv. in X-Ray Anal. 23:313-316 (1980).

3. W. Parrish, Advances in Computerized X-Ray Diffractometry and X-Ray Analysis, Adv. in X-Ray Chemical Anal. Japan 12:33-50 (1980).

4. G. L. Ayers, T. C. Huang and W. Parrish, High-Speed X-Ray Analysis, Jour. Appl. Cryst. 11:229-233 (1978).

5. W. Parrish and T. C. Huang, Accuracy of the Profile Fitting Method for X-Ray Polycrystalline Diffraction, Proc. Symp. on Accuracy in Powder Diffraction, NBS Special Publ. 567:95-110 (1980).

AUTOMATICALLY CORRECTING FOR SPECIMEN DISPLACEMENT ERROR DURING

XRD SEARCH/MATCH IDENTIFICATION

Walter N. Schreiner

Philips Laboratories
Briarcliff Manor, NY

Ronald Jenkins

Philips Electronic Instruments Inc.
Mahwah, NJ

Over the past several years there has been considerable interest in computer search/match programs for qualitative analysis
of powder diffraction patterns. This interest has been stimulated
by the availability of modern minicomputers supported by relatively
inexpensive mass storage devices capable of containing the entire
JCPDS (1) data base on line. As the traditional search/match algorithms have been reviewed for possible implementation on the
slower speed and restricted memory minicomputers being supplied
with today's automated diffractometers, new ideas have emerged for
such algorithms. One very extensive set of new algorithms has been
developed by our group and these are contained in the SANDMAN
search/match/identify program which was described at this conference last year (2). Experience has shown those algorithms to be
extremely effective, particularly in handling cases where the presence of systematic errors in the data has precluded the correct
analysis by other computerized search/match systems.

A summary of the new techniques developed for SANDMAN is seen
in Fig. 1. The first technique, which is the subject of today's
presentation, is a method of automatically correcting for unspecified amounts of a systematic error, such as specimen displacement
error. Within rather generous limits, it is not necessary to know
beforehand how large the error is since this value is determined
at search time. Once the systematic errors are removed, use of a
probability-based scoring algorithm is possible in which probabili-
ties for hits, misses, etc. are assigned based on models of the

231

SANDMAN TECHNIQUES

- **AUTOMATIC CORRECTION FOR DISPLACEMENT ERROR**

- **PROBABILITY SCORING**

- **WINDOWLESS SEARCHING**

- **ADDITIVE PHASE IDENTIFICATION**

- **ISOTYPICAL SEARCH**

FIG. 1

EXPERIMENTAL PATTERN

D	2-THETA	I
4.328	20.51	16
3.385	26.30	100
2.480	36.19	15
2.302	39.10	15
2.256	39.93	7
2.145	42.09	10
1.994	45.44	8
1.830	49.79	29
1.682	54.52	9
1.669	54.98	5
1.550	59.61	23
1.459	63.71	5
1.388	67.41	14
1.381	67.82	22
1.377	68.01	13

FIG. 2

residual random errors, which are generally smaller than the systematic errors. By modelling the random errors in terms of the statistical processes which are responsible for them, it is no longer necessary to have the user of the S/M system arbitrarily specify match windows around diffraction lines, which is the traditional method of treating both random and systematic errors in experimental patterns. Probability-based scoring also leads naturally to a method of additive phase identification in which overlapping lines are combined to build up a match to an experimental pattern, instead of subtracting reference lines from the experimental pattern and studying the residue in the traditional manner. Finally, entire isotypical series can be located by changing the form of the systematic error correction function which is sought during the search process.

How does one go about compensating for a systematic error without first knowing how much of it there is? Consider the experimental data for alpha quartz (5-490) in Fig. 2. Few people would readily accept this as a quartz pattern because, on inspection, the lines disagree significantly with the lines expected for quartz. For example, the (101) line should be around 26.66° (2θ). This pattern was acquired by displacing a quartz specimen approximately 600 microns below the focusing circle of the diffractometer. Nevertheless, the pattern is quite good because the random errors associated with it are small. This can be seen by plotting the 2θ difference between every observed line and every reference line, as a function of 2θ of the reference line. In Fig. 3 a clear trend is seen which corresponds to those cases where the observed line position is subtracted from the correctly matched reference line. The other points result when an experimental line is compared to an incorrectly matched reference line. Here, I have only plotted those differences which are less then 1°(2θ).

If we assume that all of the systematic shift exhibited by the clustered data points results from displacement error, we can de-

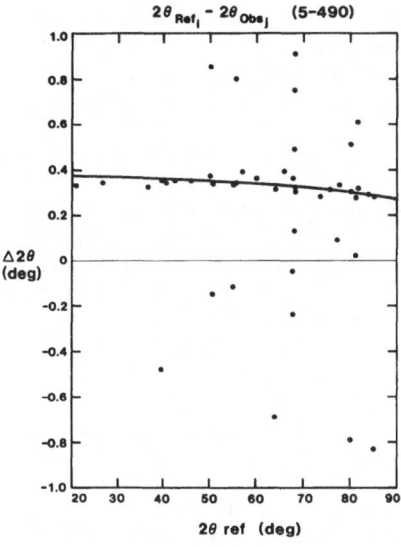

FIG. 3

termine the amount of that error by least squares fitting those points to a function of the form $K\cos(\Theta)$, where K is a free parameter. To determine which points should be included in the fit, we begin by fitting them all and then discarding the outliers on the basis of a χ^2 test until an appropriate limit is reached. The final coefficient, K, represents, of course, the previously unknown amount of displacement error.

The result of this is the solid line seen in Fig. 3 which gives the fitted value of K=580μm. It is seen that the scatter of points around this line is quite small. The RMS value of the differences is about 0.03°(2θ). Thus, aside from the large systematic error, the pattern is indeed quite good, and should be readily matched by any computer S/M program, once the systematic error is removed.

This example is one of a series of runs we made to study the effects of specimen displacement error. The run covered displacement errors from 0 to 600 microns. In each case, once the systematic error was removed, a high probability-based score resulted because residual random errors were small. The scores were independent of displacement error. Equally important, in almost every case, the second highest scoring pattern scored so low that the pattern was never even reported as a possible match candidate.

In experimental diffraction data, more than one type of systematic error is usually present. For example, a close inspection of the fit in Fig. 3 will reveal the presence of a second systematic trend probably resulting from uncorrected instrumental aberrations. In principle, one could least squares fit the data to a sum of terms, each representing unspecified amounts of various systematic errors. The fit would determine the optimum value for each of the undetermined coefficients. However, unless the terms are orthogonal to each other, the coefficients will couple, causing them to lose their physical significance. Of course, a physical interpretation for the fitting parameters is not necessary, since one is only trying to separate systematic from random errors. It is only the random component of the error that determines the score in a probability-based scoring algorithm. From a practical point of view, however, each free parameter added causes an extra burden on the least squares fitting procedure, which must be executed for every potential candidate match. In a complex mixture, it may be necessary to examine virtually every pattern in the data base, and long search times would result. Fortunately, however, the major contribution to the total systematic error commonly encountered in experimental patterns is displacement error, and thus a single free parameter has been found to suffice. In any event, the $\cos(\theta)$ term also partially compensates for the other systematic errors and so we call the coefficient, K, Equivalent Displacement Error (Eq.D.E.) to emphasize the fact that its value may not have an entirely physical interpretation.

Several interesting applications of the automatic systematic error detection algorithm can be found and we will briefly mention two. Both deal with systematic line shifts resulting from structure related effects rather than instrumental errors. The first application is to order phases which form an isotypical series. We refer to a set of phases belonging to the same space group as being isotypical, if their axial ratios are similar. We also refer to phases as being isostructural when the differences in their lattice parameters are so small that it becomes difficult to distinguish them experimentally. Isostructural and nearly isostructural phases cause difficulties for computer search/match programs because, in the past, there has been no tool to aid in distinguishing them. The Eq.D.E. parameter affords such a tool, because it provides a quantitative measure of systematic shifts. Fig. 4 shows a series of phases located by SANDMAN which are isotypical with Fe_3O_4 which was one component of an unknown mixture submitted to the program for analysis. All of these phases belong to space group $Fd\bar{3}m$ (227). It is seen that the Equivalent Displacement Error correlates closely with the lattice parameter of these phases. How does one determine which phase is correct? If the true displacement error for the specimen can be determined, the equivalent displacement error parameter can be used to select the best candidate from this series. In this multiphase example, Al_2O_3 (10-173), was unequivocally identified and its displacement error parameter was 0 microns. Since all the phases in the mixture must have the same displacement error, we conclude that Fe_3O_4 is the best, and indeed correct match.

The second example is an application to solid solutions. Solid

Phases Isotypical with Fe_3O_4

Lattice Parameters	Eq.D.E.	SANDMAN RANKING	Chemical Name	JCPDS Card
8.330	-285	10	$(Fe,Mg)(Al,Cr,Fe,Ti)_2O_4$	25-1376
8.333	-435	17	$MgCr_2O_4$	10-351
8.339	-405	9	$NiFe_2O_4$	10-325
8.364	-240	4	$FeCr_2O_4$	24-512
8.367	-240	5	$Fe_{2.43}Ni_{0.53}O_4$	23-1119
8.369	-225	11	$CuFe_2O_4$	25-283
8.375	-135	3	$MgFeO_4$	17-464
8.3967	0	6	Fe_3O_4	19-629
8.397	30	2	$MgFe_2O_4$	17-465
8.4080	60	12	Fe_2GeO_4	25-359
8.4411	270	14	$ZnFe_2O_4$	22-1012

FIG. 4

solutions can be thought of as forming a continuous isotypical series, although frequently the axial ratios are not independent of stoichiometry. Instead of looking for a $\cos(\theta)$ systematic shift, one looks for a $\cot(\theta)$ systematic shift which would characterize an isotypical series. Such a search was built into SANDMAN and a series of binary solid solutions was made (3) from Cr_2O_3 and Fe_2O_3 to test whether the pure phase end members could be identified. The cell dimensions of Cr_2O_3 and Fe_2O_3 differ by about 1.3%, whereas the isosearch command in SANDMAN, which implements the isotypical search, is designed to look for variations up to about 5%. The search successfully locates the pure phase end members in every case, and ranks them among the top five matches. In Fig. 5 we show the $\Delta d/d$ parameter fits made during search and match. $\Delta d/d$ is the amount in parts/1000 [PPT] by which the lattice dimensions of the Cr_2O_3 and Fe_2O_3 pure phases must be scaled in order to match the observed lines of the solid solutions. The solid dots are for Cr_2O_3 and the open circles for Fe_2O_3. The curves are hand drawn. The behavior of $\Delta d/d$ is typical of a binary solid solution. The curves can be thought of as calibration curves for determining the stoichiometric ratio from the $\Delta d/d$ parameter. The scatter of the data points about the curves in this example indicates a $\Delta d/d$ accuracy of about +-0.3 PPT. This represents lattice parameter sensitivities of $\sim 1/15000$. As in the case of substantial displacement errors even though systematic shifts may be large, after they are removed the residual random errors are small.

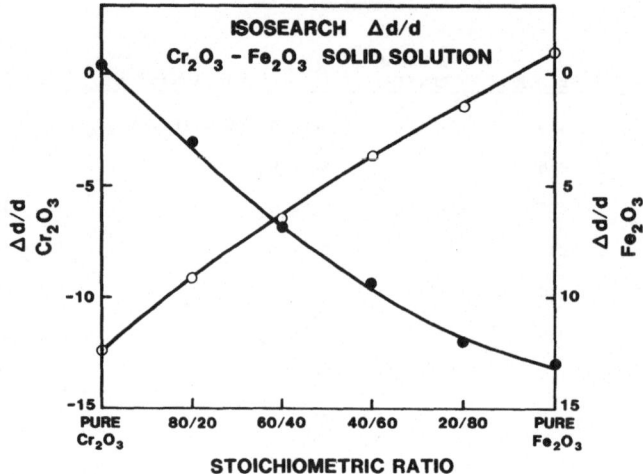

FIG. 5

To summarize, systematic errors in powder diffraction data must be treated before an unknown pattern is submitted to a computerized search/match program for phase identification. If the functional form of the systematic error is known, it is possible to remove the error during the search/match, even though the magnitude of the error may not be known beforehand. The remaining random errors are usually small enough to allow successful phase identification by computer.

References

(1) International Centre for Diffraction Data (JCPDS), Swarthmore, PA.

(2) Schreiner, W.N., Surdukowski, C., Jenkins, J., "Qualitative Phase Analysis Using An X-Ray Powder Diffractometer", Adv. X-Ray Anal., 24, 91 (1981).

(3) The specimens were prepared by Greg McCarthy, N. Dakota State University.

X-RAY DIFFRACTION PHASE ANALYSIS USING MICROCOMPUTERS

T.M. Hare, J.C. Russ, and M.J. Lanzo

Engineering Research Services Division
North Carolina State University
P.O. Box 5995 Raleigh, NC 27650

INTRODUCTION

X-ray diffraction to determine the phases present in complex specimens
generally proceeds by acquiring a pattern of D-spacings and intensities from the
sample, which is then compared to patterns from a series of known or standard
phases. The process of searching through large data bases to identify matches
is too time consuming for manual methods, and so computerized search/match programs have come into use (1,2). The magnitude of the general problem, in which
perhaps 40,000 known patterns (corresponding to the present size of the JCPDS
powder diffraction file) are involved, places it in the realm of large main-
frame computers, which are often accessible on a time-share basis by many laboratories or researchers using XRD. On the other hand, in many practical applications, the search need only be carried out over a limited range of compounds
which are expected or may possibly be present based on the known history of the
material, or on its independently determined composition. In this case, the
data base becomes much smaller, and can be accommodated in micro- and minicomputers. The system we report here utilizes a 64K-byte 8-bit microcomputer
(Apple II+) with dual floppy disk drives. One 5-1/4 inch disk can hold the
pattern information for fifty compounds, and can be searched for possible
matches in a few minutes. The same computer can control the scanning diffractometer to acquire the pattern data, and can act as an intelligent terminal to
larger time-share machines when it is necessary to access the large data bases.

INPUT OF DATA

Pattern information consists of a series of D-spacings and intensities.
For a typical known or standard phase, these can be entered from JCPDS cards
(3), by hand through the computer keyboard, for storage on disk. It is also
possible to 'download' the data from large time-share computers via telephone
lines (assuming of course that the user first establishes an account with the
service). In either case, up to 100 lines per compound, sorted in intensity
order, along with several lines of descriptive, compositional, or other comment
information are stored on disk. It is also possible to measure a pattern directly from a standard, and enter it as an unknown would be entered (a utility program converts data from the format for an unknown to that for a standard). This
allows unique patterns known to the user but not in the JCPDS file to be readily
incorporated in search/match problems.

237

For unknowns, although the search/comparison/match process actually uses the D-spacings, it is more convenient to store the angle and wavelength. This allows for simple correction for shifts in angle (offsets as are commonly encountered in film work). The conversion to D-spacing is automatically carried out using the Bragg equation, using the known and stored wavelength of the X-radiation being used (the default value of 1.5418 Å is for Cu K_α). These files are also maintained in intensity order, for up to 100 lines per compound, and can include comment lines entered by the user.

The angles and intensities can be entered by hand from the keyboard, or obtained automatically in several ways. The angles may be entered in theta, two-theta, or four-theta depending on the type of camera being used. Also, the intensities can be specified numerically (the program normalizes them with the most intense scaled to 100 before storage), or as codes of w (weak), m (medium), s (strong), etc. which the program converts arbitrarily to numbers. This is useful if manual reading of films is used.

The computer is equipped with an interface (4) that can control stepping motors, and this can be connected to a film reader or directly to a scanning diffractometer (provided they are motorized). For the film densitometer, the reading can be obtained with a simple analog-to-digital converter reading the output voltage from the photomultiplier tube. For the scanning diffractometer, the pulses from the detector(s) are amplified and sent to the same interface card that controls the stepping motor (this is in parallel to the normal counter/ratemeter and strip-chart recorder output). The pulses are counted for present lengths of time, and the values stored in the computer (and displayed on its TV screen).

When a scan is complete, it may include thousands of discrete points. Normally the user will set up the scan range (in either units of d-spacing or two-theta angle) and specify either the step size or number of steps. The complete spectrum can be stored on disk, or printed out in 'strip-chart recorder' style on the thermal printer (with peaks labelled), if desired. Most often, it will simply be processed to obtain a list of angles and intensities for storage and subsequent identification. The methods available for peak identification include purely manual designation (the user moves a cursor to the peak and presses a button), and a semi-automatic mode. In the latter case, the user specifies a minimum peak-to-background ratio and a window width in number of steps or in degrees. The program moves a sliding window of this width along the spectrum and compares the average intensity within it to that in the neighboring regions. If the ratio exceeds the specified minimum, the local maximum is found (first derivative test), and the angle is saved in an array. This method will not reliably detect multiplets of overlapped lines, and so operator interaction and assistance are still required. More complex peak-finding algorithms using higher derivatives, cross correlation (digital filter method), and so on can be readily programmed into the computer, but thus far have not been needed. The comprehensive method proposed by Mallory and Snyder (5) can also be used if necessary.

The complete files of standards or unknowns can be examined and edited. It is also practical to combine the standards onto various disks, each holding up to 50 patterns, sorted by type, composition, or in other ways useful to permit mini- search/match operations. In addition, utility programs can transfer the data in either standard or unknown files to the format required by the NIH-EPA Chemical Information System (CIS). This is a time-share program available via GTE-Telenet to users throughout the world. It includes a program to perform search/match operations in the full JCPDS database, including the use of chemistry information. Provided that the user has an established account with CIS, the Apple microcomputer can be readily equipped with a telephone connection via Modem, and a program to convert it to an

intelligent terminal. Under this program, it can transmit files up to the
host computer, download files from it, and run programs with the output being
captured for printout or display by the user. This provides an obvious and
inexpensive backup for the microcomputer, when the additional power of the
larger machine is needed, since the microcomputer can still acquire the data,
format and pre-screen it, and then control the communications with the larger
machine.

SEARCH/MATCH METHOD

A 'match' between a stored standard and an unknown pattern is defined as
occurring when at least a number N1 out of another number N2 most intense
lines in the standard pattern are present in the unknown. The values N1 and
N2 are user-selected. Typical values might, for instance, be six out of ten
or four out of seven. This means that of the ten (N2) most intense lines for
the standard, six (N1) are present in the unknown. There is no restriction
on the relative intensity of the lines at this point. It is necessary, of
course, to take into account the fact that the D-spacings of lines in the
standard and unknown may not agree exactly. The permissible difference in
D-spacing is not a constant, but varies with D, and a second degree equation
is built into the program to compute a nominal acceptable difference. This
varies from 0.01 Å at 2 Å, to 0.05 Å at 4 Å, and to 0.5 Å at 10 Å. The 'default'
mismatch can then be multiplied by any user-entered factor, greater or less
than one, to accommodate patterns with more or less trustworthy D-spacings.

Using these parameters, the program searches all of the standards files
on the data disk for matches. It also can be restricted to search a subset
of the files, for instance ones that have already survived a search/match
operation using less stringent criteria. Restricted subsets based on chemistry
criteria (elements known from independent sources to be present or absent) can
also be established with a suitable program to read the optional comment lines
in the standards files, but generally with only fifty standards patterns per disk,
the classification of standards data and establishment of subsets based on chemistry is best accomplished by collecting the appropriate standards onto specific disks.

Patterns that match the unknown are assigned a figure of merit (F.O.M.)
for subsequent ranking as to the likelihood of the compound being present in
the unknown. The present calculation is simply the sum of the intensities in
the standard for the N1 matched lines, divided by the sum of all N2 candidate
lines. This value will lie between zero and one, and weights favorably matches
for the more intense lines in the standard. It does not include any comparison
of the relative intensities of the lines in standard and unknown, but more
complex F.O.M. formulae which do take this into account can readily be added
to the Basic language program.

The list of standard compounds found by the search/match algorithm are
sorted in decreasing order of F.O.M. and printed out, with the number of lines
matched. It is also possible to print a complete listing of the lines in the
unknown, in decreasing intensity order, with the lines and intensities of each
of the matching standards. This array shows many blanks, of course, and ideally
would show just one matching line for each line in the unknown. In practice,
some lines may not be matched at all, and some will have more than one match.
Since the relative intensities of the lines for both standard and unknown are
shown, it is usually straightforward to reject obviously incorrect 'matches.'
To aid in perusing this matrix of match intersections, a 40 by 40 checkerboard
display is also shown on the video screen, and a cursor can be moved through
the array to list the unknown D-spacing and intensity, and the standard corresponding to each match. Lines of multiple matches, or extent of agreement
for a given standard, are quickly found.

COMPARISON DISPLAY

Once the search/match operation has been performed, the user will normally want to compare visually the pattern from his unknown with each candidate standard, beginning with those having a high figure of merit. This is accomplished by displaying the patterns on the video screen, using a format of vertical lines with length proportional to intensity, with the unknown below the standard. A cursor can be moved along a path between the two patterns, with its position in terms of D-spacing shown numerically. Any portion of the entire spectrum display can be expanded to fill the screen, with correspondence between unknown and standard automatically maintained. Shifts of the unknown to correct for slight angle (theta) offsets can also be entered to improve the alignment of the patterns. This may be done before or after the search/match is performed.

When a particular standard shows good agreement with the unknown, it can be 'stripped out' to leave the residual. This is done by subtracting each line of the standard from the corresponding lines of the unknown. The intensity can be scaled to the relative intensity of the strongest line. No negative intensities are allowed, of course, but it may and in fact does often occur that some intensity remains for a line in the unknown, where the line resulted from contributions of more than one standard compound. The residual pattern after stripping away of a standard can be compared to another standard (and it stripped away), or can be re-saved to disk for further search/match operations. In this way, the user can interactively and iteratively 'take apart' his pattern using the standard patterns stored on disk, to determine the compounds present in the unknown.

EXAMPLE - PHASES PRESENT IN 'SYNROC'

As an example of the use of these programs, we have been applying them to the study of 'SYNROC,' a complex, multiphase ceramic material being considered as a host matrix for radioactive waste ions for power-generation nuclear reactors. The design of the ceramic is based on at least three known host phases - hollandite, perovskite, and zirconolite - that offer attractive lattice sites for the more than 30 species of waste ions. The addition of the waste, in amounts typically 10 to 15% by weight, and subsequent sintering of the ceramic, produces an extremely fine grained and heterogeneous structure containing at least the three expected host phases, although possibly with some lattice distortion or change in size due to the waste ions. Detection of these changes, and of other minor phases that may occur because of the presence of the waste or the thermal cycle followed during treatment, is of great importance in characterizing the material and understanding its ability to resist leaching out of the waste ions in water.

A diffraction pattern from sintered, technically dense SYNROC containing 10% simulated waste (PW4B type) was compared by the search/match program against a database of likely phases (oxides containing the known host-phase elements) taken from the JCPDS file, and some patterns measured on specially prepared single-phase standards containing no waste. Figure 1 shows a portion of the printout, which required at least six of the ten most intense lines in the standard to be present for a match. The first three phases found are the hollandite, zirconolite and perovskite expected to be present. The detailed printout shows which of the lines in the unknown are matched by each standard. The figure terminates this printout before some of the low intensity minor lines are listed, and shows clearly the cases of multiple matches that occur. Most of the less likely phases, with low F.O.M., match only a few lines and do not agree well with the relative intensities.

SUMMARY:

PHASES FOUND:	NO. LINES MATCHED	FIGURE OF MERIT
HOLLANDITE (1/6 AIR)	10	1
CAO.ZRO2.TI2O4	10	1
CAO.TIO2	7	.95
TIO2.ZRO2	8	.88
TIO2 (A)	7	.73
TIO2 (R)	8	.55
ZRO2+CAO	8	.45

UNKNOWN SUBSTANCE: 26-5

STANDARD PHASES:	FIG. OF MERIT	UNKNOWN D-SPACINGS/INTENSITIES:			
		3.145/100	1.560/ 64	2.189/ 64	2.441/ 64
HOLLANDITE (1/6 AIR)	1.0	3.149/100	1.560/ 64	2.191/ 64	2.442/ 64
CAO.ZRO2.TI2O4	1.0				
CAO.TIO2	.95				
TIO2.ZRO2	.88				
TIO2 (A)	.73				2.431/ 10
TIO2 (R)	.55			2.188/ 25	
ZRO2+CAO	.45		1.548/ 25		

STANDARD PHASES:	FIG. OF MERIT	UNKNOWN D-SPACINGS/INTENSITIES:			
		2.698/ 64	2.921/ 64	1.370/ 32	1.656/ 32
HOLLANDITE (1/6 AIR)	1.0			1.371/ 64	1.659/ 16
CAO.ZRO2.TI2O4	1.0		2.929/100		
CAO.TIO2	.95	2.701/100			
TIO2.ZRO2	.88		2.920/100		
TIO2 (A)	.73			1.364/ 6	1.667/ 20
TIO2 (R)	.55			1.360/ 20	
ZRO2+CAO	.45				

STANDARD PHASES:	FIG. OF MERIT	UNKNOWN D-SPACINGS/INTENSITIES:			
		1.730/ 32	1.858/ 32	1.908/ 32	2.223/ 32
HOLLANDITE (1/6 AIR)	1.0		1.860/ 64		2.227/ 16
CAO.ZRO2.TI2O4 (CONTINUED)	1.0	1.740/ 65 1.737/ 60			
CAO.TIO2	.95			1.911/ 50	2.217/ 6
TIO2.ZRO2	.88	1.740/ 9	1.850/ 11		
TIO2 (A)	.73				
TIO2 (R)	.55				
ZRO2+CAO	.45				

STANDARD PHASES:	FIG. OF MERIT	UNKNOWN D-SPACINGS/INTENSITIES:			
		2.780/ 32	2.910/ 32	3.520/ 32	1.304/ 16
HOLLANDITE (1/6 AIR)	1.0			3.526/ 64	
CAO.ZRO2.TI2O4	1.0	2.789/ 55	2.903/ 80		
CAO.TIO2	.95				
TIO2.ZRO2	.88				
TIO2 (A)	.73			3.520/100	
TIO2 (R)	.55				
ZRO2+CAO	.45				1.284/ 4

Fig. 1. Printout from search/match program applied to SYNROC specimen, requiring 6 of 10 major lines in each standard. The first three phases are the hollandite, zirconolite, and perovskite that actually form the bulk of the matrix.

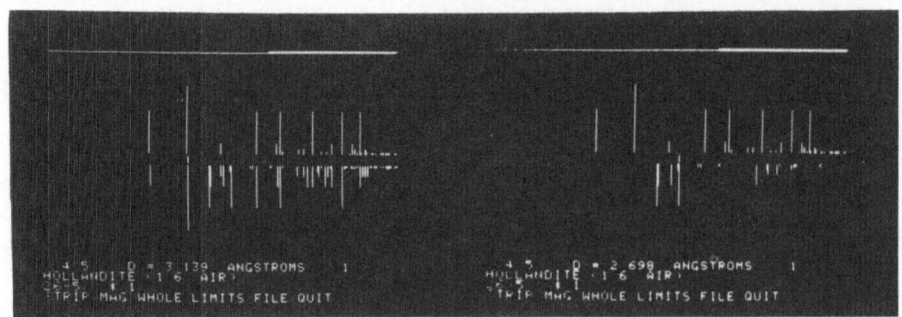

Fig. 2.   A: Comparison display of diffraction pattern from hollandite (top)
          measured by us from a single-phase standard, against the unknown
          (bottom).   B: The residual after stripping away the hollandite lines.

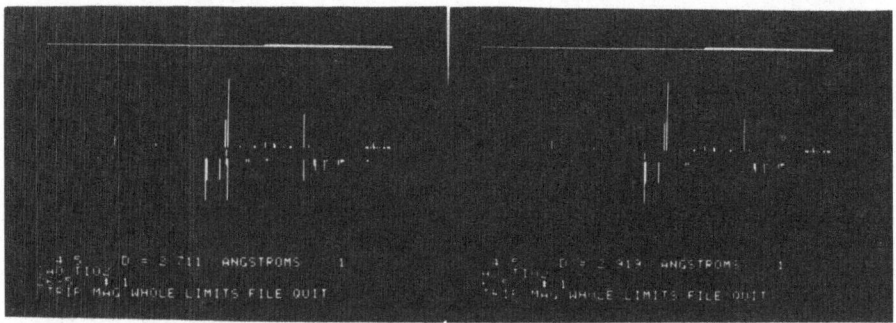

Fig. 3,   A: Comparison display of the residual from figure 2 (bottom)
          against the pattern from zirconolite taken from the JCPDS file. B:
          The residual after stripping away the zirconolite lines.

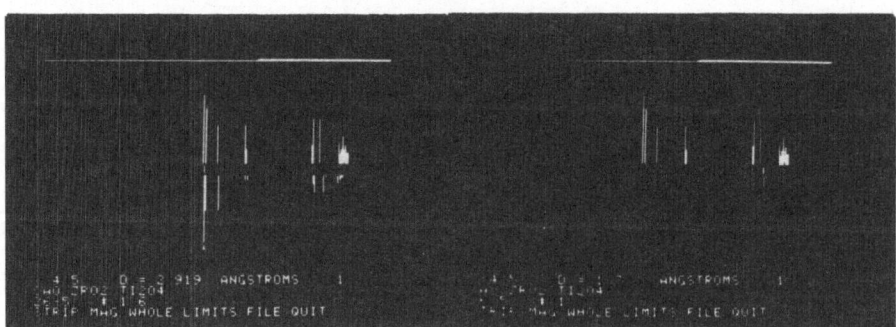

Fig. 4.   A: Comparison display of the residual from figure 3 (bottom)
          against the pattern from perovskite taken from the JCPDS file.   B:
          The residual after stripping away the perovskite lines.

Figure 2 shows the comparison display of the unknown pattern (bottom) with the first standard pattern (top), and the remainder after stripping it out. Figures 3 and 4 continue this process with the second and third standard phases. After these three phases are removed, the remainder contains only a few minor lines. Longer exposure times to record these weak line patterns are now being used to collect enough information to identify the minor phases. The usefulness of the search/match/compare method for identifying phases present is well illustrated by this sequence.

CONCLUSION

A program to perform search/match analysis of X-ray diffraction patterns against a database of standard compounds has been implemented in a microcomputer. The computer can also be used to control a scanning diffractometer to acquire the patterns. User interactive display programs aid is stripping away identified phases to reveal the residuals. The computer can readily handle applications requiring matches to a small number of potential phases, and can act as an intelligent terminal to pass data to and from, and control programs running on large time-share computers capable of searching the entire JCPDS file.

REFERENCES

1. M.C. Nichols, The Search-Match Problem, in "Advances in X-Ray Analysis, Vol. 23," Plenum Press, N.Y., 1979, p. 273-278.
2. T.C. Huang, W. Parrish, A New Computer Algorithm for Qualitative X-Ray Powder Diffraction Analysis, in "Advances in X-Ray Analysis, Vol. 25," Plenum Press, N.Y., 1982.
3. N.I.H. - E.P.A. Chemical Information System, CIS Project, 2135 Wisconsin Ave., Washington, DC.
4. Dapple Systems, P.O. Box 2160, Sunnyvale, CA.
5. C.L. Mallory, R.L. Snyder, The Alfred University Powder Diffraction Automation System, N.Y. State College of Ceramics Tech. Paper 144, Alfred Univ., 1979.

WITH REGARD TO

"X-ray Diffraction Phase Analysis using Microcomputers"

A POLEMIC

The Powder Diffraction File, accessed
through PDSM on the Chemical Information System, is a copyrighted data
base owned solely by the JCPDS--In-
ternational Centre for Diffraction Data.
Subscribers to CIS/PDSM are referred
to the copyright laws governing the
use and duplication of such data bases.
Should you have any questions, please
call or write.

JCPDS--International Centre for Diffraction Data
1601 Park Lane
Swarthmore, Pennsylvania
(215) 328-9404

A SECOND GENERATION AUTOMATED POWDER DIFFRACTOMETER CONTROL SYSTEM

Robert L. Snyder

NYS College of Ceramics
Alfred University
Alfred, NY 14802

Camden R. Hubbard

National Bureau of Standards
Washington, D.C. 20234

Nicolas C. Panagiotopoulos

JCPDS International Centre for Diffraction Data
National Bureau of Standards
Washington, D.C. 20234

ABSTRACT

The real-time x-ray powder diffractometer control system AUTO incorporates several advances in data collection and analysis. Counting procedures for selected area data collection are optimized to achieve either a preselected statistical error in minimum time or a minimum error in fixed total time. Run files are employed to greatly simplify quantitative analysis procedures and for controlling repetitive runs. External calibration curves for 2θ are used to eliminate all but sample dependent aberrations to peak positions. A generalized data file structure is used to document the instrumental variables and sample parameters.

INTRODUCTION

The introduction of computer control to powder diffractometers has opened new horizons for the powder diffraction method. However, most of the data collection algorithms in use today do

245

not take advantage of the computer's ability to bring "intelligence" to this process. The commonly used fixed angular increment move, followed by a fixed time count will be referred to here as a non-optimizing procedure ("dumb"). Optimizing data collection ("smart") methods have seen limited development (1-4).

The AUTO (5) diffractometer control system was designed to bring a certain amount of optimization to the data collection process. In selected 2θ region data collection, both the peak location and the distribution of counting time between background and the peak are optimized. However, in full powder pattern data collection, the conventional "dumb" move and count method is still employed. When peak height measurements are to be made, the algorithm first uses a Savitsky-Golay (6) smoothing procedure to locate the peak maximum. This eliminates intensity errors arising from a shift in peak position caused by solid solution or specimen displacement.

OPTIMIZATION

Four types of selected area intensity measurement are permitted by AUTO. In each type, the program optimizes the distribution of counting time between the diffraction profile and background regions. Either integrated areas or peak heights may be collected in either an optimized fixed error or fixed time counting mode. The method employed in integrated area data collection involves a step and count procedure with computer determined count time at each point. The background is collected over a small range of 2θ below the peak (NBG points) and over an equal range of 2θ above. Intensity values are recorded for each of the NPK points in the peak profile; the $\Delta 2\theta$ between data points (background and peak) are equal and preferably equal to the smallest step permitted by the instrument (typically 0.005 deg). The collection of many points over the profile and for background minimizes the effect of large sample grains or electrical noise spikes. The integrated intensity corrected for background is given by:

$$I(int) = S \cdot \sum_{i=1}^{NPK} RATE(i) - S \cdot NPK \cdot \frac{\displaystyle\sum_{j=1}^{2NBG} RATE(j)}{2NBG} \quad , \qquad (1)$$

where S is the step size; RATE(i) is the counts per second, corrected for dead time, at profile point i; and RATE(j) is the corresponding count rate for background point j. The entire profile is recorded for future reference, reanalysis, or plotting.

Peak heights are measured in an analogous manner, except for two differences. First, only a small number of points at the peak (again called NPK) are collected. Second, background can be estimated from measurements of a single region or from two bracketing regions, as for integrated intensities. In this case, the net peak height is given by:

$$I(pk) = \frac{\sum_{i=1}^{NPK} RATE(i)}{NPK} - \frac{\sum_{j=1}^{(2)NBG} RATE(j)}{(2)NBG} \quad , \qquad (2)$$

where RATE(i) and RATE(j) are as before, the (2)NBG term is either NBG or 2*NBG depending on whether one or two background regions, each of NBG points, were measured. We find for step sizes of $0.005°$ 2θ, an NPK of nine $(0.045°)$ and NBG of 10 $(0.050°)$ are appropriate.

The fixed error optimization mode of data collection aims at achieving a preselected estimated standard deviation, based on counting statistics, in I(int) or I(pk) in minimum time. The time to be spent at each NPK profile or peak point is given by:

$$T(pk) = \frac{RO(pk) + RO(bg) \cdot R}{E^2 \cdot NPK \cdot [RO(pk) - RO(bg)]^2} \quad , \qquad (3)$$

where RO(pk) is the observed average count rate over the entire profile region (or at the peak); RO(bg) is the observed count rate at a background point; $R = SQRT[RO(pk)/RO(bg)]$; and E is the fractional counting statistical error desired in the background corrected intensity. The time to be spent at each NBG background point is,

$$T(bg) = \frac{NPK \cdot T(pk)}{(2)NBG \cdot R} \qquad (4)$$

The total data collection time is simply given as,

$$T(tot) = NPK \cdot T(pk) + (2)NBG \cdot T(bg) \quad . \qquad (5)$$

If T(tot) exceeds a user selected maximum count time, T(max), then a fixed time optimization is performed. Here, the counting statistical error is minimized by adjusting the peak and background measuring time. In this case,

$$T(pk) = \frac{T(max)}{NPK} \cdot \frac{1}{1 + R} \quad , \qquad (6)$$

and T(bg) is calculated by equation 4, as before.* To apply
these optimizing equations, both RO(pk) and RO(bg) are required.
AUTO obtains these values by very rapid estimates of the intensities. First, a background point is measured for one second.
Next, a rapid (30 deg/min) scan over the profile or a one second count at the peak is performed.

FEATURES OF THE AUTO CONTROL SYSTEM

     AUTO, written in FORTRAN, was designed to allow real time
control of multiple diffractometers on a single user 16 bit
minicomputer. In its current form, it controls two diffractometers, while simultaneously allowing the user to carry out
status enquiries, file manipulations, run set up, run initiation,
and/or run termination. The program can readily be reconfigured
to control only one diffractometer. The real-time feature is
achieved by frequently polling each interface card to see if its
function (counting or driving to a selected 2θ) is completed.
When a function has been completed, a branch from the polling
subroutine to the next appropriate subroutine is performed. The
calling sequence is stored in a last-in-first-out stack. This
stack is simply an array containing integers, which are used as
arguments of a FORTRAN computed GO TO statement. The user can
interrupt the polling subroutine by typing an escape character.
This initiates a branch to the console handling subroutine.

     The various features of the AUTO system are selected from a
menu as shown in Table 1. In this example, option 0 (status) was
requested. The operator learned that diffractometer number 1 was
inactive, while diffractometer number 2 was collecting a powder
pattern. The data from unit 2 was being written to file CHO53D
and the run would be completed at 9:22 a.m. Other options in the
menu allow pattern collection, selected region data collection,
instrument control, and file manipulations.

     The selective region (menu option 2) data collection mode
is used for three methods of quantitative analysis, for profile
determination, for reference intensity ratio measurement, and for
relative intensity measurement. Because selected region data
collection is usually repeated for a number of similar samples,
the control information is stored in a run file. Creation of
this run file requires input of the beginning and ending angles
of each region and a label for the region. To collect data, the
user is required to input the name of the run file and labels for

---

*Equation 6 is equivalent to that of Jenkin and DeVries' fixed
time optimization equation (7), where NPK and NBG are 1.

the sample. The collected profile data points and pertinent
control information are written to a data file. The data file
and run file structure are very similar. Both are stored as
ASCII formatted records of 128 characters to facilitate copying,
editing, and transferring of these files.

An example of the use of run files for quantitative analysis
is shown in Figure 1. The first step involves determination of
the quantitative constants (e.g., Ipure or I/Ic) for each phase.
To collect this data, a run file is built, and the pure phase or
phase + internal standard is mounted on the diffractometer. AUTO
proceeds to collect the data and to write a data file for each
standard. Data files for all phases of the mixture to be analyzed are read by program RUNFIL, which determines the appropriate
analytical constants and writes a run file for controlling data
collection for any mixture of these phases. AUTO is run to
measure the intensity at each of the regions required for analysis of the mixture and writes a data file. The data file for an
unknown is analyzed by program QUANT for the weight fraction of
each unknown phase, using non-overlapped lines. If overlaps
occur, QUANT writes a file for further analysis by program COBRAG.

Table 1
AUTO CONTROL MENU

Enter one of the following options:

        0 - Status
        1 - Powder Pattern Collection
        2 - Quantitative Analysis
        4 - Set 2-Theta
        5 - Slew to 2-Theta
        6 - Terminate Run on a Unit
        7 - Panic Stop on Both Units
        A - Create a Run File
        C - List Disk Contents
        D - Debug
        E - Delete a File
        F - Transfer File
    <u>0</u>

            Status on 80/11/21 at 9:01:26

Unit:	1	2
File:		CH053D
Run:	INACTV	POWDER
To End at:	0/ 0 0: 0	11/21 9:22
2-Theta:	79.995	16.140

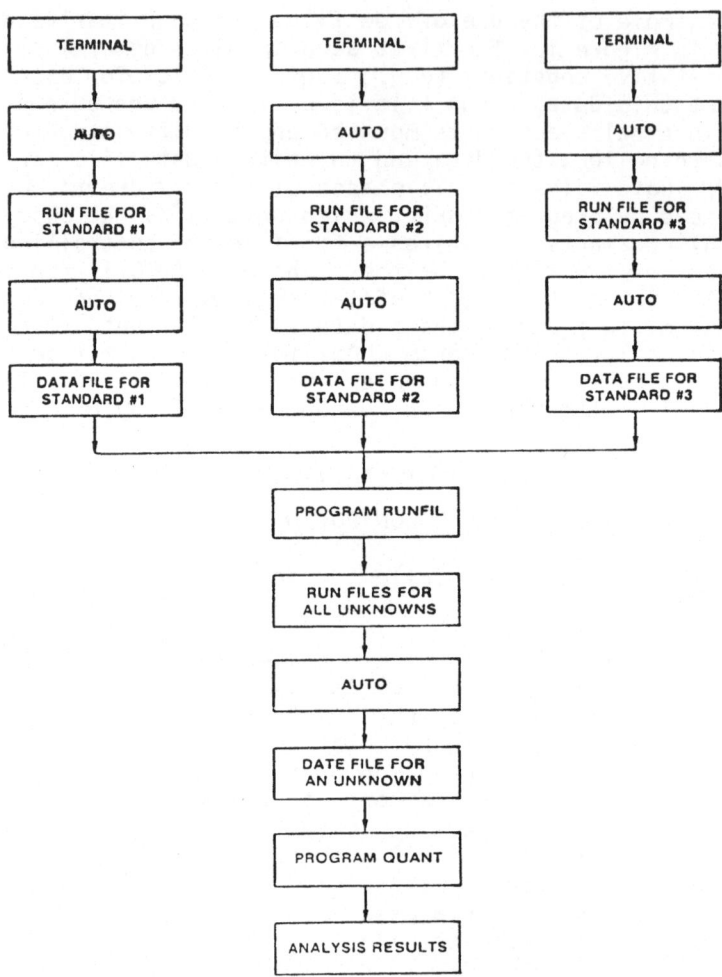

Fig. 1. Example of the use of run files in quantitative analysis.

DATA/RUN FILE STRUCTURE

The data file fully documents all aspects of the data collection and contains all flags and data required for analysis.
Appendix 1 contains the definition of each record in the selected
region data file. A run file contains record type 1 and pairs of
record types 7 and 8, one pair for each selected region.

Record type 1 provides sample, operator, and data documentation, as well as flags for file type, data purpose, analysis
method, whether elemental analysis was entered, and how many
complete repetitions and remountings of this sample were measured.
Record type 2 documents all aspects of the instrument, such as x-
ray target, take-off angle, slits, monochromator, and detector.
This data can be used to make corrections, such as for dead time
or for the shift in peak position due to physical and geometrical
aberrations.

External calibration constants for correction of instrumental and physical aberrations affecting peak positions are
contained in record type 3. Constants recorded in record 3 are
determined from analysis of the observed and expected peak positions of various standards by program CALIB. This program determines, by a least squares procedure, the constants of a polynom-
mial, up to fourth order, which describes $\Delta(2\theta)$ vs. 2θ. The
$\Delta(2\theta)$ to be added to an observed 2θ is given by:

$$\Delta(2\theta) = a_0 + a_1E + a_2E^2 + a_3E^3 + a_4E^4 ,$$

where E is 2θ minus the 2θ(offset) value (columns 7-16). The
2θ(offset) is employed in order to reduce correlations between
the polynomial coefficients. The value of using an external
calibration is shown in the next section. Record types 4 and 5
are used for propagation of error treatment.

The partial or complete elemental composition of the sample
can be saved in record type 6. These data are useful not only as
a sample identifier but can be used in phase analysis (8). The
program COBRAG implements the equations for internal standard
analysis with possible line overlap (9), as generalized for use
of reference intensity ratios (10,11). COBRAG also allows for
introduction of any elemental analysis observations into the list
of linear equations to be solved. The diffraction and elemental
analysis data are then solved for the weight fraction, X, of each
phase by least squares techniques with inequality constraints
($X_i \geq 0$, $X_i \leq 1$, $\Sigma X_i \leq 1$) and an equality constraint ($\Sigma X_i = 1$)
optional.

Record type 7 contains the control information for collection of data in the selected 2θ region. Columns 34-45 contain the maximum time and desired percent error for the optimization procedures. Record 7 also contains values needed for data analysis, such as the peak and background counting time, relative intensity (columns 70-75) and quantitative constant (columns 80-88) for the phase, and region label (columns 51-62). Estimated standard deviations of these values are also present for propagation of error analysis.

If several lines overlap within a given selected region, then name, hkl, and relative intensity values of each overlapping line are stored in record type 8. Up to three overlapping phases can be handled at present. Record 8 is also used for recording the theoretical elemental composition of the major phase (column 51-56, record 7).

Record type 9 contains the observed intensity data, as well as 2θ angle, time, and sample temperature information.

EXTERNAL CALIBRATION

Application of the external calibration $\Delta(2\theta)$ to observed peak 2θ values removes many of the systematic errors present in diffractometer data, such as due to α_1-α_2 splitting, axial divergence, and flat specimen. The sample displacement is the primary source of error uncorrected after external calibration. Records 4 and 5 contain the variance-covariance matrix elements for the five polynomial coefficients a_0-α_4. These values can be used for propagation of errors in peak positions.

Data sets for arsenic trioxide and quartz were collected using AUTO and analyzed by profile fitting a modified Lorentzian to each resolved peak. An internal standard, SRM640 Si powder (12), was used to obtain peak 2θ's free of instrumental and physical aberrations. A least-squares refinement of lattice parameters at 25 ± 1 °C, $\lambda(CuK\alpha_1)$ = 1.5405981 Å gave $\underline{a}$ = 11.0768(2) Å for AsO_3 and $\underline{a}$ = 4.9133(1) and c = 5.4052(3) Å for quartz. The Smith-Snyder figure of merit values of F_N (13) are given in Table 2 for the as observed, the external calibration corrected, and the internal standard corrected data. For both phases, the external calibration made a significant improvement in the accuracy of the data.

ANALYSIS PROGRAMS

Several data analysis programs using the run/data file structure discussed above are being written in FORTRAN 77. Included are programs for quantitative analysis (SPIKE, RUNFIL, QUANT, and

Table 2. $F_N(<|\Delta 2\theta|>, N_{poss})$ for Three 2θ
Calibration Methods

Method	Arsenic Trioxide N = 29	Quartz N = 30
None	9.9(.049,59)	16.4(.052,35)
External	15.4(.026,59)	30.0(.028,35)
Internal	42.0(.012,59)	66.1(.013,35)

COBRAG) for determination of profile shape, intensity, position, and external calibration (CALIB), for analysis of powder diffraction patterns (PDDF), for internal standard correction (PRCRFL), and for particle size and strain analysis (SIZE). Currently, each program is operational on a Univac 1108 computer. Status and details of these programs, as well as AUTO can be obtained from the authors.

DISCUSSION

The role of a minicomputer controlled diffractometer should be more than simple control of the stepping motor and counter. To minimize time and maximize statistical accuracy optimization procedures should be employed. To simplify operator interaction, run files and option menus are of great value. To permit complete documentation and subsequent analysis, a data file structure, which includes complete instrument configuration, external calibration, quantitative constants, and sample information, is required. Each of these have been incorporated in the second generation automated powder diffractometer control system, AUTO.

REFERENCES

1. Goehner, R. P. and Hatfield, W. T., Peak Search--A Program to Find Diffraction Peaks, Private Communication (1981).
2. Mallory, C. L. and Snyder, R. L., The Alfred University X-Ray Powder Diffraction Automation System, New York State College of Ceramics, Technical Paper No. 144.
3. Szabo, P., Optimization of Quantitative X-Ray Diffraction Analysis, J. Appl. Cryst., 13:479 (1980).
4. Cohen, J. B. and Schlosberg, W. H., Interpreting X-Ray Scattering, Industrial R/D, p. 105 (1978).

5. Snyder, R. L., Hubbard, C. R., and Pantiotopolous, N. C.,
    AUTO: A Real Time Diffractometer Control System,
    NBSIR 81-2229, NBS, Washington, D.C. (1981).

6. Savitsky, A. and Golay, M. J. E., Smoothing and Differen-
    tiation of Data by Simplified Least Squares Procedures,
    Anal. Chem. 36:1627 (1964).

7. Jenkins, R. and DeVries, J. L., "Practical X-Ray Spectro-
    metry", 2nd Ed., Springer-Verlag, NY (1972 p. 188.

8. Hubbard, C. R., Multicomponent Quantitative Analysis From X-
    Ray Diffraction and Fluorescence Data, Paper JN4, ACA
    Winter Mtg., Asilomar, CA (1977).

9. Copeland, L. E. and Bragg, R. H., Quantitative X-Ray Diffrac-
    tion Analysis, Anal. Chem.., 30:196 (1958).

10. Chung, F. H., Quantitative Interpretation of X-Ray Diffraction
    Patterns of Mixtures. I. Matrix-Flushing Method of Quan-
    titative Multicomponent Analysis, J. Appl. Cryst., 7:519
    (1974).

11. Hubbard, C. R., Evans, E. H., and Smith, D. K., The Reference
    Intensity Ratio, I/Ic, for Computer Simulated Powder
    Patterns, J. Appl. Cryst., 9:169 (1976).

12. Hubbard, C. R., Swanson, H. E., and Mauer, F. A., A Silicon
    Powder Diffraction Standard Reference Material, J. Appl.
    Cryst. 8:45 (1975).

13. Smith, G. S and Snyder, R. L., F_N: A Criterion for Rating
    Powder Diffraction Patterns and Evaluating the Relia-
    bility of Powder Pattern Indexing, J. Appl. Cryst.
    12:60 (1979).

Appendix 1

AUTO Run File and Output Data File Format
for Quantitative Analysis

Record 1: Problem Specification

```
Col   1- 80 Title of run (40A2)*
Col  81-100 Name of operator (10A2)
Col 101-106 Date of file creation YYMMDD (3I2)
Col 107-112 Time of file creation HHMMSS (3I2)
Col 113 IFLAG (1) = 0 Output data file from pattern collection
                  = 1 Output data file from quantitative analysis
                  = 2 Output data file from QUANT for a standard
                  = 3 Run file output from the data reduction
                      program on the off line computer, directing
                      the quantitative analysis of an unknown
                  = 4 Run file directing the determination of a
                      standard for quantitative analysis
                  = 5 Run file for profile data collection
                  = 6 Output data file of profiles
Col 114 IFLAG (2) = 1 Run to analyze an unknown
                  = 2 Run to measure standards
                  = 3 Run to measure relative intensities
Col 115 IFLAG (3) = 1 Internal standard method of analysis
                  = 2 Intensity ratio method of analysis
                  = 3 Spiking method of analysis
                  = 4 Relative intensity measurement
                  = 5 Profile measurement
Col 116 IFLAG (4) = 0 Elemental composition is not in file
                  = 1 Elemental composition is in file
Col 117 IFLAG (5) = N The number of repetitive runs for this
                      sample (all to be output in the same data
                      file)
                      (Default = 3)
Col 118 IFLAG (6) = N The number of mountings of each sample
Col 123-128 Name of run file controlling data collection
```

Record 2: Instrument Specification Parameters

```
Col  1- 4 The name of the diffractometer used for the run
Col  5-10 The date that the instrument specification (IS) file
          was created (e.g., 073080)
Col 11-12 Initials of person who created the IS file
Col 13-14 Element symbol of x-ray tube target (e.g., CU)
```

---

* Standard FORTRAN format specifications are given in parentheses.

Col 15- 16 Characteristic x-ray line used (e.g., KA = Kα)
Col 17- 22 X-ray wavelength (e.g., 1.5406)
Col 23- 24 Focal spot type (e.g., FF = fine focus, SF = standard focus)
Col 25- 28 Take off angle in degrees (e.g., 5.50)
Col 29- 30 Divergent slit type (TC = Theta compensating, FX = Fixed)
Col 31- 36 Divergent slit angle if Fixed slit or irradiated length in mm if Theta compensating (e.g., 12.500)
Col 37- 42 Goniometer circle radius in mm (e.g., 172.00)
Col 43- 46 Incident soller slit divergence angle in degrees
Col 47- 50 Receiving soller slit divergence in degrees
Col 51- 54 Receiving slit divergence in degrees (e.g., 0.10)
Col 55- 56 Monochromator code (GR = graphite, LF = lithium fluoride, blank = none)
Col 57- 58 Detector code (SC = scintillation)
Col 59- 62 Dead time in microseconds (e.g., 2.00)
Col 63- 66 Instrument intensity stability factor in percent
Col 67- 70 Time base for scalar (e.g., 0.01) in seconds
Col 71-128 Not used

Record 3:   Instrument Calibration Information

Col 1   Flat specimen correction (0 = not applied, 1 = applied)
Col 2   Axial divergence correction (0 = not applied, 1 = applied)
Col 3   Lorentz correction (0 = not applied, 1 = applied)
Col 7- 16 2θ-offset to be subtracted from the 2θ values before applying the polynomial correction (F10.3)
Col 17- 32 a_0 Y intercept of calibration polynomial (E16.6)
Col 33- 48 a_1 first order polynomial coefficient (E16.6)
Col 49- 64 a_2 second order polynomial coefficient (E16.6)
Col 65- 80 a_3 third order polynomial coefficient (E16.6)
Col 81- 96 a_4 fourth order polynomial coefficient (E16.6)

Records 4 and 5: Inverse Matrix from Instrument Calibration Run (8E16.6). The order is:

Record 4: (1,1), (1,2), (1,3), (1,4), (1,5), (2,2), (2,3), (2,4)
Record 5: (2,5), (3,3), (3,4), (3,5), (4,4), (4,5), (5,5)

Record 6: Elemental Composition. (Blank if IFLAG(4) = 0.)

Col 1- 2 Element symbol (A2)
Col 3- 7 Percent of element in unknown (F5.2)
Col 8- 12 Standard deviation of percentage (F5.2)
Col 13- 14 Element symbol number 2
Col 15- 19 Percent of element 2

Col   20- 24 Standard deviation of percentage

   .   .   .

   .   .   .

   .   .   .

Col 109-110 Element symbol number 10
Col 111-115 Percent of element 10
Col 116-120 Standard deviation of percentage

Record 7:   Line Parameters

Col   1- 7 2θ of low angle background of peak (F7.3)
Col   8- 14 2θ of peak maximum--used in peak count mode (F7.3)
Col   15- 21 2θ of high angle background of peak to be recorded
             (F7.3). In peak count mode this may be zero if only
             one background is desired.
Col   22- 28 Step width to be used in moving over the peak. This
             defaults to 0.005° (F7.3)
Col   29- 32 Rate of movement between points in deg/min (F4.1)
Col   33   Step and count method: 0 = independent, 1 = simul-
             taneous
Col   34- 40 Maximum time in minutes to be spent counting a 2θ
             region. In fixed time mode, this value will be opti-
             mally divided between background time and peak time.
             In fixed error count mode, this will be treated as an
             upper limit on the count time (F7.3)
Col   41- 45 The percent error to which a peak should be counted.
             If zero, fixed time count mode will be used (F5.2)
Col   46- 50 Maximum number of counts allowed in scaler-usually
             32700 (F5)
Col   51- 56 Name of the phase (3A2)
Col   57- 62 hkℓ of the line (3A2)
Col   63   = 1 This line belongs to the unknown phase being
                 analyzed
             = 2 This line belongs to the internal standard
             = 3 This line is an intensity reference belonging to a
                 special narrow slotted sample holder
             = 4 This area is from an amorphous halo
Col   64   = 0 This is a fully resolved line
             = N There are N more lines overlapped with this line
                 whose phase name, hkℓ and I(rel) values will be
                 supplied on Record 8
Col   65   = 0 Integrated step scan of the peak
             = 1 Peak height mode, locate position
             = 2 Peak height mode, do not locate position. Instead
                 use previous 2θ(obs)-2θ(input)
Col   66- 67 The number of points to be measured at the top of a
             peak in peak count mode or the number of points to be
             used in averaging at the peak to estimate peak inten-
             sity in peak scan mode (I2)

Col 68- 69 The number of points to count for BKGTIM seconds at
             the beginning and end of a peak (I2). (Zero if
             Col 63 = 4)
Col 70- 75 Relative intensity of the line (F6.1)
Col 76- 79 Standard deviation of the relative intensity (F4.1)
Col 80- 88 The standard intensity to be used in analyzing an un-
             known line. This is I(pure) or I(pure)/I(ref) or
             I(unk)/I(std), depending on the analysis method.
             This value usually computed by the analysis program
             QUANT (F9.4)
Col 89- 96 ESD of standard intensity (F8.4)
Col 97-104 Concentration of the added phase if IFLAG(3) = 3.
             Weight fraction of the internal standard if IFLAG(3) =
             1. Mass absorption coefficient of this sample if
             IFLAG(3) = 2 (F8.3)
Col 105-111 Count time at each peak point in seconds (F7.2)
Col 112-118 Count time at each background point in seconds (F7.2)
Col 119   = 1 If region counted in fixed time mode
            = 2 If region counted in fixed error mode
Col 120   = 0 (Integrated mode)
            = 1 Located peak within ± 0.09° 2θ of expected 2θ
            = 2 (located peak > 0.09° 2θ from expected 2θ
            = 3 Failed to locate peak
Col 121-128 Weight fraction of the standard in the sample (F8.4)

Record 8:   Line Overlap and Phase Chemistry Information
            (Record 8 will always be present but usually blank.
            Columns 1-48 will be entered only if Record 7 Col 64
            ≠ 0)

Col   1- 6 Name of second phase overlapping this line (3A2)
Col   7- 12 hkℓ of overlapping line (3A2)
Col 13- 17 I(rel) of overlapping line (F5.1)
Col 18- 24 Standard deviation of I(rel) or 2θ peak if IFLAG(3) =
             5, of second overlapping line (F7.3)
Col 25- 30 Name of third overlapping phase (3A2)
Col 31- 36 hkℓ of third overlapping line (3A2)
Col 37- 41 I(rel) of overlapping line (F5.1)
Col 42- 48 Standard deviation of I(rel) of third overlapping line

The following chemical information for the phase indicated in
Record 7 is only required if IFLAG(4) = 1 and Record 7 Col 63,
= 1.

Col 49- 50 Symbol of element 1 of phase (A2)
Col 51- 58 Percentage of element 1 in phase (F8.4)
Col 59- 60 Symbol of element 2 of phase

Col 61- 68 Percentage of element 2 in phase

. . .

. . .

. . .

Col 119-120 Symbol of element 8 of phase
Col 121-128 Percentage of element 8 in phase

Record 9: Intensity Data

Col 1- 7 The angle of the last data point in the record (F7.3)
Col 8- 11 The time the record was written (2I2)
Col 12- 16 100 times the temperature in °C (I5)
Col 17- 23 Counts at first point (I7)
Col 24- 30 Counts at second point (I7)

. . .

. . .

Col 122-128 Counts at sixteenth point (I7)

As many of these records as needed are included to record the number of counts observed at each point. A negative data point indicates that LIMCNT counts were detected, and the data value is the negative of ten times the number of seconds spent counting.

APPENDIX 2

Counting Time Optimization
(Derivation of Equations 3 and 4)

Let the estimated count rate averaged over the peak and at background be RO(pk) and RO(bg), respectively. Then the number of counts expected from integration is $NPK \cdot T(pk) \cdot RO(pk)$, to be measured in time $NPK \cdot T(pk)$, and the number of background counts expected is $(2)NBG \cdot T(bg) \cdot RO(bg)$ in time $(2)NBG \cdot T(bg)$. The total time is then given by eqn. 5. Let the ratio of peak to background time be:

$$R = \frac{NPK \cdot T(pk)}{(2)NBG \cdot T(bg)} \ . \qquad A2.1$$

The approximate net intensity and approximate estimated standard deviation based on RO(pk) and RO(bg) are:

$$I(net) = [NPK \cdot T(pk) \cdot RO(pk)] - R \cdot [(2)NBG \cdot T(bg) \cdot RO(bg)] \qquad A2.2$$

$$\sigma^2_{I(net)} = [NPK \cdot T(pk) \cdot RO(pk) + R^2[(2)NBG \cdot T(bg) \cdot RO(bg)]. \qquad A2.3$$

Two cases are of special interest. One is achieving a fixed relative error, E, in minimum counting time. The other is obtaining the lowest relative error in a fixed total time. In either case, one wishes to optimize the counting times T(pk) and T(bg) given estimates of the integrated and background count rates. The solution to the second case was presented by Jenkins and DeVries (7). The derivation of eqns. 3 and 4 for the first case are given below.

From eqns. 5 and A2.1, we see that:

$$NPK \cdot T(pk) = R \cdot T(tot)/(1 + R) \quad , \qquad \text{A2.4a}$$

and

$$(2)NBG \cdot T(bg) = T(tot)/(1 + R) \quad . \qquad \text{A2.4b}$$

Substituting these identities into the expression for the square of the relative error gives,

$$\frac{\sigma^2_{I(net)}}{I^2(net)} \equiv E^2 = \frac{[RO(pk) + R \cdot RO(bg)]}{\dfrac{R \cdot T(tot)}{1 + R} \cdot [RO(pk) - RO(bg)]^2} \quad . \qquad \text{A2.5}$$

Rearranging eqn. A2.5 yields T(tot) as a function of R and E. The optimum division of time between peak and background regions, in order to minimize T(tot), is when,

$$\frac{\partial T(tot)}{\partial R} = 0 = \frac{1 + R}{R} \cdot \left[\frac{RO(bg)}{E^2 \cdot [RO(pk) - RO(bg)]^2} \right]$$

$$- \frac{1}{R^2} \cdot \frac{RO(pk) + R \cdot RO(bg)}{E^2 \cdot [RO(pk) - RO(bg)]} \quad ;$$

which yields,

$$R^2 = RO(pk)/RO(bg) \quad . \qquad \text{A2.6}$$

Replacing $R \cdot T(tot)/(1 + R)$ in eqn. A2.5 with $NPK \cdot T(pk)$ and rearranging yields,

$$T(pk) = \frac{RO(pk) + RO(bg) \cdot R}{E^2 \cdot NPK \cdot [RO(pk) - RO(bg)]^2} \quad , \qquad \text{A2.7}$$

which is eqn. 3. T(pk) is the optimum time to be spent at each peak point in order to obtain the fixed relative error, E. Substituting eqn. 3 into eqn. A2.1 yields eqn. 4, the background time:

$$T(bg) = \frac{NPK \cdot T(pk)}{(2)NBG \cdot R} \quad . \qquad \text{A2.8}$$

APPLICATION OF THE MODIFIED SNYDER'S PROGRAM FOR THE DATA

PROCESSING OF AN AUTOMATED X-RAY POWDER DIFFRACTOMETER

G. Platbrood, J.M. Quitin, and H. Barten

LABORELEC (Laboratoire de l'Industrie Electrique)
BP 11
1640 Rhode-St-Genese, Belgium

INTRODUCTION

X-ray diffraction analysis is an excellent analytical tool. But if a certain quality of the results is needed or if solutions of analytical problems are to be obtained in a short time period, the X-ray diffractometer must be automated and the spectra reduced with dedicated algorithms.

In LABORELEC, three programs are principally used to solve the problems encountered in the X-ray diffraction analysis: a modified program given by R. L. Snyder;[1,10] the search/match G. G. Johnson program (last version);[2] and the POWD5 program.[3]

In 1978, R. L. Snyder and C. L. Mallory wrote a program able to process the data of digitized X-ray diffraction spectra. As they promote the software exchange between laboratories, the program has been requested and easily implemented on an SEL 32 computer (Systems Engineering Laboratories).

The program has the three following levels:

Level 1: Fortran analysis program
Level 2: Fortran interface translation subroutine
Level 3: Assembly language interface drivers

In LABORELEC, the level 1 program has been deeply modified. But the programs of levels 2 and 3 and the graphic software have not been used because the automated installation (Siemens diffractometer, Tektronix 4051, interface) and the graphic peripheric

261

(display Tektronix 4015, CALCOMP pen plotter) differ too much from the X-ray diffraction material and the computer of the N.Y.S. College of Ceramics.

AUTOMATED X-RAY POWDER DIFFRACTION

In LABORELEC, the analytical department is separated from the mathematics department. This one is also in connection with another IBM calculation center where the G. G. Johnson program is used in time sharing. The design proposed by R. L. Snyder has been followed and applied to the LABORELEC hardware installation (Fig. 1).

DATA REDUCTION

In the following paragraphs, only the points differing from the R. L. Snyder program are taken up.

The first data reduction is a smoothing using a least square moving average published by Savitsky and Golay (1967).[4] R. P. Goehner has followed the same procedure.[9] The subroutine allows

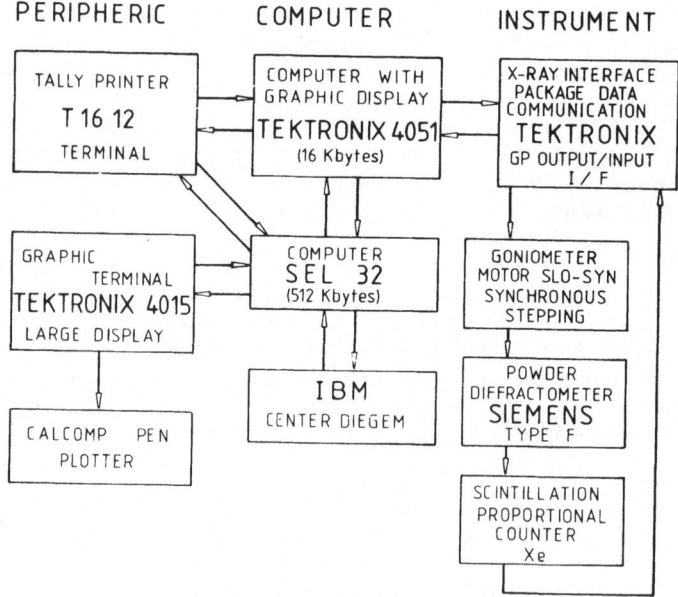

Fig. 1. Automated X-ray powder diffraction system

smoothing using 5, 7, 9, 11, 15, 17 or 25 points. When the intensities are recorded after a complete scan, they are displayed on the local Tektronix 4051 graphic screen. The analyst decides which smoothing shall be chosen later (7 for the usual spectra and 9-15 for the spectra with many background fluctuations).

The background level is calculated in each quarter degree span of a five degree section on the basis that the condition $(I_a + Ysigma$ sigma $(I_a) \geqslant I_b - Ysig$ sigma $(I_b))$ which defines background, must be satisfied for each point. The data I_a, I_b and I_x are the intensities at angles a, b and x. Ysig is a constant defined by the analyst and sigma $(I_x) = (I_x/counting\ time)^{1/2}$. When a second degree curve plot (also valid for a 1th degree function) fits the background, extrapolation outside the data cluster can be a source of problems when the accepted points are not regularly distributed on the 5 degree span. Thus, the background is subtracted only up to the last accepted point.

PRINCIPLE OF THE Kα2 ELIMINATION

The principle, but not the calculation, of the Kα2 elimination method developed by J. Ladell, A. Zagofsky and S. Pearlman[5] has been applied for the Kα2 elimination for the three targets: Cr, Co, Cu.

Experience has shown that if we want to eliminate completely the Kα2 line, the arms and weights have to be determined for the X-ray diffractometer working in the laboratory. Table 1 shows for the three targets the arms and weights necessary for a Kα2 elimination and calculated for the LABORELEC diffractometer. A 5 arms and weights selection seems to be a good choice.

Table 1

	ARMS			– WEIGHTS	
Cr	Co	Cu	Cr	Co	Cu
1.00136185	1.00201225	1.00238609	0.076038380	0.086960205	0.051996831
1.00147724	1.00209141	1.00243568	0.115941191	0.124022149	0.13128700
1.00170612	1.00270156	1.00248623	0.160371200	0.141401561	0.173987782
1.00182152	1.00225067	1.00253582	0.148888128	0.131615009	0.138863584
1.00193596	1.00232983	1.00258541	0.116641102	0.097448987	0.056705799

Two programs are necessary to obtain these parameters: one for the calculation of the peak parameters of a perfect separated doublet: SN.PARDI and another one for the arms and the weights calculation: HV.FOUR1 (Fourier transform program).

The shapes of the Kα1, Kα2 peaks are different. To calculate the Kα2 function from the Kα1 peak points, the half widths at half maximum for the two sides of the Kα1 and Kα2 doublet have been calculated, as well as the peak heights.

The SN.PARDI program calculates the four half widths at half maximum (HWHM) of the peaks of a good resolved Kα1, Kα2 alpha-quartz doublet.

A convolution product exists between the two observed line profiles; u_1 for the Kα1 and u_2 for the Kα2 line. δ is the α2, α1 displacement on the s scale ($s = 2\sin\theta$). $u_2(s) = h(s) * u_1(s-\delta)$.

The arms and weights calculation is based on a classical Fast Fourier Transform (FFT) algorithm using 64 points. In the Fourier transform procedure, a period is defined to calculate the Fourier coefficients of the Kα1 from a function synthesized with the experimental peak parameters (half width at half height, peak height).

The defined period starts from an abscissa equal to 5 times the time half width at half height value up to a point at which intensity equals the intensity of the starting point on the other peak side.

This period is divided into 63 parts and 64 intensity points which are directly used for the FFT calculation. The same period is kept for the Kα2 line. Thereafter, the Fourier coefficients ($U(k)$ = transform of $u(s)$) are calculated up to a fixed number of harmonics. The $U_2(k)/U_1(k)$ ratio is determined to calculate later the function $h(s)$ that shall be used in the convolution product.

In the program, the function $h = T(H)$ is calculated and plotted as a histogram. The histogram is normalized such that the sum of the ordinates is set equal to $\int u_2(s)\, ds\, /\, \int u_1(s-\delta)\, ds$. Now, 3 up to 7 pairs are selected in the histogram and used in the convolution product.

PEAK REFINEMENT

When the peak parameters of a peak group have been roughly evaluated by the second derivative of the polynomial fit using the Savitzky Golay smoothing[4] process, they have to be adjusted. Therefore a suitable algorithm Marquardt[6,7] determines 30 parameters (position, height, full width half height) of a 10 peak group. This slope-following method evaluates the first derivatives of the error function and uses this information to indicate how the parameters should be changed to minimize the error.

The function fitting the experimental curve and their derivatives has to be determined. Some limits can be fixed to discard meaningless parameter values after the iteration procedure. The limits of variation can so be defined: position (peak ± 1 degree), full width half maximum (maximum 1 degree), height (no negative peak height).

. Moreover, the Marquardt algorithm has been improved in order to optimize the length of the step for the variation of the parameters. There is termination of the subroutine calculation when the relative change in each parameter is less than $\varepsilon 1$,(0.05%) or when the relative change of the residual sum of squares is less than $\varepsilon 2$,(0.01%). When more than ten peaks constitute the peak group, only the ten greatest are taken into account for the peak parameter refinement. The cpu time is diminished by a factor three to four in comparison with the simplex algorithm.[1,8,10]

The program used in the diffraction analysis has been also modified and applied successfully for X-ray fluorescence spectra reduction. As the quality of the results has increased, the search/match Johnson program finds 80% of the phases encountered in LABORELEC after one year's experience and statistics.

REFERENCES

1. Chester L. Mallory and Robert L. Snyder, The control and processing of data from an automated X-ray powder diffractometer, in "Advances in X-ray Analysis," 22:121 (1978).
2. G. G. Johnson, "User Guide, Data Base and Search Manual," Pub. by JCPDS-International Centre for Diffraction Data, Swarthmore, PA.
3. Connie M. Clark, Deane K. Smith and Gerald G. Johnson, "A FORTRAN 4 program for calculating X-ray powder diffraction patterns." Version 5, The Pennsylvania State University.
4. A Savitzky and M. J. E. Golay, Smoothing and differentiation of data by simplified least squares procedures, Anal. Chem. 36:1622 (1964).
5. J. Ladell, A. Zagofsky and S. Pearlman, Cu Kα2 elimination algorithm, J. Appl. Cryst. 8:499 (1975).
6. D. W. Marquardt, An algorithm for least-square estimation of nonlinear parameters, J. Soc. Ind. Appl. Math. 11:431 (1963).
7. N. Janssens, Identification de parametres, LABORELEC (1980).
8. Richard W. Daniels, "An Introduction to Numerical Methods and Optimization Techniques," Elsevier North-Holland (1978).
9. Raymond P. Goehner, Specplot-an interactive data reduction and display program for spectral data, in "Advances in X-ray Analysis," 23:305 (1980).
10. Chester L. Mallory and Robert L. Snyder, New York College of Ceramics, Technical paper No. 144 (1979).

INDEX, A PROGRAM TO RECONCILE POWDER DIFFRACTOGRAMS

Tommy Hom, Ron Jenkins
Philips Electronic Instruments, Inc., 85 McKee Drive,
Mahwah, N. J. 07430

and

Joshua Ladell
Philips Laboratories, Briarcliff Manor, N.Y. 10510

ABSTRACT

This paper describes an interactive computer program, INDEX, which was written at Philips Laboratories as part of the software development effort for the APD 3600: Automated Powder Diffractometer. This program reconciles the differences between calculated and observed powder diffractogram "d-spacings" by unit cell parameters refinement and/or correction of observed lines for "significant" systematic errors. The reconciliation procedures, described below, yield a set of residual differences which may be due to x-ray photon statistics, random gear imperfections, or minor instrumental aberrations. Lattice parameters accuracies of 4-5 parts per 10^4 can be obtained routinely on powder diffractometers when systematic effects have been taken into account. The interactive nature of INDEX permits intervention at critical stages where it is most beneficial with a minimal number of keystrokes.

INTRODUCTION

Indexing of powder diagrams has gained added interest in recent years from the increased application of powder diffraction to the refinement or determination of crystal structures (1, 2, 3); especially for compounds whose thermodynamical properties prevent the growth of suitable sized crystals. Indexing is a preliminary procedure for the determination of crystal structures from powder diffraction; in order to eliminate non-analyte lines and correlate structure factors and diffraction lines.

The earliest suggested method of indexing (4) successively select low angle reflections to form powder zones; the intersection

of unique powder zones yield evaluated angles from which reciprocal
lattices are determined. Ito (5) rediscovered this "zone-indexing"
approach, and refinements were introduced later by deWolff (6) and
Visser (7). The success rate of the zone-indexing method is highest
for low symmetry materials. Other indexing techniques have been
introduced by Werner (8), Loüer & Loüer (9) and Shirley (10).
Shirley (11) has reviewed available indexing computer programs and has
indicated which indexing technique each employ. The above techniques assume no prior information concerning cell lattice parameters, and as such execution times in many incidences are prohibitive
even for modern high-speed main-frame computers. In many situations
some information is available; for example, in the areas of high/
low temperature studies, solid solution phases, isostructural
materials and residual strain measurements. INDEX performs most
efficiently in such indicated areas regardless of crystal symmetry.
INDEX is written in Fortran and executes on a Data General series
minicomputer.

RECONCILIATION OF POWDER DIFFRACTOGRAMS

    Measured angular positions of powder diffraction lines are
subject to random and systematic errors. In cases where lattice
parameters have been reported, the accuracy may be inferior to that
obtainable from powder diffractometry. Reconciliation of calculated
and experimental x-ray powder diffractograms may be achieved through
several analytical approaches. These procedures can be characterized as either (i) systematic error evaluation and correction, (ii)
least-squares cell parameters refinement or (iii) iterative combination of (i) and (ii), see Figure 1. The most significant errors

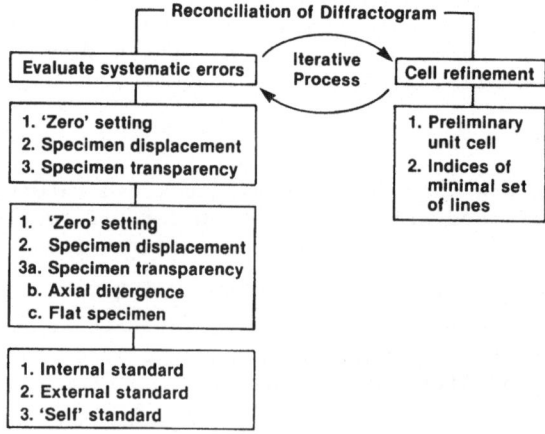

Fig. 1

encountered in a modern Bragg-Brentano parafocusing diffractometer
are 1) instrumental zero missetting, 2) specimen displacement from
the focusing circle, and 3) axial divergence, specimen transparency
and flat specimen (12). Parameters representing systematic errors
due to 1) and 2) can be refined as part of the reconciliation effort.

Calibration from external or internal standards is conveniently
handled by INDEX. The magnitudes of the systematic errors obtained
from an external calibration pattern can be imposed. For internal
standards, the analyte lines can be temporarily deleted and a
regression analysis performed on the remaining pattern to evaluate
the zero missetting , [DEL]; specimen displacement, [SD]; and a
slope parameter ($\sim$1.0), [EM], to take into account residual systematic errors. True two-theta is related to experimental two-theta by

$$2\theta(\text{true}) = [EM] * (2\theta\ (\text{exp.}) + K * [SD] * \cos\theta\ (\text{exp.}) + [DEL])$$

where,

K = constant inversely proportional to the diffractometer radius.

Least-squares refinement of lattices parameters for an experimental pattern free of systematic errors can start with either preliminary unit cell parameters or the indices of a minimal set of
reflections (six or less lines depending on the crystal class) as
input information. If the indices of a minimal set were entered, the
lattice parameters are evaluated for each successfully indexed line.
Candidate indices are selected when the difference between calculated and observed values falls within an error window. The error
window is enlarged for low intensity reflections to account for the
high uncertainty associated with poor counting statistics. The
best effective two-theta error window determined empirically is
given by:

eff. err. window = PA * (Io / I(i)) * exp (1 / NN)

where,

    PA   = experimental two-theta error window
    Io   = intensity of most intense line
    I(i) = intensity of i-th line
    NN   = Io / 1000

In case of multiple candidate indices, the best fitting index is
selected and the top ten (or less) candidates are listed.

Each indexed line is weighted according to the following
scheme, which assigns greatest weight to high intensity and high

angle reflections:

$$Weight(i) = I(i) / d(i)$$

The choice of weighting scheme is based upon the assumption that high two-theta lines are less sensitive to systematic errors and that lines of greater intensity are less subjected to random errors due to the photon statistics. Evaluation of the magnitudes of systematic errors should be limited to low angle lines to benefit from the high sensitivity to systematic effects and minimize the influence of "d" on the weighting scheme.

A regression analysis of the residuals is carried out by the program INDEX. The weighted least-squares procedure determines the parameters [EM], [SD], and [DEL] which minimizes the residual differences between calculated and observed two-thetas and maximizes the figure of merit (13) defined by:

$$FIGURE\ OF\ MERIT = n(obs.) / n(poss.) / \langle \Delta\ 2\theta \rangle$$

n(obs.)   = number of lines observed
n(poss.) = number of lines which could possibly have been observed
             consistent with the assigned space group
$\langle \Delta\ 2\theta \rangle$ = average difference between calculated and observed
             two-thetas.

When calibration standards are not available the analyte can serve as a "self" standard and through successive iterations alternating between systematic error analysis (after imposing the lattice) and least-squares refinement of lattice parameters a self-consistent set of parameters is obtained.

DATA ENTRY AND OUTPUT

Figure 2 illustrates an input dialogue with INDEX, where input responses have been underlined. The input data set, GOODQTZ, is an APD 3600 file generated by a peak search algorithm APDPEAK. Weak lines can be excluded from the analysis by imposing an intensity threshold. Corrections to the experimental pattern for axial divergence, specimen transparency and flat specimen are applied for an affirmative response to "default corrections?". A working list reports the line number, d star, d-spacing, experimental two-theta, two-theta corrected and peak height for the remaining lines to be used in the reconciliation. Any number of diffraction lines not necessary in a particular stage of the reconciliation may be deleted temporarily, whether the lines are due to the analyte, internal standard or impurity. The % error bias sets an upper limit in terms of Δ d/d for inclusion of a line in the regression analysis. Space

```
INDEX REV. 1.00
LEGEND: QUARTZ
DATE:  7/29/81   TIME: 15:13:15
WAVELENGTH= 1.540598
ANGULAR ERROR: DEGREES 2-THETA, PA=0.10
ENTER CRYSTAL CLASS: HEXAGONAL
ENTER DATA FILE NAME: GOODQTZ
DEFAULT CORRECTIONS? YES
CALIBRATED ZERO ANGLE= 0.00
LINEAR ABSORP. COEF.= 148.5

*SECF* CORRECTION: TRUE=[EM]*(2*THETA)+[DEL])
  DEL = 0.0
  EM = 1.0

SPEC DISPLACEMENT (MICRONS)= 100.0

MAX PEAK= 55743.  MIN PEAK=    30.
IGNORE PEAKS LESS THAN: 220.
WORK LIST? NO
DELETIONS? YES
SPECIFY LINE NUMBERS 53 61

THANKS
WORK LIST? NO
% ERROR BIAS= 0.5
IMPOSE LATTICE? NO

INPUT #1:  LINE # = 2
  H,K,L= 1 0 1

INPUT #2:  LINE # = 8
  H,K,L= 1 1 2

IMPOSE SPACE-GROUP LIMITATIONS? YES
ENTER SPACE-GROUP NUMBER =152
```

Fig. 2

group limitations of reflection types may be imposed by either entering the space group number (14) or the non-extinction conditions after entering a zero, if the space group symmetry is not standard.

Results of each analysis cycle include a record of input parameters, a summary of the reconciliation effort, indices found, apparent reciprocal lattice parameters and a report of the values for the figure of merit, [DEL], [SD], and [EM].

Test samples of well-crystallized materials giving rise to "complex" diffraction patterns show that accuracies of 4-5 parts per 10^4 in the measurement of lattice parameters can be routinely obtained on powder diffractometers when corrections for systematic effects have been made.

REFERENCES

(1) Rietveld, H.M., "A Profile Refinement Method for Nuclear and Magnetic structures", J. Appl. Cryst. 2, 65-71 (1969)

(2) Malmros, G. and Thomas, J.O., "Least-Squares Structure Refinement Based on Profile Analysis of Powder Film Intensity Data Measured on an Automatic Microdensitometer", J. Appl. Cryst. 10, 7-11 (1977)

(3) Young, R.A., Mackie, P.E. and von Dreele, R.B., "Application of the Pattern-Fitting Structure-Refinement Method to X-ray Powder Diffractometer Patterns", J. Appl. Cryst. 10, 262-269 (1977)

(4) Runge, C., "Analysis of Crystal Structures by X-rays", Z. Zeits 18, 509-513 (1917)

(5) Ito, T., "A General Powder X-ray Photography", Nature, 164, 755 (1949); "X-ray Studies on Polymorphism" Maruzen, Tokyo p. 187 (1950)

(6) deWolff, P.M., "A Simplified Criterion for the Reliability of a Powder Pattern Indexing", J. Appl. Cryst., 1, 108-113 (1968).

(7) Visser, J. W., "A Fully Automated Program for Finding the Unit Cell from Powder Data", J. Appl. Cryst., 2, 89-95 (1969)

(8) Werner, P.E., "Trial and Error Computer Method for Indexing of Unknown Powder Patterns", Z, Kristallogr., 120, 375-387 (1964)

(9) Louer, D. and Louer, M., "Methode d' Essais et Erreurs pour I'Idexation Automatique des Diagrammes de Poudre", J. Appl. Cryst., 5, 271-275 (1972).

(10) Shirley, R., "Recent Advances in Determining Unknown Cells from Powder Diffractometer Data", Acta Cryst., A31, S197 (1975)

(11) Shirley, R., "Indexing Powder Diagrams", Crystallographic Computing, Proceedings of 1978 Summer School, Delft Univ. Press and Oosthoeks.

(12) Schreiner, W. N., Surdukowski, C., Jenkins, R. and Villamizar, C., "Systematic and Random Diffractometer Errors Relevant to Phase Identification", in press (1981)

(13) Smith, G.S. and Snyder, R.L., "A Criterion for Rating Powder Diffraction Patterns and Evaluating the Reliability of Powder Pattern Indexing", J. Appl. Cryst. 12, 60 (1969)

(14) International Tables for X-ray Crystallography, Vol. 1, Kynoch Press

IDENT - A VERSATILE MICROFILE-BASED SYSTEM FOR FAST INTERACTIVE XRPD PHASE ANALYSIS

Barbara A. Jobst and Herbert E. Göbel

Forschungslaboratorien der Siemens AG

D 8000 München 83, West Germany

ABSTRACT

A minicomputer program system for phase identifications in x-ray powder diffraction is described. The routines written in FORTRAN comprise the users' special purpose standard-file management (creation of microfiles from the complete JCPDS-data base or own standards, as well as utility routines for corrections, supplements and bookkeeping) and the measured data processing, including raw data acceptance, data reduction by two selectable peak-search routines (trendoriented or 2nd derivative) and an exhaustive search/match procedure through selected microfiles. The data reduction and analysis programs have manifold graphical displays allowing manual interference. In one run up to five phases can be matched. The weight portion of every phase is estimated, using I/I (corundum)-values. The whole analysis (in automatic operation) consumes about 1 or 2 minutes of the processor time on a pdp 11/34 minicomputer. This matches well to the data collection times accessible for fast powder diffraction systems /1/.

INTRODUCTION

Crystalline phase analysis is one of the major applications of x-ray powder diffraction (XRPD). The capability of modern mini-computers in combination with an automated diffractometer allows the solution of this problem in a short time. Two approaches have been successfully applied: a Hanawalt SEARCH strategy in the full JCPDS-standard file of about 35000 compounds (program written by R. L. Snyder /2/) or an exhaustive IDENTification through specific users' microfiles. Both methods were incorporated into the program

273

system DIFFRAC /2/ for XRPD applications that is distributed by the Analytical X-Ray Division of the SIEMENS AG. While SEARCH is mainly useful for totally unknown samples as a "batch" program consuming up to half an hour of processing time, IDENT was developed as an interactive tool for identifications based on a reduced number of eligible standards. The answers are instantaneous and the inter-active mode, besides allowing adjustments and own proposals, mediates a detailed in-situ understanding of the analysis to the operator. In routine analyses IDENT can also operate automaticly. A general view of the functions of IDENT is shown in figure 1.

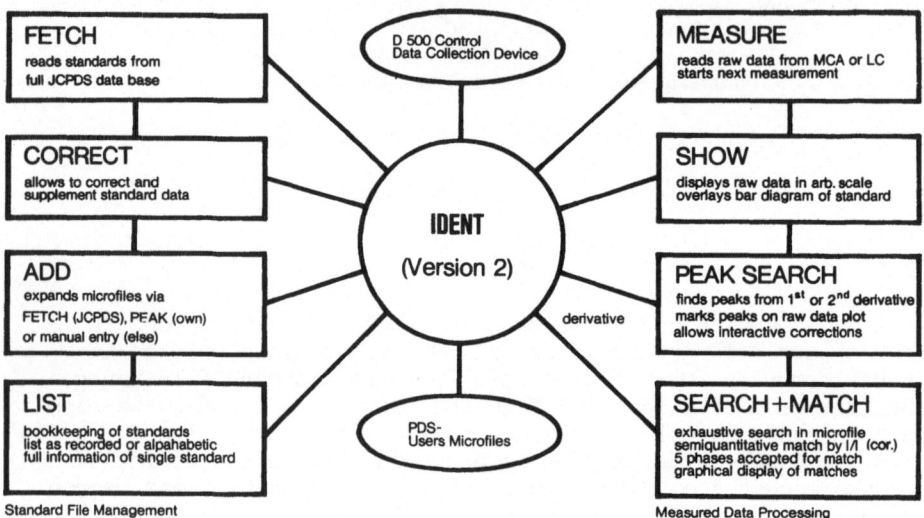

Fig. 1: The functions of IDENT

In this work the prototype program of IDENT is described as it was developed at the Materials Research Laboratory of the Siemens AG Munich. In the official version, distributed in DIFFRAC /2/, names of subroutines or parts of the handling may be altered, leaving the principles unchanged.

HARDWARE DESCRIPTION

The XRPD system installed at our laboratory is outlined in figure 2. The raw data are collected with a continuously scanning position-sensitive proportional counter mounted on an automated Siemens D 500 diffractometer. The method of data collection was described in /3/. From the 4096 channels of a multichannel counter (Canberra 8100) the patterns are transferred to a pdp 11/34 mini-computer (Digital Equipment) with 64 kwords of memory working

under RT 11 (V. 3b) operating system. As peripheral mass memories
two RL 01 hard discs with 5 MBytes of storage volume for each are
installed. Drive 1 contains the complete JCPDS standard data base
in a packed and rearranged form /2/ occupying about 2/3 of its
capacity for both the catalogue and the data. Drive 0 contains the
program files, raw data, d/I files and the users' microfiles.

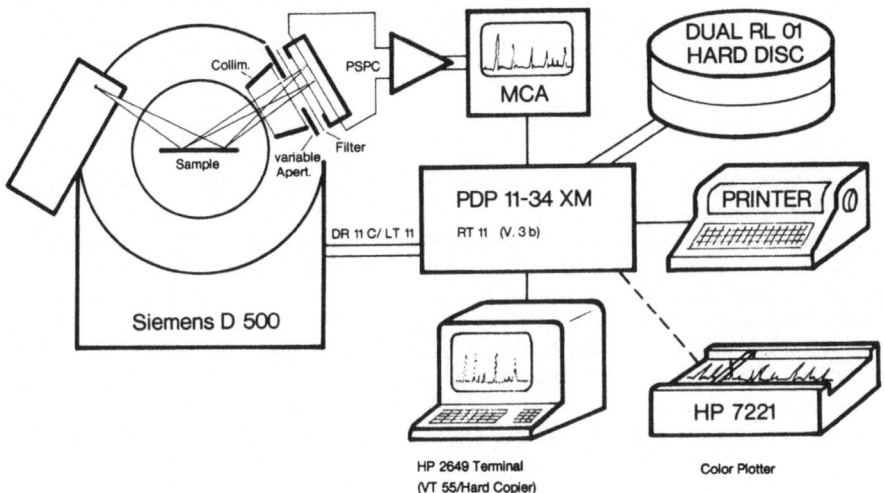

Fig. 2: Computer-aided XRDP

The system allows an easy separation between data collection
and data processing, i.e. while collecting a new pattern in the
MCA the computer is free for evaluating previous samples. In this
way a throughput of 20-30 samples per hour is routinely possible
with the automatic operation of IDENT.

MEASURE AND SHOW

The data acquisition system with a PSPC and multichannel
counter (MC) requests a low amount of software controls. Those are
the commands for clearing the MC-memory and start data collection
at the beginning, and readout the data at the end of a measurement.
Due to the high measuring speed of a position-sensitive detector
the diffractometer control can be reduced to a continuous $\vartheta/2\vartheta$-
scan at a fix speed over an angular range which corresponds to
the address range of 4096 channels of the MC and the selected
channel width of 0.01°, 0.02° or 0.05°. This operation can be
verified using a general-purpose DR11-C parallel interface from
DEC or the standard logic controller available for the D 500
diffractometer.

In addition to the raw data the MEASUR subroutine has to store
the set of experimental conditions like x-ray wavelength and
calibration curves for 2 Theta adjustment or intensity corrections.

The SHOW facility is mainly thought as a graphical display of
raw data or selected parts of a pattern in arbitrary linear scales
or a logarithmic intensity scale. The measured points can be overlayed by bar-diagrams of standards to monitor the line positions of
a questionable phase. On an extended 2 Theta scale also the line
positions of unknown and standard can be accurately compared.

PEAK SEARCH

For data reduction two different peak-search subroutines can be
linked to IDENT: A second-derivative peak-search program (SDRIP,
100 raw data points/second) finds peaks from minima of the second-
derivative of a set of raw data, applying the Savitzki-Golay /4/
digital filter technique for data smoothing and calculation the
second-derivative function. It strips the $K\alpha$ -contributions from
the raw data using a modification of the Rachinger-method /5/. The
subroutine SDRIP can be regarded as a condensed modification of the
automatic data reduction program ADR from C. Mallory and
R.L. Snyder /6/ and will therefore not be described here in detail.

Ten times faster than SDRIP is a trend-oriented peak-search
strategy (1000 raw data points per second). The program TOPS senses
peak areas from a changing trend of the background function. An
increasing trend stronger than a selected maximum background-slope
will mark the beginning of a peak area. A crest is expected exceed-
ing the background by at least three times its statical error.
After crest the end of the peak area is indicated if the decreasing
trend falls below the slope threshold.

A prior check allowing the preset maximum background slope has
to decide whether the peak ends on the brackground level or on an
elevated valley between two partly resolved peaks. If this is the
case the background is extrapolated horizontally. To recapture
the measured background level the slope test is continued until
crossing the measured data. This guarantees an uncritical behaviour
even in heavily overlapped peak areas.

Changes in the trend are easier to detect if a low raw data
density (about 2-3 points on the fwhm of a peak) is selected. Such
a low point density, however, is not acceptable if high accuracy
or details of the line profiles are requested. TOPS therefore uses
a reduced point density (weighted average of original data) for
the peak detection and the original point density to calculate the
peak parameters. The position and height are determined from a
regressive fit of a parabola through the upper half of the peak

area, the width is twice the smaller half width at half maximum (hwhm) on both sides of the crest. The integrated intensity is the sum of all measured net counts over the peak area. If Kα-radiation is used, the $K\alpha_2$-contributions are recognized from their relations (position, height) to the corresponding $K\alpha_1$-lines and can be eliminated from the d/I-list.

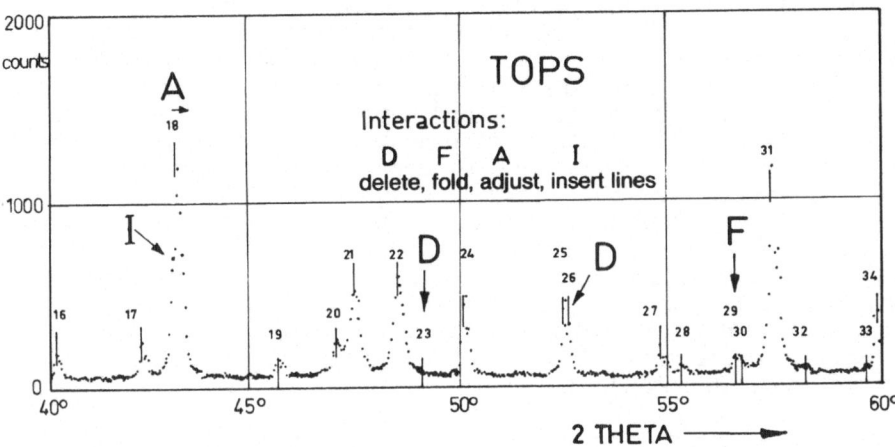

Fig. 3: Interactions to the results of the peak-search subroutines of IDENT

In the interactive operation of IDENT the results of the peak-search subroutines are monitored by markers on the graphical display of the raw data set. Manual corrections like deleting unwanted peaks (D), overlaying wrongly splitted peaks (F), adjusting peak positions (A) or inserting lines (I) missed by the peak-search routine can be made before further processing (fig. 3). The list of peaks found is finally corrected using an external-standard correction curve for the line positions and an intensity correction curve taking the geometrical beam conditions into account (variable detector window or secondary graphite mono-chromator). Both curves are established by the program CALIB from the DIFFRAC software system.

Both TOPS and SDRIP perform an accuracy of about 0.01° of 2 Theta for the location of unbroadened peaks routinely, which is about five times better than the truncation errors made by storing the standard patterns as integers in terms of 1000 Å/d on the JCPDS data base. So the results are accurate enough for phase analysis. If a higher accuracy is necessary a least-squares-profile fitting program FIT can be chained after IDENT using the peak-search results as start parameters. Further chainings of peak-search results are possible for unit cell refinements.

STANDARD FILE MANAGEMENT

A reduced data base has to be taylored for the problems under analysis. The powder diffraction standards manangement subroutine (PDSMAN) of IDENT provides a convenient transfer of standards from the full JCPDS-file (entered by the JCPDS number) to a selectable user file as well as adding standards from own measurements or literature data. Also calculated patterns can be entered. A check routine allows correction of every date of a standard and to add specific informations that are not contained in the JCPDS file like functional groups, or I/I(corundum) for quantitative estimates. The following sequence of information is stored for every standard: standard number, -name, up to 10 elements or functional groups, 3 strongest peaks, I/I(corundum), up to 50 d/I-pairs in the order of decreasing d-values. The d/I-pairs are written as two integers, d in terms of 10 000 Å/d and I in % relative to the strongest peak. Storing the d-values in terms of 10 000 Å/d, i.e. one digit more than in the JCPDS data file /7/ takes the high accuracy of computerized diffractometers into account. It is, however, meaningless if the files are created exclusively from the JCPDS data base. The length of a standard microfile may be arbitrarily expanded, a recommended limit was set to about 500 standards.

For the bookkeeping of the microfiles a directory of all compounds can be listed in alphabetic order of characteristic elements. Also, directories for compounds containing a selected element or combinations of elements can be requested as well as the full information of a single standard.

SEARCH/MATCH PHASE IDENTIFICATION

While data collection was mainly a technical problem, data reduction mainly a mathematical one, so the identification of

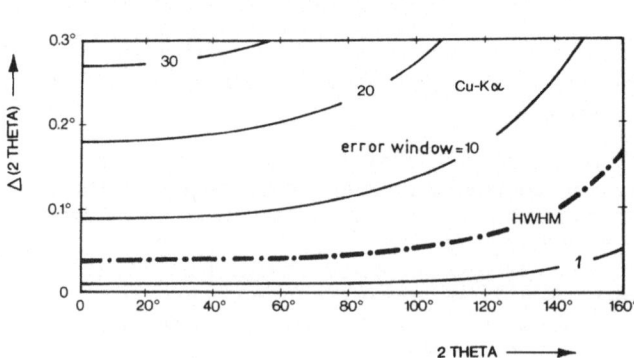

Fig. 4: Error windows for the line position match. The windows are in terms of 10 000 Å/d. The broken line, showing the half width at half maximum of un-broaded lines is about the envelope (error window 5) of truncation errors in the JCPDS file.

components in the unkown sample is based on experience. The problem
is that XRPD data are influenced by a number of circumstances like
line shifts by the formation of solid solutions, line broadenings,
different ways of sample preparation and data collection, preferred
orientations etc., altogether making no diffraction pattern look
exactly the same as a synthetic pattern from standard data. From
this reason an error window has to be tolerated within which line
positions should match. Still larger aberrations have to be
allowed for the intensity match making the definition of the
residual pattern after substraction of identified components
complicated.

The phase identification in IDENT is a two stage process. In
the search-part the strongest (up to 3) lines of every standard
are compared to the measured lines within the preset error window.
From the match of at least one of them a first estimate of the
portion is made giving an answer to the question of how many lines
could be detectable relative to the smallest measured peak. A
standard matching this first check and not being forbidden from
other reasons will be considered as a possible match and a figure-
of-merit is calculated on the base of all standard lines. All
possible matches are ranked after their figures-of-merit. The 25
best matches are saved. The lines of the best matching standard
are substracted from the measured pattern and a new order of the
remaining matches is calculated for the residual pattern. This
cycle is repeated four times maximum and is discontinued either
after the fifth accepted match or if the proposals of matches are
exhausted. The remaining residual pattern is listed in addition to
the identified phases. The weight portions of every phase are
estimated from the integrated intensities using I/I(corundum)
values. In standards with unknown I/I(corundum) a default value of
1 is used. Quantitative results therefore can be accepted only
seriously if all identified phases have known I/I(corundum) values.

The interactive operation of IDENT allows to refuse the
acceptance of proposed matches, make own proposals and adjust the
intensity match. For this purpose the measured raw data are
displayed on the graphics terminal and overlayed with a bar
diagram of the standard on trial. To visualize minor components
easier the raw data are plotted on a nonlinear intensity scale
(square-root of the counts). The example of an identification of
a 3-phase mixture is shown in figure 5.

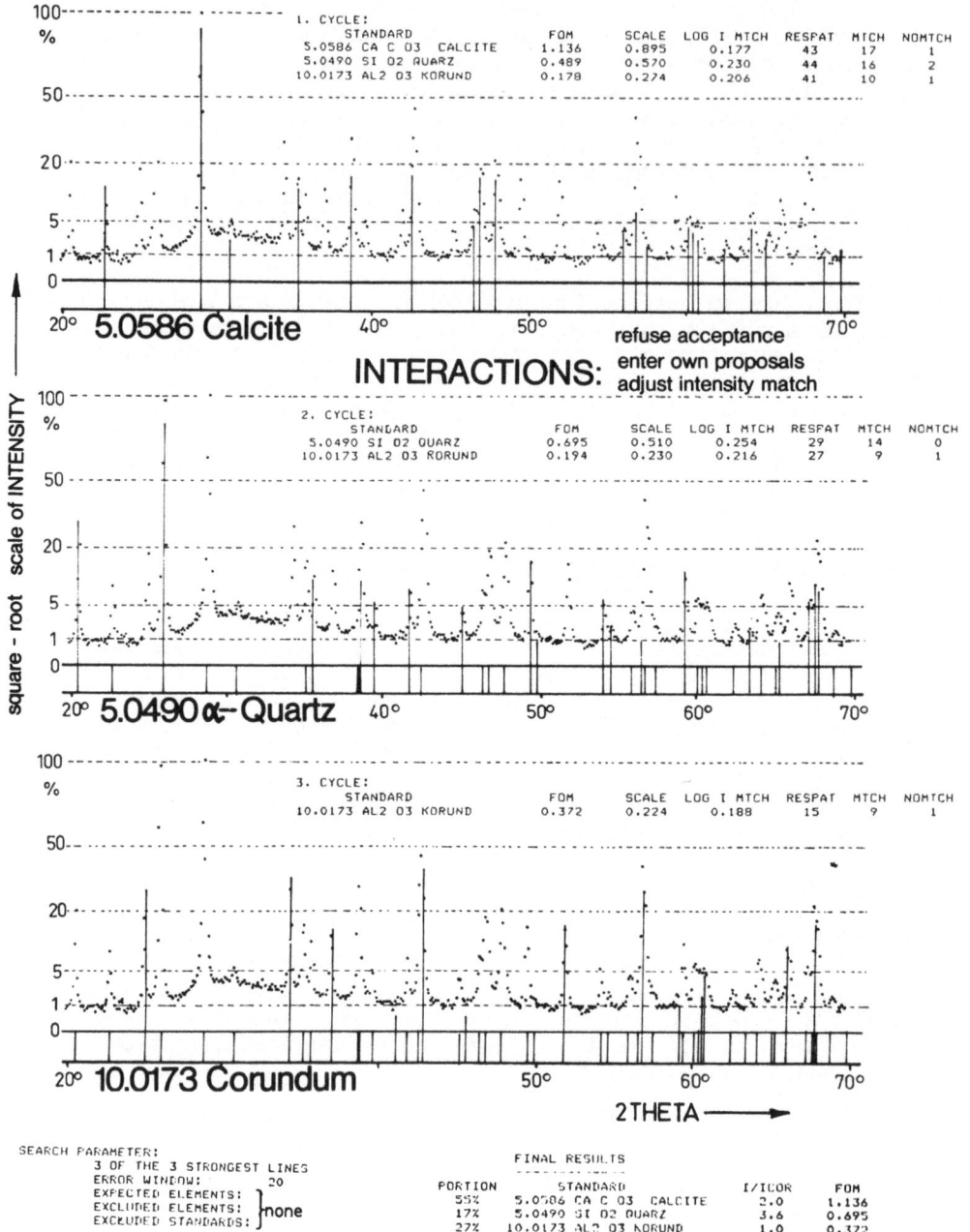

Fig. 5: Search/match analysis

The central point of the match process is the calculation of the figure-of-merit (FOM). The formula used in IDENT is:

$$FOM = CHEM \cdot \frac{SCALE}{0.1 + LOGIM} \cdot \frac{MTCH}{RESPAT} \cdot \frac{MTCH-NOMTCH}{MTCH+NOMTCH} > 0.01$$

with CHEM
= 1 (without any information about chemistry),
= 2 (at least one chemical element of the sample is known)
= 3 (more than one chemical element is known)

SCALE = concentration of the unknown, calculated from the comparison of the intensities of all matches lines.

$$LOGIM = \frac{1}{MTCH+NOMTCH} \sum_{1}^{MTCH+NOMTCH} \log \frac{SCALE \cdot I(stand.)}{I(meas.)}$$

MTCH = number of matching lines

NOMTCH = number of unmatches lines

RESPAT = number of lines in the unknown (residual) pattern.

The strongest factor in this formula is the last one, requiring more than 50 % of the detectable standard lines to match. The FOM range between 0 and more than 1. The threshold of 0.01 was only set as a figure larger than 0. It can be raised if with growing microfiles the number of proposals will become too large. The other factors have the following purpose: CHEM shall give a bonus to standards with the right chemistry. SCALE and LOGIM are measures for the portion of a standard in the unknown and of the correspondence of intensities between standard and measurement. To attenuate the influence of overlaps and preferred orientations or to compensate different apparatus functions, a regressive linear approach of I_{meas}/I_{stand} against 2 Theta is built up from the I_{meas}/I_{stand} values of all matched lines. The value of SCALE is taken at the position of the strongest line to provide the correlation with $I/I_{corundum}$ for the final estimate of weight percentage. LOGIM is a measure for the intensity match. An ideal match would be 0, so 0.1 is added to prevent a division through zero. LOGIM = 0.1 characterizes a good match. The factor MTCH/RESPAT should prefer those standards with the largest number of lines matching the unknown.

This way of calculating the figure-of-merit proved a high reliability in finding the right, and only the right answers to an analytical problem. The practical performance at many places and in many applications will have to demonstrate its success for the future.

CONCLUSIONS

The program IDENT represents a reliable tool for a no-delay phase analysis on the base of preselected users' standard files. Together with fast raw-data collection systems /1/ samples can be analyzed in a two to five minutes timing using the complete angular range. The interactive operation does not only provide a manual interference but, with the extensive graphical displays, also helps the operator to understand and accept the results in real time of the measurements.

The higher accuracy of standard data storage (10 000 Å/d) is a first step for identifying patterns with systematic line shifts caused, for example, by the formation of solid solutions.

ACKNOWLEDGEMENTS

The authors want to thank D. W. Beard and C. Harkins (Siemens Corporation) as well as users of DIFFRAC for critical comments on version 1 of IDENT. Stimulating discussions with G. G. Johnson from Pennsylvania State University and R. L. Snyder from Alfred University helped to accomplish the success of version 2.

REFERENCES

/1/ H. E. Goebel: Adv. in X-Ray Anal. 24, 123–138 (1981)

/2/ Manual of the Siemens DIFFRAC System (1980), edited by D. W. Beard and C. Harkins

/3/ H. E. Goebel: Adv. in X-Ray Anal. 22, 255–265 (1981)

/4/ A. Savitzki, M. J. E. Golay: Anal. Chem. 36, 1627–1639 (1964)

/5/ R. Delhez, E. J. Mittemeijer: J. Appl. Cryst. 8, 609–611 (1975)

/6/ C. L. Mallory, R. L. Snyder: Adv. in X-ray Analysis 22, 121–131 (1979)

/7/ G. G. Johnson, V. Vand: Industr. and Eng. Chem. 59/8, 18 ff. (1967)

COMPLETE QUANTITATIVE ANALYSIS USING BOTH X-RAY FLUORESCENCE AND

X-RAY DIFFRACTION

M. F. Garbauskas and R. P. Goehner

General Electric Company
Corporate Research and Development Center
Schenectady, New York   12301

The advantages of using elemental information to assist in the interpretation of x-ray diffraction (XRD) data are well known and many of the modern search/match algorithms utilize chemical information to make the procedure faster and more reliable (1). Elemental data can also be used to facilitate the quantitation of phases identified by x-ray diffraction, in some cases eliminating the need for performing the quantitation based upon the diffraction pattern. There are certain advantages to this approach. The first is that certain difficulties encountered in XRD quantitation such as line overlap, particle statistics, preferred orientation, microabsorption, extinction, and lack of pure standards are alleviated. The second is that the elemental analysis is, as a rule, easier and more sensitive than the diffraction technique. It therefore should be possible to increase the ease and the accuracy of quantitative phase analysis by utilizing the chemical data available. Since most materials submitted for XRD analysis are in form acceptable for x-ray fluorescence (XRF) analysis, this is the quantitative elemental technique of choice. However, it should be noted that other elemental analysis techniques can be used instead of XRF to furnish the chemical analysis.

The general outline of the quantitative phase analysis scheme is currently in the form of a series of interactive computer programs. Qualitative elemental data is combined with digitized XRD data to provide the basis for a search of the JCPDS file and the identification of the phases present in a material. In the case of elementally distinguishable phases, the qualitative phase identification is combined with quantitative elemental data obtained from either energy-dispersive or wavelength-dispersive

XRF. By using the fundamental parameters correction program NRLXRF
(2), quantitative phase analysis can be obtained directly. In
the case where some of the phases cannot be distinguished elementally, relative ratios of these phases can be determined using
appropriate XRD quantitation techniques (3) and the results combined with the NRLXRF results to provide the complete quantitative
analysis of the material. This approach has been applied to several
cases where complete phase analysis was desired.

Synthetic Corrosion Product

As part of an internal round-robin on corrosion product analy-
sis, a mixture of Fe_2O_3, Fe_3O_4, MnO_2, NiO, Cr_2O_3, CuO, ZnO, and
Co_3O_4 was analyzed. The diffraction pattern obtained from this
mixture is presented in Figure 1 and illustrates clearly the line
overlap problem that can occur in XRD with complex mixtures.
The relative positions of the diffraction lines from some of the
phases present are indicated below the diffraction pattern. The
lines from the minor phases are difficult to detect because of
the overlaps and their presence in the mixture is difficult to
confirm using only XRD. The (104) line of Fe_2O_3 and the (220)
line of Fe_3O_4 are separated well enough to obtain a relative ratio
for these two phases. The (104) Fe_2O_3 line overlaps the (104)
line of the Cr_2O_3, but in this case, the error introduced by the
overlap is minor due to the low concentration of Cr_2O_3 in the
mixture.

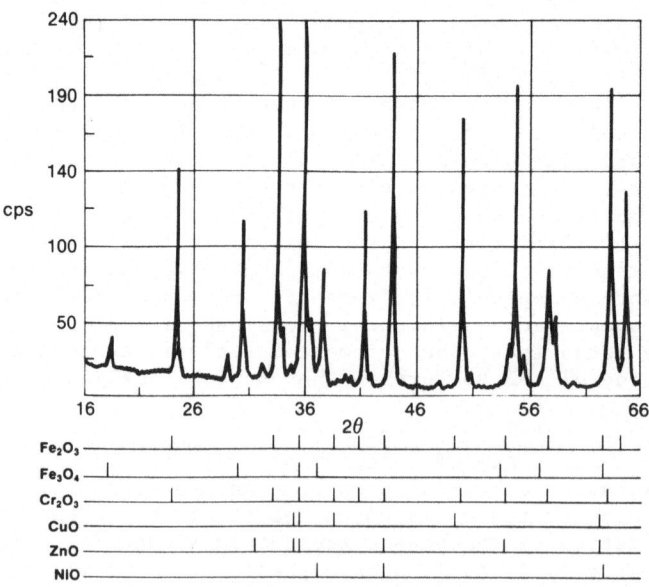

Figure 1: XRD Pattern of Synthetic Corrosion Product

Quantitative XRF analysis of the mixture was performed using an energy-dispersive system with pure element standards. The K-ratios obtained in this way could be used with NRLXRF for the final analysis.

When no phase information was used, very poor elemental analysis was obtained, illustrating that even oxygen can have a significant effect when present in the matrix (see Table 1). However, when the phase information was available, NRLXRF plus the Fe_2O_3/Fe_3O_4 ratio obtained from XRD gave results in relatively good agreement with the expected values (see Table 2). It was later discovered that sample inhomogeneity was present in the mixture and could account for some of the variation observed in the final analysis. This example illustrates the way in which the elemental analysis is dependent upon good phase information and how it in turn can be used to provide quantitative phase analysis.

Table 1: XRF ANALYSIS OF CORROSION PRODUCT WITH NO
PHASE INFORMATION AVAILABLE

	Calculated	Expected
Ni	4.49%	3.27%
Cr	2.52	1.64
Cu	0.62	1.00
Co	1.35	0.74
Mn	0.90	0.82
Zn	1.89	1.25
Fe	81.21	62.6
	92.99	71.3

Table 2: XRF ANALYSIS OF CORROSION PRODUCT WITH
PHASE INFORMATION AVAILABLE

	Calculated	Expected
NiO	4.25%	4.16%
Cr_2O_3	2.76	2.40
CuO	0.58	1.25
Co_3O_4	1.36	1.02
MnO_2	1.08	1.30
ZnO	1.74	1.6
Fe_2O_3*	52.97	53.54
Fe_3O_4*	34.40	34.78

*Fe_2O_3/Fe_3O_4 ratio obtained from XRD

NBS Limestone

As a second case, an attempt was made to characterize the mineral content in a sample of limestone which had been certified elementally by NBS. Qualitative elemental analysis indicated the presence of large amounts of Ca, Si, Al, K, and Fe. XRD analysis indicated that the major phases present are calcite, quartz, muscovite, and clinochlore ferroan. In this mixture, Si is present in every phase except calcite. However, the presence of K in the muscovite phase and Fe in the clinochlore ferroan phase enables the quantitation of this mixture without the need for any quantitative XRD measurements. A quantitative energy-dispersive XRF analysis with pure element standards using NRLXRF provided the analysis presented in Table 3. This analysis was elementally consistent with the certified elemental data (see Table 4). Errors present in the elemental data may be due to trace phases present that cannot be easily detected in the diffraction pattern. The results for the calcite and quartz phases were verified using XRD and an internal standard technique.

Table 3:   QUANTITATIVE PHASE ANALYSIS OF NBS LIMESTONE

Calcite 73%
Quartz 6%
Muscovite 11%
Clinochlore Ferroan 10%

Table 4:   ELEMENTAL CONSISTENCY OF QUANTITATIVE PHASE
ANALYSIS AND CERTIFIED COMPOSITION OF NBS LIMESTONE

	From Phase Analysis	NBS Certified
Ca	29.3	29.53
Si	6.5	6.60
Al	2.7	2.20
Fe	1.7	1.14
Mg	1.5	1.32
K	1.1	0.59

Mullite

The final example illustrates how a coupling of the XRD and XRF data can be used to yield an analysis for a solid solution. Mullite is a ceramic material that is nominally 60 mole percent Al_2O_3 and 40 mole percent SiO_2. However, the material can easily vary from 60-68% Al_2O_3. This change in composition can be observed in XRD as a change in the â lattice parameter of the material. By measuring the â lattice parameter for a series of mullites and then analyzing the material using XRF for Al_2O_3 and SiO_2,

a linear calibration curve of lattice parameter with composition
has been obtained (see Figure 2). Quantitation of the amount
of mullite present in a mixture can now be obtained without having
to make assumptions about the behavior of the solid solution relative to its end members. Once the stoichiometry of the phase
is determined from the â lattice parameter, XRF can be used to
provide the quantitation.

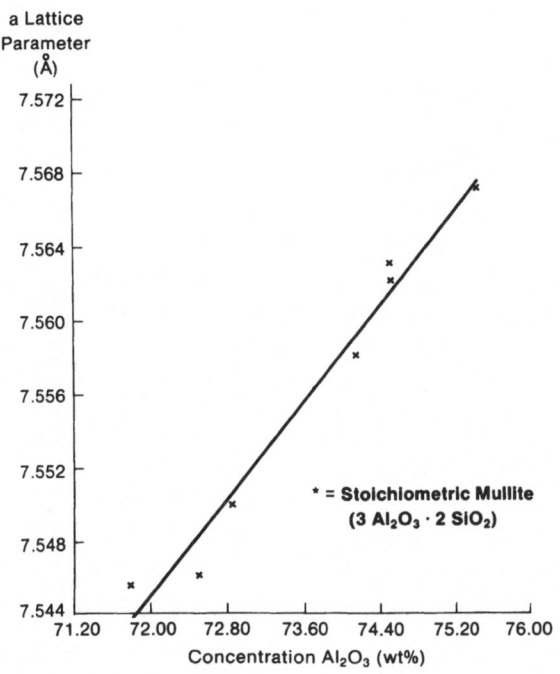

Figure 2:   Calibration of Mullite â Lattice
Parameter with Composition

Conclusions

It has been demonstrated that elemental information can be
coupled with XRD to provide quantitative phase identification.
Currently, the method employs a series of interactive computer
programs to acquire and process the XRD and XRF data. It is hoped
that this type of analysis will be available in a single program
in the near future. With both automated diffractometers and spectrometers interfaced to a single computer system, the potential
for completely automated phase quantitation is present today.

References

1.   G. J. McCarthy and G. G. Johnson, Jr., Adv. in X-Ray Analy-
      sis, 22, 109 (1979).
2.   J. W. Criss, L. S. Birks, J. V. Gilfrich, Anal. Chem. 50,
      33 (1978).
3.   H. P. Klug and L. E. Alexander, X-Ray Diffraction Procedures
      for Polycrystalline and Amorphous Materials, 2nd Ed., Wiley
      Interscience, New York, (1974).

CALIBRATION OF THE DIFFRACTOMETER AT LOW VALUES OF TWO THETA

R. Jenkins, T. Hom, and C. Villamizar
Philips Electronic Instruments, Inc., Mahwah, N.J. 07456

and

W. N. Schreiner
Philips Laboratories, Briarcliff Manor, N.Y. 10510

INTRODUCTION

A set of experimentally obtained "d" values is subject to a variety of random and systematic errors, some of which are inherent to a given diffractometer configuration and some of which may result from incorrect alignment of the diffractometer or technique in establishing peak positions and subsequent calculation of "d" values. In a previous paper (1) we have discussed the problems involved in the identification and control of errors in the computer controlled diffractometer and in that paper we indicated that it is useful to differentiate between different types of two-theta values. Firstly the Theoretical 2θ values are those values which are dependant only on the size and distribution of atoms in the unit cell of the phase. The Practical 2θ values on the other hand result when aberrations inherent in a given diffractometer geometry are convoluted with the Theoretical 2θ values. In terms of their typical magnitudes the most important of these aberrations in decreasing order are: Specimen Displacement, Axial Divergence, Flat Specimen and Transparency Errors. Other inherent errors include refraction and the effects of focal line and receiving slit widths etc., but these latter effects are generally considered small. At the second level, Experimental 2θ values also depend on misalignment errors. The third level errors accrue in the conversion of the experimental 2θ value to "d" spacing. These errors may arise simply from round-off errors or from more fundamental reasons arising from the polychromatic nature of the source and uncertainties in the wavelength of the diffracted radiation.

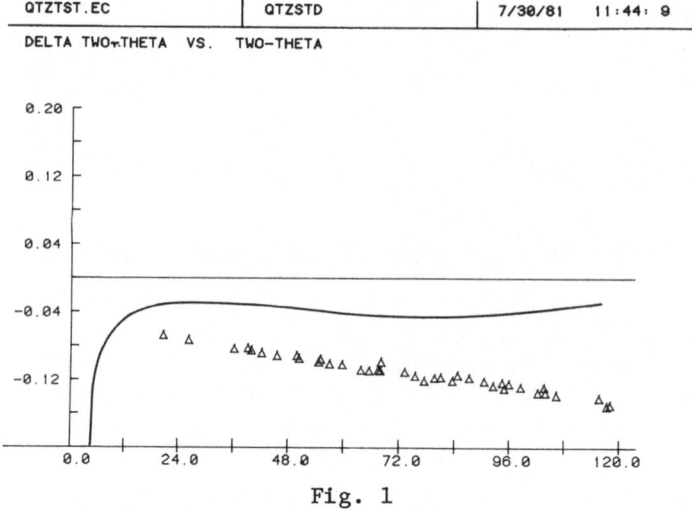

Fig. 1

        As we have indicated previously, (2) we believe that the
currently used correction algorithms do not reflect the geometry of
the modern diffractometer. Generally, the error functions currently
employed refer to the center of gravity of the intensity distribution
but in practice most modern computer controlled diffractometers now
utilize a second derivative algorithm and report peak locations
instead. As shown in Figure 1, the currently employed correction
algorithms predict an "S" shaped curve which has a sharp fall-off at
low two-theta values. Also indicated in the figure, however, are
experimental data points taken on a surface ground novaculite
specimen and clearly, the experimental peak data are different. We
have recalculated the contribution of the axial divergence term to
the theoretical errors and found a shape that is more similar to the
data shown. Note especially that the error at low angles is in the
opposite sense to that previously predicted.

        Two problems occur in practice, the first concerns the
measurement of an error curve on the diffractometer to establish
the integrity of alignment, and the second relates to the need to
correct one's experimental data for geometrical errors, plus additional errors which may relate to the sample itself, e.g. a specimen
displacement error. The first of these two problems is generally
managed with an external instrument standard and the second with an
internal standard. In our own laboratories we typically use a surface ground novaculite (α-quartz) specimen as the external instrument
standard and the NBS SRM-640 Silicon powder as the internal standard.
Such a procedure is quite successful for diffraction angles down to
20 degrees or so but is not good in the low angle regions. Difficulties often occur in recording calibration data at low 2θ values
because of the problem in finding material of suitably large "d"

value. As an example, the silicon SRM gives its first line at about
28 degrees. This is unfortunate because the low angle region is
one in which the systematic errors are large and in which if uncorrected, result in very poor "d" values. Reasonably accurate
values of "d" are, of course, required both for computer search
matching and cell indexing. Misalignment errors usually occur in the
form of a zero angle calibration error or a missetting of the $2\theta/\theta$
axes. The zero angle error introduces an error in "d" in terms of
$Cot\theta$ and hence tends to be large at low 2θ i.e. large "d" values.
Unless it is particularly bad, the $2\theta/\theta$ misalignment does not
markedly affect the accuracy of the measured "d", since the major
effect of this aberration causes asymmetric broadening of the diffraction profile which has only a minor effect on shifting the peak
position. In practice, either or both of these effects can be sufficiently minimized by careful alignment of the diffractometer.

CALIBRATION OF THE POWDER DIFFRACTOMETER AT LOW TWO-THETA VALUES

      We have recently been investigating various classes of
materials which might be used as low angle calibration standards and
two types seem to offer some promise.   One class of materials that
has been suggested as internal standards is the group one acid
phthalates (e.g. salts of NH_4+, K+, Rb+ and Tl+). These materials
are orthorhombic and give an intense (010) reflection at about 26A.
Figure 2 shows the diffraction diagram of thallium acid phthalate

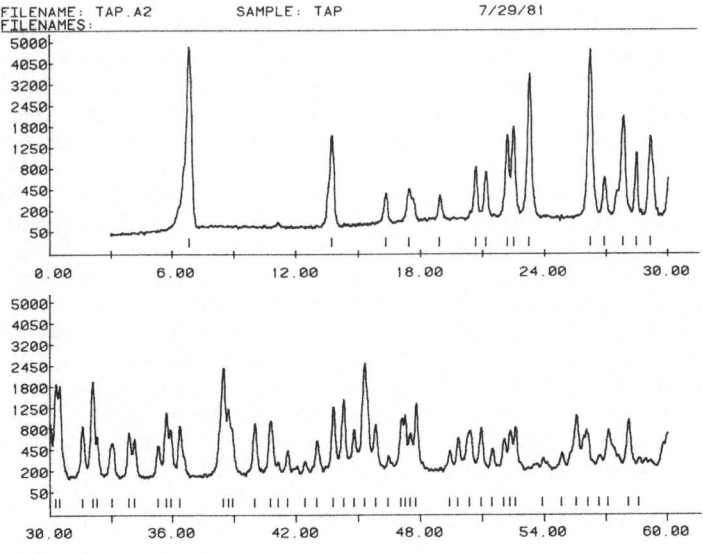

Fig. 2

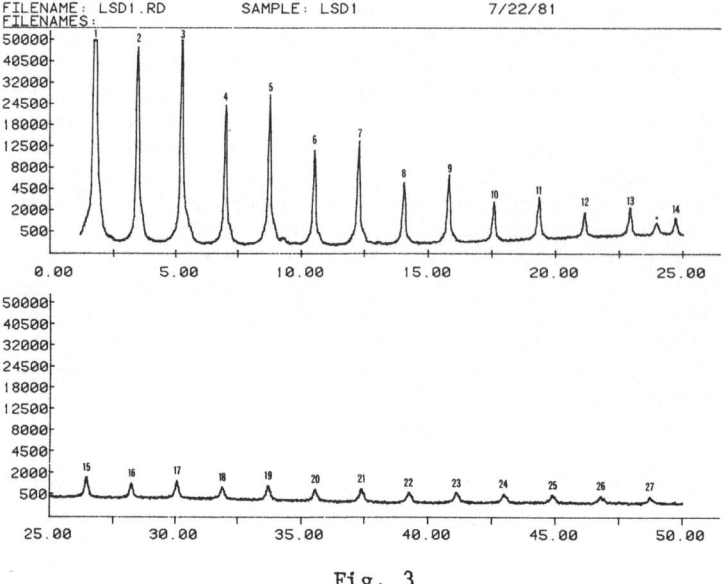

Fig. 3

and indicates that indeed the diagram does include a couple of strong reflections in the low angle region. Thus, in principle, it could be used as a wide range instrument calibration standard. Unfortunately it gives rather too many lines in the mid-angular range to be of much use as an internal standard.

A second, and what looks like a particularly promising group of calibration standard materials is the heavy metal stearates typified by lead stearate. These materials have been used for a number of years as analyzing crystals for long wavelength x-ray spectrometry and are available as thin layers on glass substrates. As such they are ideally suited as external calibration standards since they can be mounted directly in the diffractometer. The "d" spacing of lead stearate is about 50A and with a well aligned diffractometer one is able to observe about 30 harmonics. As shown in Figure 3 reflections range from about 2-50 degrees. Just one additional reflection is also seen this occurring at about 23.5 degrees. We have used these stearates to check diffractometer calibration down to very low two-theta values and a delta two-theta curve which includes both stearate and quartz data confirms our earlier postulate about the shape of the error curve. In practice one has to employ two separate standards and two separate measurements to establish this curve and this is rather inconvenient.

One possible approach to avoid this problem is to fabricate a composite standard of both silicon and lead stearate by deposition of a stearate layer onto the surface of the pressed silicon powder.

The art of deposition of stearate layers is well known in the field
of x-ray analysis since the classic work of Henke (3) in the late
1960's.   In trial runs we have successfully deposited about 250 layers
of lead stearate decanoate onto silicon using the Langmuir-Bloggett
dipping method (4) and have obtained very stable composite crystals.
Figure 4 shows a diffractogram of such a composite crystal and
clearly shows both lead stearate and silicon lines.   The additional
displacement of the silicon specimen below the focusing circle
because of the stearate layer is a relatively insignificant effect
since this displacement is only of the order of 0.5 microns.   The
relative intensity of the two contributions can be controlled by the
dipping process and the particular example was obtained with about
100 monolayers.   The usefulness of the NBS silicon SRM as an internal standard could be greatly enhanced if one could overcome the
lack of low angle information, the problem being that one is never
sure of the intensity of the calibration curve at low two-theta
values.   If there were some way to ensure that an extrapolation of
the calibration curve for silicon down to very small angles was a
reasonable assumption this problem would be minimized.   It appears
that use of the composite silicon/lead stearate crystal does allow
a judgment to be made as to the reliability of such an extrapolation.
In conclusion we would say that the use of pseudo crystals of lead
stearate either alone or deposited on pressed silicon powder specimens offers an attractive means of checking low angle calibration
as well as establishing the viability of extrapolation of silicon
internal standard data down into the low two-theta region.

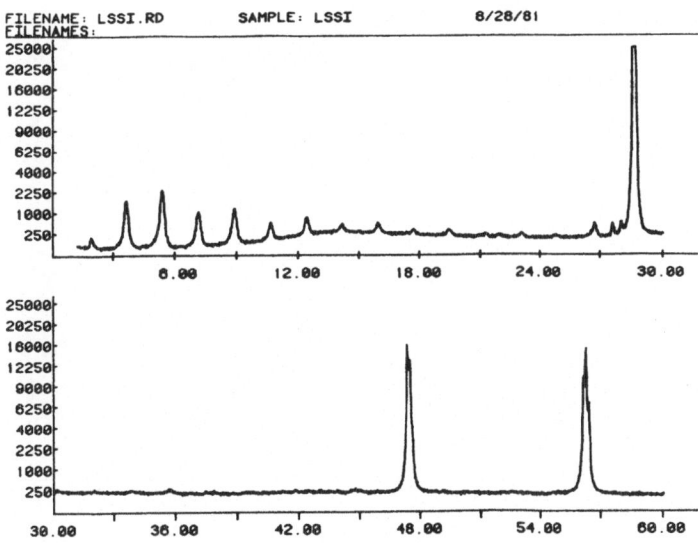

Fig. 4

REFERENCES

(1)   Jenkins, R., Hahm, Y., Villamizar, C., Schreiner, W.N. and
      Surdukowski, C., "Control of Systematic Errors in the
      Computer Controlled Diffractometer" Norelco Reporter,
      in press
(2)   W. N. Schreiner, C. Surdukowski, R. Jenkins and C. Villamizar,
      "Systematic and Random Diffractometer Errors Relevant to
      Phase Identification", in press
(3)   Henke, B. L., "An Introduction to Low Energy X-Ray and
      Electron Analysis" Adv. X-ray Anal., 13 (1969) 1
(4)   e.g. Langmuir, F., Proc. Royal Soc., 170A (1939) 1.

SAMPLE PREPARATION AND METHODOLOGY FOR X-RAY QUANTITATIVE ANALYSIS

OF THIN AEROSOL LAYERS DEPOSITED ON GLASS FIBER AND MEMBRANE FILTERS

Briant L. Davis and L. Ronald Johnson

Institute of Atmospheric Sciences
South Dakota School of Mines and Technology
Rapid City, South Dakota 57701

INTRODUCTION

The purpose of this paper is to summarize the theoretical
basis and experimental techniques for application of the reference
intensity method to quantitative, multi-component analysis by
x-ray diffraction (XRD). Detailed descriptions of the technique
and formal error analysis are discussed by Davis (1978, 1980,
1981a, 1981b, 1981c).

THEORETICAL BASIS

The reference intensity method of quantitative, multi-component
analysis is based on the "adiabatic" equation presented by Chung
(1974);

$$W_i = \left[\frac{k_i}{I_i^\infty} \sum_{i=1}^{n} \frac{I_i^\infty}{k_i} \right]^{-1} \tag{1}$$

where I_i is a component intensity corrected for matrix and
absorption effects, k_i is the component reference intensity
constant, and W_i is the weight fraction of component i in the
mixture. The reference intensity constant is defined as the
ratio of the intensity of the strong peak of component i to the
strong peak of the reference component (usually Al_2O_3) when
both exist in a 50-50 weight mixture. If the aerosol layer is
less than "infinite" thickness, the raw intensities must be
corrected according to the equation (Davis, 1981c):

295

$$I_1^\infty = \frac{I_1^\alpha \mu_H^*}{W_B^H \mu_B^* \left[1 - e^{-2\mu_H^* \breve{M}_B / W_B^H \sin\theta} \right]} \qquad (2)$$

where I_1^α is the raw integrated intensity for component i in the mixture, W_B^H is the weight fraction of the sample material in the sample-matrix composite, $\breve{M}_B$ is the mass per unit area of sample on the substrate, μ_H^* is the mass absorption coefficient of sample material, and μ_B^* is the mass absorption coefficient of the composite sample-matrix taken down to the depth of sample penetration in the matrix (filter substrate). If t_f and ρ_f are the effective thickness (pore space omitted) and density of the matrix material, respectively, the parameter W_B^H is given by $\breve{M}_B / (\breve{M}_B + t_f \rho_f)$.

In the event that all components are crystalline the calculation of the weight fraction of each component is a straightforward application of Eqs. 1 and 2. When amorphous components have been identified (through optical examination, for example) then additional information on sample mass absorption characteristics is required to obtain a quantitative evaluation of both crystalline and amorphous components. The observed mass absorption coefficient of the sample, μ_{BO}^*, is obtained by use of the relation

$$\mu_{BO}^* = - \frac{1}{\breve{M}_B} \ln \left[\frac{I_{fB}/I_o}{I_f/I_o^\prime} \right] \qquad . \qquad (3)$$

where I_{fB}/I_o is the (normal-beam) transmission ratio through the loaded filter and $I_f/I_o^\prime$ is the transmission measured through the blank filter substrate alone. Once the crystalline component XRD analysis has been completed the mass absorption coefficient of the crystalline portion of the sample, μ_{BC}^*, can be calculated from standard elemental or oxide mass absorption values. This value is then combined with corresponding values for the amorphous components in the relation

$$\mu_{BO}^* = \mu_{BC}^* W_{BC} + \sum \mu_{ai}^* W_{ai} \qquad (4)$$

where W_{BC} is the weight fraction of the crystalline portion of the sample, μ_{ai}^* are the respective mass absorption coefficients of amorphous components in the sample, and W_{ai} their corresponding weight fractions. Since the conservation relation $\sum W_i = 1$ must hold, solution of Eq. 4 may be carried out in a straightforward manner for one amorphous component without additional information. For analyses of atmospheric aerosols, however, two amorphous components are commonly observed, one a carbonaceous component and the other a silicious (fly ash) component. The mass absorption coefficients μ^*_{ai} must of course be known or estimated in order to complete the analysis.

When two amorphous components are present in the sample additional information is required, such as the determination of W_{BC} independently, by polarizing optical microscopy, or by use of the specific diffracting power of the sample, defined as

$$\text{SDP} = \mu^*_{BO} \sum \frac{I_i^\infty}{\mu^*_i k_i} \quad . \tag{5}$$

The SDP is experimentally calibrated against W_{BC} for those components observed in the sample being analyzed. It is very important that the reference intensity constants of these components be accurately known. The SDP method appears to have general applicability for all components, providing the variation of reference intensity constants with differing petrogenesis can be determined.

SAMPLE PREPARATION

The key to successful application of the reference intensity method is the manner of sample preparation. It has long been known that the accuracy of x-ray quantitative analysis is highly sensitive to preferred orientation and primary extinction. Assuming that the latter can be eliminated by proper particle size reduction, the former can be eliminated (or at least be made a consistent and reproducible characteristic of the sample) very simply by reducing all samples to a fine aerosol and loading the aerosol onto a filter substrate. A simple method for carrying this out has been discussed by Davis (1981c). Experience suggests that aerosol loads on the order of 300 to 1,000 $\mu g \ cm^{-2}$ are needed when using a glass fiber (such as Whatman GF/C filter) substrate, whereas loads of 100 to 300 $\mu g \ cm^{-2}$ are adequate for teflon substrates. The random particle orientation, initiated in the first few particle layers within the glass fibers of the substrate, appears to be carried out into the free aerosol surface with subsequently heavier loads, resulting in a uniform random aggregate providing the surface is not disturbed after loading. The two substrate filters mentioned above provide the best surface for loading and analysis. Millipore, nuclepore, or other substrates have been found to be less desirable because of their poor particle retention and background scatter characteristics (Davis and Johnson, 1981).

EXAMPLES OF ANALYSES

We conclude by presenting a number of analyses completed by the reference intensity method using bulk materials such as rocks as well as ambient collections of atmospheric aerosol particles.

Bulk Sample Containing No Amorphous Component

Table 1 presents two analyses obtained from the peninsular batholith of southern California. Rock samples were collected which were suitable for both x-ray diffraction and polarizing optical particle count analyses. The results of the analysis using both techniques and showing the variance errors in the case of XRD are presented in Table 1. In this case, because of no uncertainty with regard to amorphous components and because of the heavy loads obtainable on the filter, the analyses shows excellent agreement with the point count by polarizing optical microscopy. The good agreement is also in part the result of the determination of specific reference intensity constants for two of the ferromagnesian constituents.

Table 1. Analyses of Two Igneous Rocks from the
Peninsular Batholith, California
(Weight Percent)

Mineral Component	Quartz Diorite			Mafic Quartz Diorite		
	Optical	X-ray	(Error)	Optical	X-ray	(Error)
Biotite	12.2	14.1	(1.6)	7.0	5.7	(2.2)
Hornblende	2.6	2.7	(0.4)	10.4	13.3	(3.9)
Chlorite	2.0	2.0	(0.4)	-	-	-
Diopside	-	-	-	15.4	13.4	(5.8)
Magnetite	-	-	-	-	1.3	(0.4)
Quartz	22.7	26.1	(2.7)	9.6	8.4	(1.2)
Plagioclase	60.4*	55.1*	(3.2)	57.6[†]	57.9[†]	(5.3)
Sphene	Tr	-	-	-	-	-
Apatite	Tr	-	-	-	-	-

*Oligoclase   [†]Labradorite

Aerosol Containing One Amorphous Component

Table 2 presents an analysis of ambient aerosols observed at Missoula, Montana, where a known large carbonaceous component was observed. The analysis was thus completed using μ^*_{a1} = 8.0 ±2 cm^2 g^{-1}, a value which is estimated to be characteristic of carbonaceous material having adsorbed impurities (primarily water). The somewhat lower accuracy (as indicated by the variance error

Table 2.   Ambient Aerosol Analyses by X-ray
Diffraction (Weight Percent)

Mineral/Amorphous Component	Missoula, MT		Provo, UT	
	$100\ W_1$	Error	$100\ W_1$	Error
Muscovite (Illite)	56.7	8.2	8.3	1.5
Gypsum	2.0	0.4	6.1	1.1
Kaolinite	-	-	1.2	0.4
Chlorite	3.5	0.9	-	-
Quartz	12.0	2.3	12.8	2.0
Microcline	1.7	0.4	-	-
Andesine	0.1	0.02	-	-
Calcite	0.3	0.2	9.0	3.6
Hematite	-	-	2.7	1.6
Carbonaceous matter (a1)	23.7	10.6	28.0	22.6
Silicious fly ash (a2)	-	-	32.0	23.4

values) is the result of uncertainties in the various reference
intensity constants and the relatively large uncertainty assigned
to μ^*_{a1}.

Aerosols Containing Two Amorphous Components

In Table 2 we present a typical analysis of ambient aerosol
collected at Provo, Utah (Davis, 1981a). The two amorphous components present were carbonaceous and silicious fly ash. For μ^*_{a1}
(carbonaceous) we used the value of 8 ±2 $cm^2\ g^{-1}$ as before,
whereas for the silicious fly ash (a2), we calculated a value of
μ^*_{a2} = 52 ±1.5 $cm^2\ g^{-1}$ based on the known elemental composition
of the fly ash. The value of W_{BC} was estimated to be 0.4 ±0.05 by
polarizing optical microscopy. The somewhat large variance errors
with regard to the amorphous components come about primarily from
the uncertainties in W_{BC} and μ^*_{a1}.

CONCLUSIONS

A combination of x-ray diffraction, x-ray transmission, and
polarizing optical microscopy may be used to complete quantitative
multi-component analysis of both crystalline and amorphous components.
The key ingredients required for high level of accuracy in such
analyses include: (a) the use of reference intensity constants
for specific sample components where possible, (b) the correction
of all intensities for matrix and absorption effects, and (c) the
use of aerosol suspension sample preparation in order to eliminate

effects of preferred orientation. It is especially important that
all samples be reduced to a fine aerosol and suspended onto a glass
fiber filter or teflon filter surface, the former being preferred
for heavy loadings and the latter where limited amount of sample
is available. Optical polarizing microscopy is an essential
supportative tool for both confirmation of compounds identified
from the XRD spectra as well as in the estimation of the crystalline
fraction of the total aerosol.

ACKNOWLEDGMENT

     This research was supported by the U.S. Environmental
Protection Agency under Cooperative Agreement CR806769-02. We
are grateful to Dr. Michael Walawender, Department of Geological
Sciences, San Diego State University, for providing the optical
point count analysis of the peninsular batholith samples.

REFERENCES

Chung, F. H., 1974, Quantitative interpretation of x-ray diffraction
     patterns of mixtures II. Adiabatic principle of x-ray diffrac-
     tion analysis of mixtures. J. Appl. Crystal., 7, 526-531.
Davis, B. L., 1978, Additional suggestions for x-ray quantitative
     analysis of high-volume filters, Atmos. Environ., 12, 2403-2406.
     , 1980, "Standardless" x-ray diffraction quantitative
     analysis compounds, Atmos. Environ., 14, 1206-1207.
     , 1981a, Quantitative analysis of crystalline and amorphous
     airborne particulates in the Provo-Orem vicinity, Utah, Atmos.
     Environ., 15, 613-618.
     , 1981b, A study of the errors in x-ray diffraction quanti-
     tative analysis procedures for aerosols collected on filter
     media, Atmos. Environ., 15, 291-296.
     , 1981c, Use of x-ray diffraction quantitative analysis in
     air quality source studies, Proc. 3rd Symposium on Electron
     Microscopy and X-ray Applications to Environ. and Occupational
     Health Analysis, P. A. Russell (Ed.), Ann Arbor Science Press.
     , and L. R. Johnson, 1981, On the use of various filter
     substrates for quantitative particulate analysis by x-ray
     diffraction. Atmos. Environ. (In Press).

DIFFERENTIAL X-RAY DIFFRACTION BY WAVELENGTH VARIATION:

A PRELIMINARY INVESTIGATION

M. C. Nichols,   D. K. Smith,   Quintin Johnson

Materials Science Department, Sandia National Laboratory, Livermore, CA 94550;   Department of Geosciences, Pennsylvania State University, University Park, PA 16802;   Chemistry Department, Lawrence Livermore National Laboratory, Livermore, CA 94550

INTRODUCTION

In powder diffraction experiments involving mixtures of compounds, identification of each individual phase is complicated by the presence of other phases. The interpretation of such complex patterns is often very difficult, and much effort has gone into computational search-match algorithims which attempt to identify individual phases (Nichols, 1966; Johnson, 1977; Frevel, 1976). The success achieved by these programs and by manual search-match methods accounts for the fact that X-ray diffractionists have, in the past, not actively searched for other techniques that could be used to simplify such complex problems. In a recent work (Nichols & Johnson, 1980), a comparison was made of the search-match methodologies employed by several similar technologies (mass spectroscopy, fingerprint identification, X-ray diffraction etc.). A significant observation was made that only in the X-ray method was there so much emphasis on analysis of phases in their "as received" condition. In the other cases, emphasis was placed on the separation of phases before obtaining the spectra. A classical example is the GC–MS instrument which employs a gas chromatograph to separate phases before the mass spectrographic analysis is carried out on what are, by that time, essentially pure phases. This "divide and conquer" strategy suggests an obvious question—can analogous separation methods be devised to solve mixture-type problems in X-ray powder diffraction analysis? If the X-ray diffraction pattern of a mixture of phases could have its major peaks labeled such that they could be attributed to individual phases, or classes of phases containing certain elements, the identification of the unknown crystalline compounds making up the mixture would be greatly facilitated.

One way to allow the tagging of an individual phase is to physically separate it from the mixture prior to any X-ray examination. This approach has been used for many years, especially by mineralogists. Another approach that allows the tagging of individual phases is to change the ratio of phases present using chemical or physical techniques. The intensity differences between the X-ray diffraction patterns made before and after the treatment

301

can then be used to identify partially or completely segregated phases (Nichols & Johnson, 1980). Such treatment usually requires several extra steps and may destroy part or all of the remaining sample. At best, such differential methods raise questions as to the possibility that some of the differences observed were due to the need for multiple handling of the sample.

Because of the difficulty of preparing reproducible samples using the various chemical and physical separation procedures, it would be advantageous to develop a method where the sample does not need to be altered or even removed from the instrument during the course of the analysis. If a technique requiring only changes in X-ray system parameters were devised such that diffraction peaks could be labeled with elemental or other information, powerful new methods of x-ray phase characterization would soon follow. Such a technique would be especially valuable for problems where preferred orientation would produce added complications if the sample had to be remounted during some part of the analysis.

The methods of phase characterization to be described in this paper are conceptually linked to the more obvious physical and chemical separation methods. We refer to all such methods as differential X-ray diffraction (DXD). This acronym is particularly appropriate for the methods described here in which the results are the difference between two normal patterns. The proposed new methods exploit absorption differences that exist between the phases of the mixture due to their different elemental compositions. Two absorption effects must be considered: macroabsorption and microabsorption. Macroabsorption is the beam attenuation due to its passage through a completely homogeneous sample where the phases present are dispersed as "grains" or clusters of crystallites which are generally 1μm or less. Microabsorption is the modification of the beam attenuation where the "grain" size is relatively coarse, and/or the sample is not homogeneous.

The effect of macroabsorption in terms of the differences in mass absorption coefficients is shown in Figure 1. The fraction of intensity contributed by each compound is dependent on the amount present and the ratio of the mass absorption coefficients of the compounds in the mixture. The intensities of the lines from the compound having high absorption will be significantly enhanced compared to the rest of the mixture. As a result, intensities from the remainder of the sample will be diminished. These absorption ratios would be enhanced if two different radiations, selected to have wavelengths on opposite sides of the absorption edge of one of the major elements in the sample, were used to produce two experimental patterns. This wavelength selection could be accomplished by adjusting a diffracted beam monochromator or by selecting a specific channel in an energy dispersive detector mounted on a scanning diffractometer. In either technique the sample remains undisturbed, and the same crystallites contribute to the experimental diffraction pattern. By placing the data on the same intensity and d scale, the two patterns can be subtracted. The difference data that results becomes a subset of the original diffraction data and can be related to the compounds containing the element whose absorption edge is bridged.

Microabsorption effects can significantly modify intensity values in patterns of mixtures. This phenomenon becomes important for coarse-grained samples in which the component phases have very different absorption coefficients. The predicted effect of microabsorption is to increase the intensities of the low absorbing phases with respect to the high absorbing ones. This

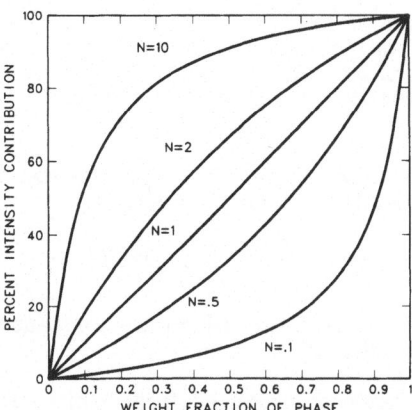

Fig. 1. The effect of different ratios of mass absorption coefficients in a binary mixture on the intensity contribution of any peak. $N = (\frac{\mu}{\rho})_\beta / (\frac{\mu}{\rho})_\alpha$ (modified from Klug and Alexander, 1974, page 535)

effect is opposite to that of the macroabsorption effects described above and in many real samples can totally dominate. Microabsorption is not easily treated theoretically, but estimates of its effects can be made for special sample configurations.

Calculations have been made to determine the magnitude of the intensity differences which could be expected. Due to the experimental convenience of using CuK_α and CuK_β radiations, and Ni and Cu compounds, initial calculations have been made using the Ni K edge. The K absorption edge for Ni lies between CuK_α and CuK_β, hence compounds containing Ni should show marked differences in scattering behavior for these wavelengths, and the ratio of mass absorption coefficients of Ni-bearing phases and non Ni-bearing phases should differ significantly. The higher absorbing phase would be accentuated in mixtures where the mass absorption ratio deviates significantly from unity.

This paper describes the preliminary development of the DXD method by absorption effects. The theoretical basis and some experimental applications are described elsewhere (Nichols, Smith and Johnson, 1982).

CALCULATIONS

The intensities of a powder diffraction pattern for a given compound may be calculated if the crystal structure is known. The expression for the integrated diffracted beam absolute intensity (I^{abs}) for a flat, randomly-oriented diffractometer sample (modified from Warren 1969, page 49) is

$$I^{abs}_{ix\lambda} = P_{ix\lambda} = \left(\frac{P_p}{16\pi R}\right) \cdot \left(\frac{\lambda^3 e^4}{m^2 c^4}\right) \cdot \left(\frac{M_i |F_T|^2_i}{2\mu V^2}\right) \cdot \left(\frac{1 + \cos^2 2\theta_i}{\sin \theta_i \sin 2\theta_i}\right) \tag{1}$$

where the subscripts i, x, and λ represent the ith diffraction maxima from phase x using a wavelength λ, P_p is the power of the primary beam, R is the distance from the sample to the detector slit, M_i is the multiplicity factor, $(F_T)_i$ is the structure factor including thermal effects, μ is the linear absorption factor, V is the volume of the unit cell and π, θ, e, m, and c have their usual meanings. The trigonometric term is the Lorentz–polarization factor usually abbreviated $(L_p)_i$.

For experimental diffraction purposes, intensity comparisons are usually made on a "relative–absolute" scale (I^{ra}), where

$$I_{ix\lambda}^{ra} = \frac{I_{ix\lambda}^{abs}}{K} = \frac{M_i(L_p)_i|F_T|_i^2}{2\mu V^2} \ . \tag{2}$$

The quantity K contains those values which are constant for a given diffractometer system and validation.

X-ray diffraction patterns produced from a mixture of phases will be composites of the patterns from all the individual components. The relative weighting factor for each of the individual component patterns will depend on absorption effects and on the relative amounts of each component present.

Using the DXD concept, a comparison of two patterns obtained using wavelengths α and β utilizes the expression

$$\Delta I_{ix(\alpha-\beta)} = \lambda_\alpha^3 (I_{ix\alpha}^{ra})_{mix} - \lambda_\beta^3 (I_{ix\beta}^{ra})_{mix} \ . \tag{3}$$

The effect of macroabsorption is illustrated in Figure 1. The higher the value of N for a component of a mixture, the more the intensity of the lines of that component will be enhanced in the composite pattern.

Because of the absorption factors, the level of the intensities of the composite α pattern is significantly higher on the relative-absolute scale than is the β pattern, and the differences will always have the same sign. It has proved more useful to scale the α and β data differently using (I/I_c) reference intensity ratios. Each pattern is scaled to an αAl_2O_3 pattern for the appropriate radiation. This scaling has the effect of using a reference compound which is only slightly affected by absorption variations to define the working level to be used for pattern comparisons. When the data are scaled in this manner, experience has shown that the DXD pattern yields mostly negative differences for the lines from compounds containing the elements whose absorption edge was bridged, and mostly positive deviations for the remainder of the pattern. This scaling has been used for the calculated examples in this paper.

The scaling of experimental data from unknown samples is more of a problem because it is not known which lines belong to which compounds or even which compounds are present. Scaling using intensity ratios from a standard sample results in α and β patterns at different levels on the relative absolute scale. Although scaling can be accomplished by adding αAl_2O_3 to the mixture and using its peaks as a reference, it is often not a convenient procedure. However, successful scaling can usually be accomplished by examining the two patterns for a moderate to strong line whose intensity ratios to several nearby lines do not show marked differences in the α and β patterns. The intensities of this line can then be scaled to be equal in the α and β patterns. Several trials may be required, regardless of the choice of reference line, although the strongly negative peaks usually stand out on inspection.

In experimental patterns the nature of the sample also becomes important, and the effect of microabsorption must be evaluated carefully. Microabsorption occurs when particles of a highly absorbing phase only partially screen the diffracted intensity from low absorbing phases in a mixture. It is extremely sensitive to the particle size and size differences when two or more phases are present. The approximate theory of Brindley (1945) enables an evaluation of the nature and magnitude of this effect on experimentally measured intensities. According to the theory, intensity ratios are ·modified by a correction factor based on absorption and size effects. It is important to note that the effect of microabsorption is to enhance the intensity of the low absorbing phase in a mixture causing the intensities from the higher absorbing phases to appear reduced on a relative basis. The limiting effect of microabsorption can be illustrated by considering the extreme case in which a sample is composed of stripes of each phase oriented parallel to the diffraction plane. In this situation no phase influences any other phase. The macroabsorption effect of a mixture is totally eliminated, and the fractional intensity contribution of each phase on the relative-absolute scale will be proportional to the volume fraction of that phase. In the absence of intergrain absorption effects, intensities will vary inversely with μ.

RESULTS

Several different approaches could be employed whereby the variations in intensity around an absorption edge might be exploited to obtain phase and/or chemical information about individual diffraction peaks. In the first of these techniques, a diffraction peak would be scanned maintaining a constant d by varying the scan angle as the wavelength is changed. The resulting trace would represent a single diffraction peak scanned over the absorption edge. A significant change in peak intensity at the edge would indicate that the peak contains the element associated with that edge. A significant discontinuity in the intensity should occur, the magnitude of which would depend on the amount of the element present in the compound. As many strong peaks as necessary could be scanned in this way until one of the phases is identified and its pattern subtracted from the composite pattern. In the second, perhaps more straight-forward approach, two complete patterns would be obtained using fixed wavelengths on opposite sides of the absorption edge for a given element. By placing the intensities on the same d and intensity scales, differences between the two patterns would indicate effects due to the absorption edge. This latter approach is more amenable to currently available instrumentation and is outlined below in greater detail.

For the purposes of the calculations performed here, compounds of copper and nickel have been used. These elements were chosen for initial study because CuK_α and CuK_β radiations having reasonable intensity are readily available using standard laboratory generators and because compounds of copper and nickel used with these radiations should show strong contrast. The variation in absorption of both the CuK_α and CuK_β radiations by copper compounds will be very small, but this is not true for Ni compounds. The CuK_β is on the strongly absorbing side of the absorption edge for Ni compounds while CuK_α is on the weakly absorbing side. This large variation in absorption accounts for a significant difference in the absolute intensity scale of the CuK_α vs. CuK_β patterns for nickel-bearing compounds and markedly affects the intensities in the patterns of mixtures. The anomalous dispersion coefficients also show significant differences which affect the scattering powers of the individual atoms. The effect is sufficient to cause perceptible differences between the patterns produced by the two wavelengths.

Intensities were calculated for rhombohedral NiO and monoclinic CuO using the POWD10 program (Smith & Nichols, 1981). Composite patterns were prepared for both the $Cu K_\alpha$ and $Cu K_\beta$ by calculating individual patterns, scaling the patterns by multiplying each data set by its appropriate I/I_c value and adding the patterns. These patterns simulate composite experimental data which could be obtained for each radiation. The problem is to reverse the procedure and to use two different experimental patterns to indicate the component phases.

One way to separate a composite pattern into its components is to compare the patterns collected using different wavelengths. Scaling using experimental I/I_c ratios and d values compensates for differences in P_θ for $Cu K_\alpha$ and $Cu K_\beta$. The intensity correction ratio I/I_c should be measured experimentally to avoid system dependent parameters such as detector quantum counting efficiencies. Patterns whose intensities have been properly scaled and plotted on the same d scale can be graphically subtracted as shown in Figure 2. The plot is similar to that which can be accomplished on split screen video terminals commonly used with modern diffraction systems. It is immediately apparent when looking at the difference pattern that certain peaks may be attributed to the compound containing the element whose absorption edge was bridged. Those differences, which are strongly negative, are all due to the NiO component. The peaks attributable to CuO have differences near zero or distinctly positive. The variations from zero are mostly due to the effect of anomalous dispersion and partly due to the Lorentz-Polarization (L_p) effect. Some of the weaker high-angle peaks of NiO are not sufficiently distinct from the CuO lines to allow tagging. Nonetheless, there are sufficient clearly-tagged peaks to allow an identification of the NiO phase.

DISCUSSION

It has been shown in the theoretical development of the DXD concept that differences will occur between diffraction patterns obtained with different radiations. These differences are enhanced when marked absorption differences exist for the different radiations employed. In powdered samples the effect of microabsorption also becomes significant when absorption differences become large. To illustrate the potential magnitude of the effect of microabsorption from non-ideal samples on the data presented above, consider a sample of CuO–NiO with particle sizes $1\mu m$ and $10\mu m$ respectively. In such a case, the experimental ratios for intensities in the $Cu K_\alpha$ pattern are little affected, but in the $Cu K_\beta$ pattern, the CuO intensities will be enhanced by a factor of 3 with respect to the intensities of the NiO. This change is opposite to the macroabsorption effect which has been discussed above. In the extreme case where there is no dependence on intergrain absorption, the magnitudes are further enhanced.

It is evident that it may prove more advantageous to accentuate the microabsorption by deliberately using coarse-grained samples rather than preparing an ideal, fine-grained sample. The magnitude of the microabsorption effect is so large that micronized samples would be required in order to observe idealized macroabsorption effects. A coarse-grained specimen also has the advantage that it will show minimum peak broadening and overlap.

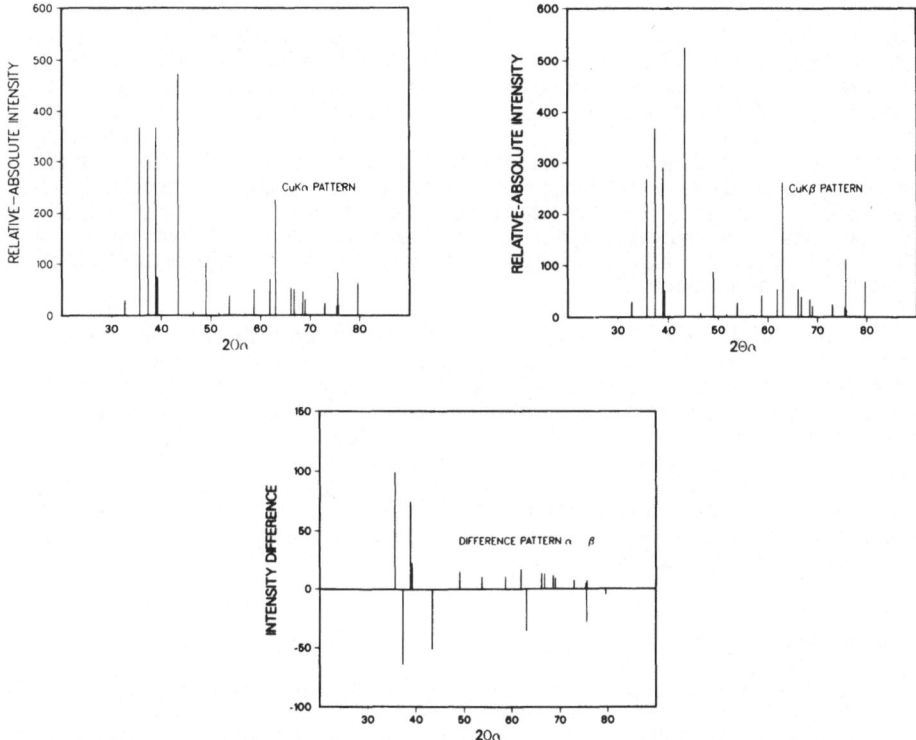

Fig. 2. Graphical representation of the X-ray powder diffraction patterns for a CuO and NiO mixture taken with CuK_α and CuK_β radiations along with the difference pattern that reflects the variation in absorption and anomalous dispersion effects for the two radiations used.

As long as the grains are small enough to assure adequate particle statistics, these "coarser" samples will thus result in sharper peaks which can be more accurately "tagged".

To utilize the DXD approach for actual samples, either with or without microabsorption effects, it will be necessary to modify a diffractometer or X-ray source to be able to select different wavelengths. A diffractometer can be easily modified to switch between K_α and K_β radiation without removing the sample by using an energy dispersive detector or by adjusting a diffraction beam monochromator. Using CuK_α and CuK_β radiation limits experiments to those involving Ni compounds. Sealed X-ray tubes employing other targets would also be limited to detecting one or two elements. There are, however, several other experimental approaches which could extend the utility of the concept to many other elements.

One way to allow elemental labeling of diffraction peaks from compounds containing other elements would be to employ a modifiable X-ray source. Although X-ray systems exist which have dial-a-target X-ray tubes, these systems have relatively low beam powers and therefore may not be suitable for most applications. An alternative solution would be to use a synchrotron source. Such a source of X-rays would allow the selection of an intense beam of almost any wavelength desired, but would not, in general, be routinely available. A more practical solution might be to use the continuum radiation from a high-intensity rotating anode generator. The continuum radiation from such a source may be intense enough to allow "individual" wavelengths to be selected using a monochromator. If not, the rotating target could be changed to allow the production of the desired high intensity characteristic wavelength, or the anode could be fabricated from strips of different target materials, and its position shifted to expose the desired target material to the electron beam.

Although the equipment requirements seem to be elaborate, the method may prove to be very useful. Using this technique, two patterns can be obtained from a multiphase sample without removing the sample from the diffractometer. The "same" crystalline grains would then contribute to each pattern and the patterns would be directly comparable. Even if preferred orientation exists, differences observed may be sufficient to allow labeling the diffraction lines as to the elements present in the compounds which gave rise to them. Such information could greatly facilitate the identification of phases for such systems. Computer graphical display and stripping algorithms already exist which, with modification, could be used to analyze such data.

ACKNOWLEDGMENTS

This work was supported by the U. S. Department of Energy. We would also like to thank J. Mansfield and C. L. Bisson for computational assistance and advice, H. L. Willyard for help in the preparation of the illustrations, and B. G. Nichols for proofreading and other assistance.

REFERENCES

Frevel, L. K., 1976, Quantitative Matching of Powder Diffraction Patterns, in: "Advances in X-ray Analysis, Vol. 20", Plenum Press, NY

Johnson, G. G.,Jr., 1977, Resolution of Powder Patterns, in: "Laboratory Systems and Spectroscopy", Marcell Dekker, NY

Klug, H. P. and Alexander, L. E., 1974, "X-ray Diffraction Procedures", Second Edition, John Wiley and Sons, NY

Nichols, M. C., 1966, A FORTRAN II Program for the Identification of X-ray Powder Diffraction Patterns, UCRL-70078, Lawrence Livermore Laboratory

Nichols, M. C. and Johnson, Quintin, 1980, The Search–Match Problem, in: "Advances In X-ray Analysis", Plenum Press, NY, Vol 23, p273

Nichols, M. C., Smith, D. K. and Johnson, Quintin, 1982, Differential X-ray Diffraction by Wavelength Variation: A Theoretical Basis, Submitted to J. Appl. Cryst.

Smith, D. K. and Nichols, M. C., 1981, A FORTRAN IV Program for Calculating X-ray Powder Diffraction Patterns — Version 9/10

Warren, B. E., 1969, "X-ray Diffraction", Addison-Wesley, NY

X-RAY DIFFRACTION QUANTITATIVE ANALYSIS USING INTENSITY RATIOS

AND EXTERNAL STANDARDS

Raymond P. Goehner

General Electric Company
Corporate Research and Development
Schenectady, N.Y.

INTRODUCTION

The use of quantitative x-ray diffraction (XRD) as an analytical technique has recently become increasingly popular. There are primarily two reasons for this increasing interest in a relatively old discipline[1]. The first is simply the need for quantitative phase analysis. This need arises from several sources such as government regulations on respiratory quartz, industrial quality control, and material research. X-ray diffraction provides a readily available technique for bulk phase analysis on chemically similiar phases. For phases which are elementally distinct, our sister science, x-ray fluorescence, can more easily provide the quantitation needed[2]. The second reason for the increasing interest in quantitative XRD is the ready availability of automated powder diffractometers. These instruments remove much of the tedium involved in the collection and reduction of the data.

There are five experimental techniques used in the reduction of data in XRD quantitative phase analysis. The first measures the intensity from the unknown and compares it against a pure standard. In order to calculate the weight fraction of the phase of interest, the mass absorption coefficients of the unknown mixture as well as the standard has to be known or measured[3]. The second method is the spiking or dilution method[4]. The spiking method adds fixed amounts of the phase of interest, recording the intensity of a line from this phase at each amount. The original amount of the unknown is then extrapolated from this data. The dilution method is exactly the same except an amorphous diluent is added. In this case the intensity of the diffraction lines decreases. The third method is the internal standard method. In this technique a fixed amount of a standard material not found in the sample is added. The intensity from a line of this standard is ratioed against the intensity of a line from the unknown and compared to a previously obtained calibration curve. The fourth method is the standardless method[5]. This method allows the analyst to determine quantitatively the amount of crystalline phase in the sample by mechanically or chemically treating the sample in order to reduce the amounts of the phase found in it. By measuring the intensity of diffraction lines from each phase and comparing these intensities against the treated sample, a quantitative phase analysis can be obtained. The fifth quantitative XRD method is the external standard method[6]. This technique is actually a variation of the internal standard method. Instead of mixing an internal standard, the external standard procedure uses the ratio of diffraction lines from the phases found in the sample and compares them to previously mixed standards. The reference intensity ratio can be used instead of an external standard. Chung has popularized this procedure, for binary mixtures calling it auto-flushing[8] and for multiphase mixtures the adiabatic principle of auto-flushing[9].

The external standard intensity ratio method as it is applied in our laboratory was formulated by Copeland and Bragg[6]. The following equation expresses it simply as:

$$\frac{X_i}{X_j} = K \frac{I_i}{I_j} \qquad (Eq. 1)$$

where X_i is the weight fraction of phase i and X_j is the weight fraction of phase j. The I_i is a diffraction intensity from phase i and I_j is a diffraction intensity from phase j. The K is a constant that can be obtained from a single standard of a one to one mixture by weight of phase i and j.

$$K = (\frac{I_j}{I_j}) \qquad 1:1 \text{ mixture}$$

Thus the weight ratio of any two phases can be obtained. If these weight ratios are calculated for each phase, then the quantitative amount can be obtained by the application of the following equation

$$\sum_{k=1}^{n} X_k = 1$$

where n is the number of phases in the sample.

By measuring intensity ratios from the unknown and using
these ratios in the quantitative procedure, the diffractionist
can minimize many problems. These include changes in barometric
pressure, changes in KeV or Ma setting, long term drift in x-ray
tube intensity, and matching the absorption of the matrix and
internal standard. If the diffraction lines that are ratioed
are close together then many other errors are reduced. These
include compactness of the powder in the sample holder, sample
transparency, sample size, sample position on the focusing circle,
slight changes in alignment, and small changes in the slits used
for the analysis. These errors can be very significant and tend
to bias the results in ways which are not obvious. The errors
which are not minimized include all of the classical problems
such as particle statistics, preferred orientation, microabsorption, extinction, and the serious problem of obtaining a standard
that is always similiar to the unknown[10].

EXPERIMENTAL

In order to demonstrate the usefulness of the external standard technique, binary mixtures of hematite (Fe_2O_3) and corundum
(α-Al_2O_3) were prepared. Figure 1 shows the relative intensity
of a peak from hematite versus the weight % of hematite. This
curve deviates radically from a linear relationship due to the
large differences in mass absorption coefficients between Fe_2O_3
and Al_2O_3 for CuKα radiation. Figure 2 is a plot of the intensity of the 1.31Å line of Fe_2O_3 divided by the intensity of the
1.374Å line of Al_2O_3 versus the ratio of the weight fraction of
Fe_2O_3 and Al_2O_3. In this case, the relationship is linear as
expected from equation 1. The K value of .642 is the slope of
the line. Table 1 is a tabulation of these results. The first
column is compositions based on the weighed amount of Fe_2O_3 and
the last column is the calculated amount using K = 0.642. Note
that the agreement is within a few tenths of a weight percent.
A major use of this technique in this laboratory is the quan-
titative analysis of mixtures of α alumina, β alumina, and β" alumina.
This represents an interesting application of quantitative XRD
since these phases cannot be easily measured by any other technique. One of the first problems encountered when doing this
analysis is that β alumina, and to a greater extent β" alumina,
picks up water from the air rapidly if the powders have high surface areas. This water pickup causes an expansion between the
basal planes. Thus the diffraction lines move as a function of
time and humidity. A BASIC program was written to control our
Siemens D500 Diffractometer[11] and provide the quantitative analysis
needed. This program scans over the region of interest, finds
the highest intensity and adds to this the intensities of a fixed
number of data points either side of the highest point. A total
integration is not done because some peak overlap may occur as

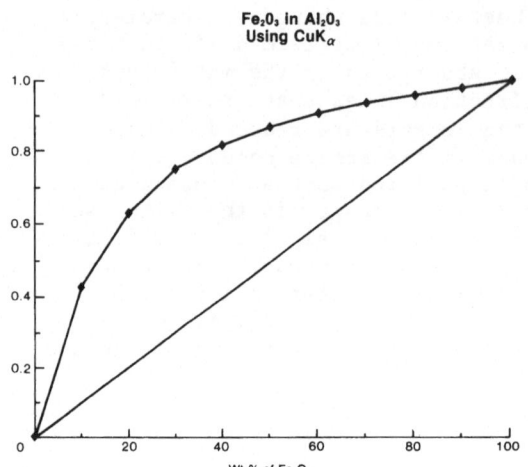

Figure 1.   Relative intensity of a peak
from Hematite versus weight % Hematite.

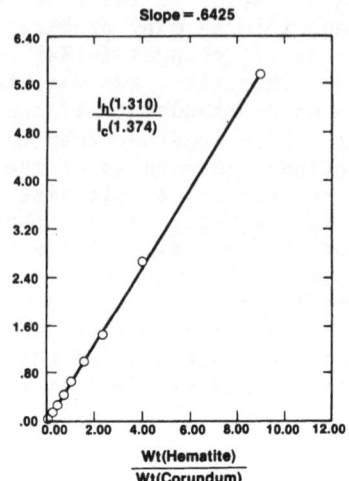

Figure 2.

INTENSITY RATIO METHOD
Fe₂O₃ — Al₂O₃ MIXTURES

% Fe₂O₃	I_h	I_c	I_c/I_h	$X_h\%$
0	0	71949	—	0
10	2386	33176	13.9	10.1
20	3827	24148	6.31	19.8
30	5097	18523	3.63	30.0
40	6587	15053	2.29	40.5
50	7157	10997	1.54	50.3
60	8155	8274	1.01	60.5
70	8217	5656	.69	69.3
80	8504	3188	.37	80.6
90	7844	1362	.17	90.0
100	8043	0	0	100.0

$$X_h = \frac{1}{1 + .6425\,\frac{I_c}{I_h}}$$

Table 1.   %Fe₂O₃ is the as
weighed standards while
X_h% is the calculated
values.

QUANTITATIVE ANALYSIS
OF ALPHA, BETA, AND
BETA" ALUMINA

Nominal			Measured		
α	β	β''	α	β	β''
0	10	90	1	9	90
5	25	70	5	26	69
5	80	15	6	78	16
10	10	80	10	12	78
20	40	40	20	39	41
33	33	33	33	32	35
40	35	25	43	32	25
50	10	40	50	10	40
60	5	35	59	6	35

Table 2.   Nominal is as
weighed standard. Measured
is the calculated weight %.

relative peak positions change. Standard mixtures were prepared
and a parabolic calibration curve was calculated for each set
of intensity ratios. A linear curve was not acceptable simply
because regions of pure background were not available on these
samples. By picking the background too high, the linearity of
the calibration curve is lost. Table 2 shows some of the calibration standards used and the agreement between the weighed standards (nominal) and the values as calculated by the program (measured). The agreement is within a few weight percent over the
entire range. The program will also automatically check if a
sodium aluminate phase is contained in the sample. This technique
has been applied to determining the ratio of β to β " alumina on
the inside or outside diameters of solid ceramic tubes as well
as powdered ceramic tubes. Since the peaks ratioed are reasonably
close in 2θ the geometrical distortion of the diffraction peaks
due to the curvature of the tubes does not seriously effect the
results. This technique has also been applied to determining
hematite and magnetite ratios, as well as the amount of Cu, CuO,
and Cu_2O in powder samples.

CONCLUSION

The external standard technique using intensity ratios has
been the quantitative diffraction technique of choice in this
laboratory, because it minimizes many common errors. A problem
can occur with this technique if a large number of phases needs
to be measured or only one phase out of many has to be determined.
In these cases, the internal standard method would yield results
easier.

REFERENCES

1. G.L. Clark and D.H. Reynolds, Ind. Eng. Chem., 8, 36-40 (1936).
2. M.F. Garbauskas and R.P. Goehner, To be Published in Adv.
    X-Ray Analysis, (1982)
3. H.P. Klug and L.E. Alexander, "X-Ray Diffraction Procedures
    for Polycrystalline and Amorphous Materials," Wiley Inter-
    sciences, New York (1974).
4. S. Popovic and B. Grzeta-Phenhovic, J. Appl. Crys., 12, 205-
    208 (1979).
5. L.S. Zevin, J. Appl. Crys., 10, 147-150 (1977).
6. L.E. Copeland and R.H. Bragg, Anal. Chem, 30, 196-208 (1958).
7. "Alphabetic Index of Inorganic Phases," JCPDS, p. XV (1980).
8. F.H. Chung, J. Appl. Cryst., 7, 519-525 (1974).
9. F.H. Chung, J. Appl. Cryst., 8, 17-19 (1975).
10. G.J. McCarthy, R.C. Gehringer, D.K. Smith, V.,. Injaian,
    D.E. Pfoertsch and R.L. Kabel, Adv. in X-Ray Analysis, 24,
    253-264 (1981).
11. R.P. Goehner and W.T. Hatfield, Adv. in X-Ray Analysis, 22,
    165-167 (1979).

A GUINIER DIFFRACTOMETER WITH A SCANNING POSITION SENSITIVE DETECTOR

Herbert E. Göbel

Forschungslaboratorien der Siemens AG

D 8000 München 83, West Germany

ABSTRACT

A strictly focussing Guinier diffractometer using a linear position-sensitive proportional counter (PSPC) to detect the diffracted x-rays is described. The data collection time for a complete pattern can so be reduced to minutes instead of hours as it used to be in conventional film- or counter-Guinier systems. The PSPC collects all diffracted x-rays over several degrees of 2 Theta in parallel and composes the full pattern by a continuous scan over the whole 2 Theta range. This principle was described in Adv. in X-Ray Anal. Vol. 22, 255 ff and 24, 123 ff. for Bragg-Brentano diffractometers.

The PSPC Guinier system allows high-quality patterns from a minor amount of sample material (comparable to the amount necessary for the Debye-Scherrer method) with peaks having a resolution of 0.06° (fwhm) and an absolute angular accuracy of a few thousandths of a degree of 2 Theta, especially at low angles. The spectral purity (no doublets) performed by a Ge (1,1,1) primary-beam monochromator reduces the background and facilitates the separation of overlapping peaks. A movement of the sample, which is inherent to the method described here, brings a high spacial angle of crystallite orientations into reflection position thus providing representative crystallite statistics and reliable intensities even for minor phases of an unknown sample.

INTRODUCTION

The Guinier method is one of the established techniques in x-ray powder diffraction. Its properties are high resolution and accuracy at low background. The amount of sample material is comparable to the Debye-Scherrer method. So it is useful for many applications in industrial troubleshooting analyses like corrosion etc. where only tiny particles of the specimen are available. In figure 1 the mainly used

315

Guinier arrangement is sketched, showing the asymmetric transmission technique
with substractive chromatic dispersion.

This technique is mostly used with x-ray films to record the diffracted beam,
but also with single-quantum x-ray detectors like scintillation counters in combination with fine receiving slits /1/ that are rotated around the 4 Theta axis on the
focussing cylinder surface. A sliding guidance orients the detector to the diffracted
beam coming from the sample.

The counter system can use the intensity dynamics of single quantum registration and can measure the data with the mechanical accuracy of the 4 Theta
step motor drive, which ranges around 0.001°, the smallest step width. While the
film collects all diffracted x-rays over the full 4 Theta range simultaneously, the
counter system has to scan over the range sequentially utilizing only a small part
of the diffracted beam. As consequence long data collection times are necessary
with scan speeds of 0.1° to 1° per minute. For this reason the Guinier method is
mainly used with film registration and automatic densitometers for data evaluation.

The registration speed of a counter Guinier system can be increased by up to
2 orders of magnitude if a position-sensitive detector is used which collects all x-
rays over an extended region of the focussing cylinder in parallel like a multidetec-
tor array.

The resolution of such a detector should range between 0.01° to 0.02° of
2 Theta corresponding to a spacial resolution of 30 μm to 60 μm at 180 mm focussing cylinder diameter. Linear position-sensitive proportional counters (PSPC) have
a principal resolution limit of 1 in 1000 of their full length /4/.

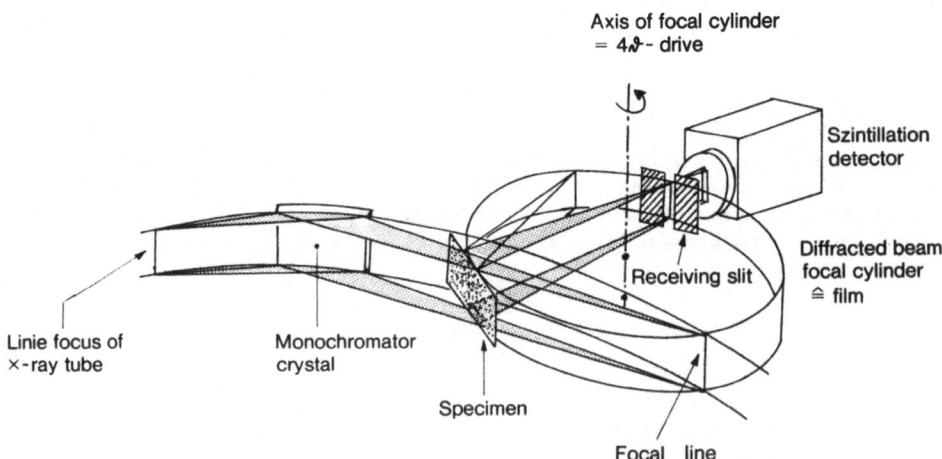

Fig. 1: The asymmetric Guinier method with film or scintillation
        counter registration

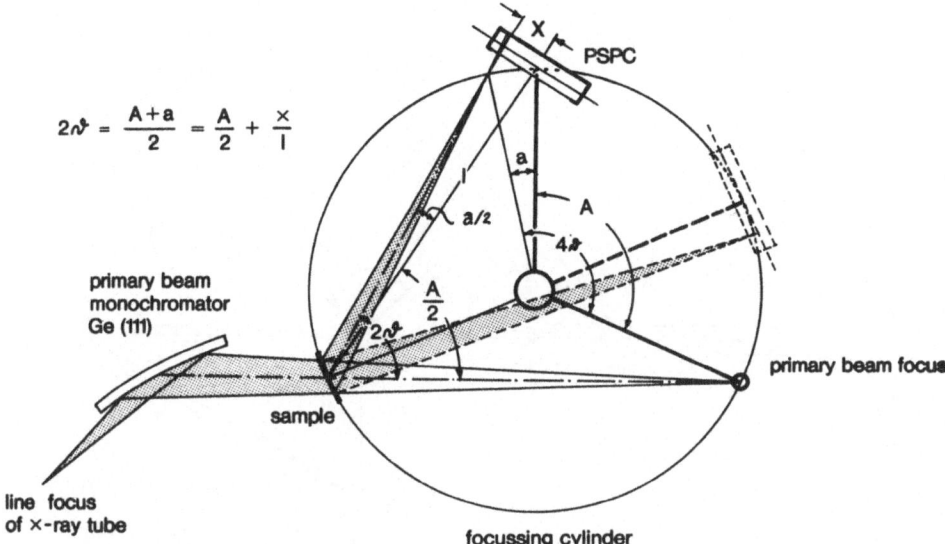

$$2\vartheta = \frac{A+a}{2} = \frac{A}{2} + \frac{x}{l}$$

Fig. 2: A Guinier diffractometer with a PSPC replacing the scintillation
counter. The broken lines show the position of strict focussing.

This means that without loss of resolution a PSPC can collect the data over a
range of 10 to 20 degrees parallel and the full pattern has to be composed by displacing the detector. A technique how this is performed in a continuous way was
described in /2/ and /3/. A further restriction for the angular range is the necessity
of nearly vertical incidence on the detector wire to avoid parallax broadening. This
comes into conflict with the focussing properties of the Guinier method (see fig. 2)
so that off-center beams have to be collected out of focus. The dashed part of
figure 2 shows the position where this effect is nearly negligible.

If fixing this diametrically opposite position of specimen and detector by a rigid
coupling, the focussing cylinder has to be rotated to scan an extended 2 Theta range.
This requests a translatory and rotary motion of the sample along the primary beam
as it was described by H. Dachs and K. Knorr /5/ in a Guinier film camera, mainly
used for single crystal investigations. The system described here applies the same
beam geometry and combines it with the continuously scanning PSPC principle /2/.

EXPERIMENTAL

The principles of the system are shown in figures 3 and 4. The focus F of the
primary beam monochromator is the only fix axis of the system. This is why the
goniometer drive (2 Theta) was positioned into this axis instead of C the center of
the focussing cylinder. With the isosceles triangle specimen(S)-center (C) of focussing circle-focus(F) the axis PSPC-specimen includes the same angle A with the
primary beam as the goniometer drive. The specimen is translated along the primary
beam. Both the sample guide running on ball bearings and the pivot bearings have
to be manufactured absolutely free from backlash.

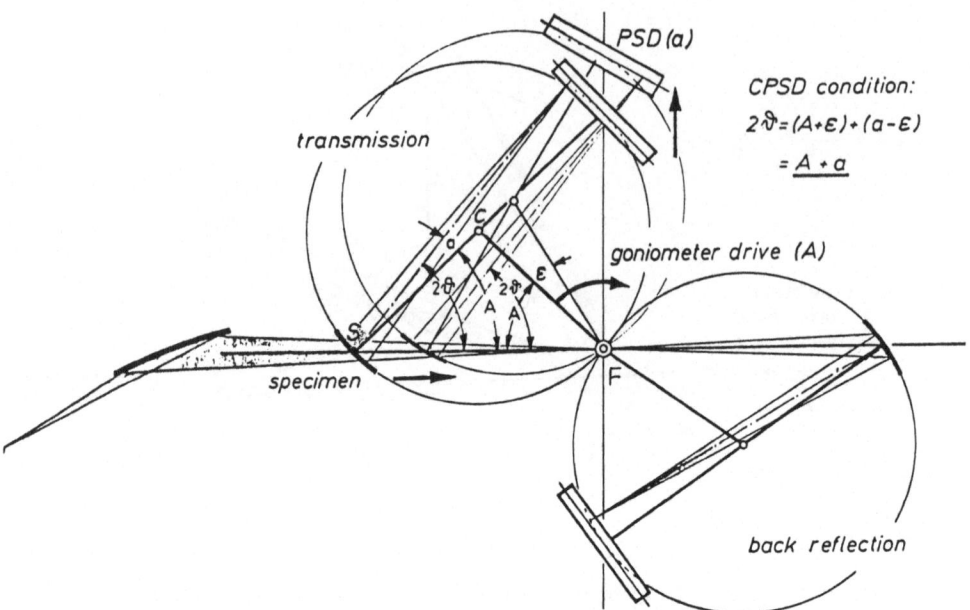

Fig. 3: Principle of the moving-sample Guinier diffractometer with PSPC

Figure 3 shows two x-ray beams diffracted by the same angle 2 Theta being re-corded by the PSPC at subsequent times. As in /2/ the digital address of 2 Theta, under which a diffracted x-ray quantum is stored in the multichannel counter, is composed as a sum of an address "A" representing the angular position of the goniometer and an address "a" representing the incidence position in the PSPC. Both addresses have to be digitized in the same scale of 2 Theta, for example in increments of 0.02 degrees. The 2 Theta address for quanta belonging to the same peak will so become independent of the goniometer movement (variation ε) be-cause "a" will decrease in the same way as "A" increases leaving 2 Theta con-stant.

Figure 4 shows a photograph of the diffractometer and a block diagram of the electronic equipment. The monochromator was a HUBER 611 with a Ge(111) crystal 615-2 with a focus length of 211 mm. As x-ray source a SIEMENS fine-focus tube FK 60-04 with Cu-target (maximum load 40kV, 37 mA) was used. The goniometer was a HUBER 410 horizontal single circle drive with a BERGER 5-phase stepper motor with 1000 steps per degree controlled by a LASER-OPTRONICS SMC 500 supply. As position-sensitive detector a high-pressure RC-line encoded proportional counter from BRAUN as described in /2/ was used. The TDC is tunable between 400 and 750 MHz. The digitization rate can so be ad-justed exactly to the step widths of the goniometer drive. The smallest selectable digital step is 0.01° of 2 Theta, which can be switched to rougher scales by fre-quency division. 0.01°, 0.02°, 0.05° and 0.1° steps are provided both from the TDC-and the MSC-modules.

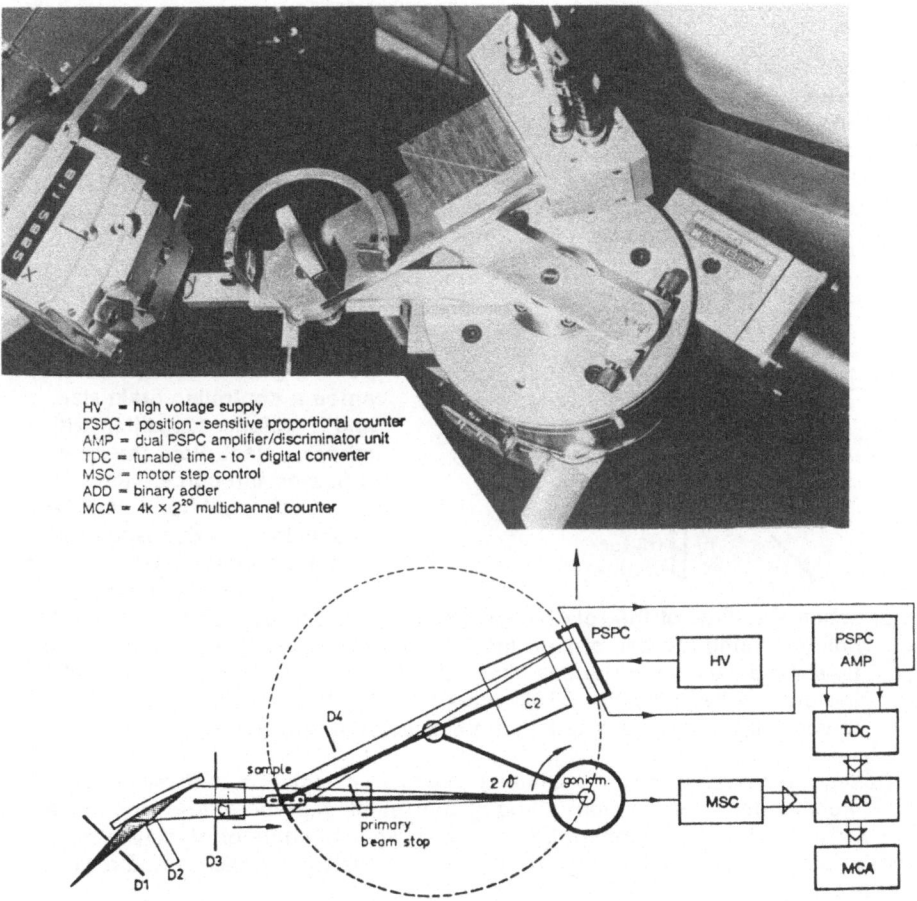

HV   = high voltage supply
PSPC = position - sensitive proportional counter
AMP   = dual PSPC amplifier/discriminator unit
TDC   = tunable time - to - digital converter
MSC   = motor step control
ADD   = binary adder
MCA   = 4k × 2^{20} multichannel counter

Fig. 4: Experimental equipment

The quality of the diffraction patterns depends strongly on the alignment of the monochromator and the exact coincidence of the beam axis and the axis of the sample translation. The goniometer axis has to be centered exactly to the focus line of the primary beam. These alignments can be performed using an adjustment fluorescent screen similar to that described in /1/.

The system of diaphragms, Soller collimators and the primary beam stopper is indicated in figure 4. The two Soller collimators in front and after the specimen avoids the broadening of the low angle tails of the peaks called "blurring effect". Together with the other diaphragms they define the beam path very effectively and allow low-background measurements. At very low angles a higher background due to air or edge scattering of the primary beam seems hardly avoidable. The useful angular range starts at about 3°–4° corresponding to d-values of about 25 Å (= 2.5 nm) which is sufficient for the most purposes of XRPD (see also figure 6). Nevertheless our aim will be to reduce the blind area to less than 2 degrees corresponding to d-values of about 50 Å, which is an interesting range for some applications like clay minerals or organic compounds.

EXAMPLES AND DISCUSSION

1) Sample Preparation

Samples for transmission measurements have to represent a good compromise for the thickness. Thick samples will offer more material to the primary beam for diffraction, however due to absorption a certain thickness will produce maximum intensities. Sample thickness will also affect the line width from geometrical reasons. So, as an average rough value, a sample should not exeed a thickness of 100 µm. As supporting film 5 µm thick mylar was used.

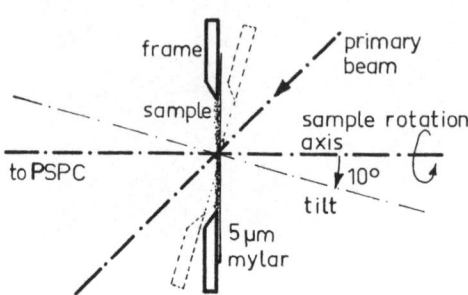

An area of about the beam diameter is coated with a thin film of vaselin as adhesive for the powder grains. In order to guarantee a controlled grain size and a uniform deposition with random orientation, the powder is passed through a micro sieve onto this film. After a plain layer of powder was deposited the holder is turned upside down removing all sample material that is not in direct contact with the adhesive. This process can be repeated several times producing thin and homogenious transmission samples.

Fig. 5: Schematic view of the rotating specimen holder. During rotation around the PSPC-to-sample axis the surface is tilted against the primary beam by the CPSD-recording thus spiralling in reciprocal lattice volume.

During the measurement the sample holder can be rotated as shown in fig. 5. This, in addition to the continuous tilting of the sample surface against the primary beam during the PSPC-recording, brings a large spacial angle of crystallite orientations into reflection position guaranteeing good crystallite statistics and reliable intensities.

2) Accuracy, Resolution and Peak/Background Ratio

For testing the diffractometer, $CaWO_4$ was used as standard specimen material. This allows a direct comparison to test measurements described in /1/.

Figure 6 and 7 show the result of these test measurements on a logarithmic intensity scale to get a better information on the background level. Figure 6 demonstrates the zero calibration by symmetric measurement around the primary beam. The reflection (101) at 2 Theta (calc) = 18.621° is plotted on an enlarged 2 Theta scale on the lower part of figure 6. By reflecting the left profile at the original zero a misalignment of 0.005° can easily be detected. After zero correction the measured position was determined at 18.622°.

The high accuracy at low angles makes the Guinier technique especially valuable for indexing of unknown materials, reducing the danger of ambiguous indexings least-squares refinements of unit-cell dimensions.

Figure 7 shows the complete pattern of $CaWO_4$ with a digitization rate of 0.02°

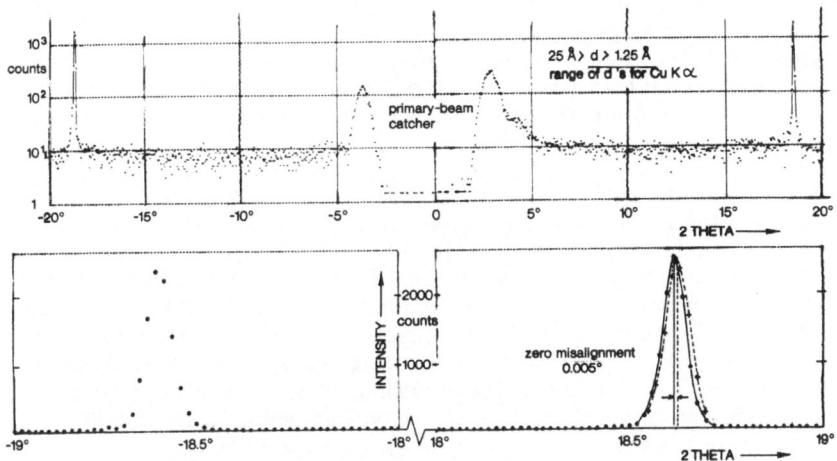

Fig. 6: Measuring range around the primary beam and zero calibration

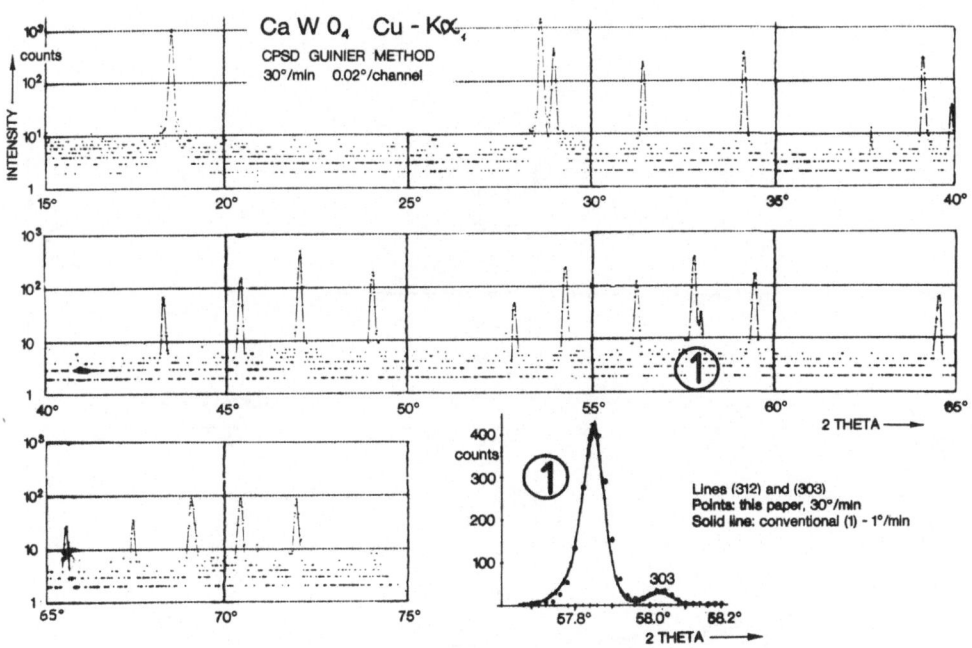

Fig. 7: Low Background pattern of an ideal CaWO₄ sample
(in logarithmic intensity scale)

per channel. The background level centers around 2-3 counts per channel while the strongest peak at 28.72° has its maximum at about 2000 counts. The detail (1) on the bottom of figure 7 compares the splitting of the closely neighbouring reflections (312) and (303) that are as clearly resolved as in /1/, however collected 30 times faster. Here also the effect of the two Soller collimators is obvious cutting the low-angle tails.

The results from the peak evaluations of the $CaWO_4$ patterns are plotted in figure 8, showing the accuracy of peak positions in the upper part and the line widths (fwhm) in the lower part. The deviations between calculated and measured peak locations become large for angles beyond 75 degrees due to geometrical effects. This is about the angular range accessible for the symmetric Guinier cameras. For $CuK\alpha$-radiation the angular range corresponds to a d-range between 1.25 and 25 Ångstroem (0.125 - 2.5 nm). The distorted S-shape of the 2 Theta calibration curve is a combination of the misalignment of the specimen guidance with respect to the primary beam axis (S - curve) and an off-center position (or thickness effect) of the specimen resulting in an U-shaped curve.

The line width has to increase with larger diffraction angles due to sample thickness and chromatic dispersion effects. The small line width of about 0.065° minimum of pure $K\alpha_1$ lines facilitates the separation of overlapping peaks considerably,

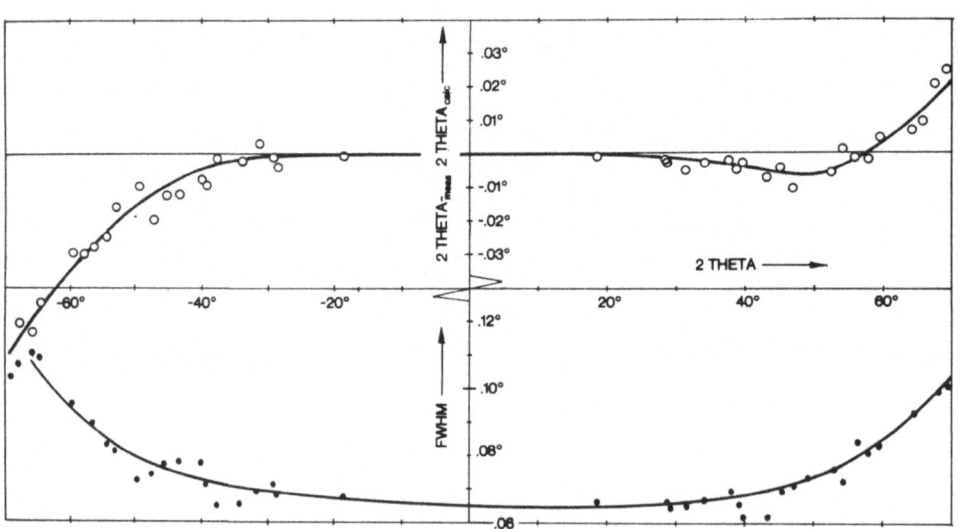

Fig. 8: 2 Theta calibration (upper part) and line width (lower part)

above all in comparison with Bragg-Brentano diffractometers without a primary beam monochromator (see figure 9).

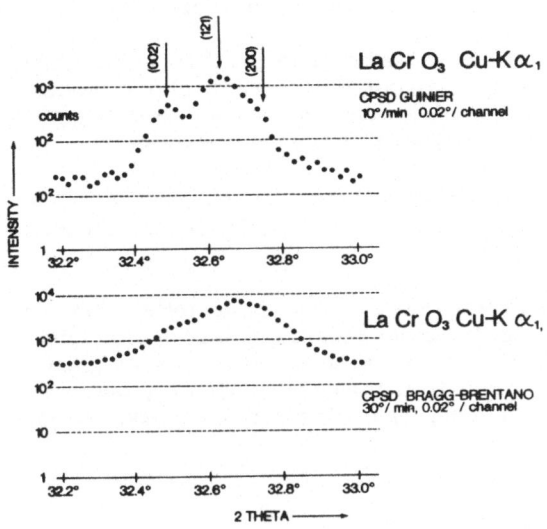

The example of LaCrO$_3$, a pseudocubic perowskite material, was also selected to show the effective reduction of fluorescence background using a primary-beam monochromator. In both cases the pulse height discriminator window of the PSPC was equal.

Fig. 9:
Effect of primary-beam monochromator:
Low background, easy recognition of overlaps.
(Intensity on logarithmic scale)

A final example was measured with the scanning-PSPC Guinier system in back-reflection mode of operation (see also figure 3). Besides the effect of simpler patterns due to pure Kα_1-lines high accuracy can be achieved by measuring symmetrically around 2 Theta = 180°, which is not practicable in Bragg-Brentano systems.

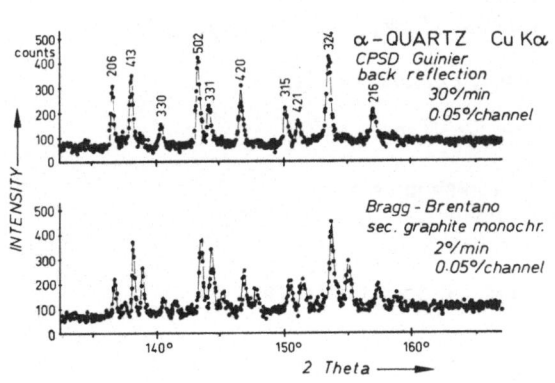

Fig. 10: Example of a back-reflection CPSD-Guinier recording (see fig. 3)

CONCLUSIONS

The use of a PSPC for data collection in a strictly focussing Guinier diffractometer reduces the measuring time considerably. Well plottable patterns can be
accumulated at scanning speeds of several tens of degrees per minute.

The necessary amount of sample material lies between about 10 % of a Bragg-
Brentano specimen and a Debye-Scherrer preparation. The quality of data makes
the system useful for XRPD-applications requiring high precision like indexing,pro-
file analysis or computer search/match analyses at narrow error windows.

As a transmission method it produces complementary information to reflection
data on preferred orientations in a sample. This may be utilized to reconcile experimental and calculated intensities.

The first laboratory model of a scanning-PSPC Guinier diffractometer was constructed for the angular range of symmetric Guinier cameras. A second prototype
having the extended range of asymmetric cameras is presently in a test program.
It also uses a longer focal distance of a new type of monochromator crystals and
can be operated with a fix sample position (fig. 2) in addition to the moving-sample
technique described here.

ACKNOWLEDGEMENTS

The author is indebted to Dr. K. L. Weiner from the Institut für Kristallographie
der Universität München for helpful discussions. Special thanks are addressed to
G. Zorn for his patience while aligning the system and to Barbara Jobst for software adaptations.

REFERENCES

/1/ K. L. Weiner: Guinier-System 600
    Manual from R. HUBER Diffraktionstechnik, 8211 Rimsting, W-Germany

/2/ H. E. Goebel: Adv. in X-Ray Anal. 22, 255-265 (1979)

/3/ H. E. Goebel: Adv. in X-Ray Anal. 24, 123-138 (1981)

/4/ M. K. Kopp: ESF-CNRS-EMBL Workshop on X-Ray
    Position-Sensitive Detectors, Hamburg, Nov. 17-21, 1980
    Proc. to be published in Nucl. Inst. & Meth.

/5/ H. Dachs, K. Knorr: J. Appl. Cryst. 5, 338-342 (1972)

OBSERVATION OF AN X-RAY BEAM OF 10 MICRORADIAN

DIVERGENCE WITHOUT USING ANY COLLIMATOR

K. Das Gupta

Radiation Research Laboratory
Department of Physics and Engineering Physics
Texas Tech University, Lubbock, Texas 79409

I report the experimental setup to obtain an x-ray beam from germanium monocrystal of unusually small angle of vertical divergence of the order of 2 sec. of arc without using any collimator. The production and the properties of the highly collimated beam of x-ray will be described. A rectangular piece of germanium 1" x 1" and of thickness 0.5 mm has been cut with specific orientation of crystallographic planes (111) and (110). The 1" x 1" surface is parallel to (111) planes and the surface of one edge is parallel to (110) planes suitable for Borrmann channelling via (220) planes.

The x-ray beam of low divergence, recorded photographically on x-ray film placed at right angle to the beam, appears more or less as a rectangular hot spot superposed on the characteristic x-ray lines from the target, via 440 Borrmann channelling.

The crystal is mounted on the tungsten plate shutter with a pin-hole at H and the 220 planes in vertical position, as shown in Fig. 1. The target line focus T_1T_2 is 12.5 mm in length and of height 1 mm. The center of the pin-hole at H, on the adjustable tungsten plate shutter, is at a distance of 8.8 mm from the point Z and covers the range of Bragg angles between 15 to 23° in Borrmann channelling from different points of the line focus target T_1T_2.

The experimental arrangement, as shown in Fig. 1, is a convenient method of precision x-ray spectroscopy by simultaneous Borrmann channelling of a convergent set of rays from various points of the line focus target to the pin-hole at H for a fixed position of the Borrmann quality germanium, silicon, diamond, and other such crystals. We have photographically recorded tungsten L-α, β, and γ lines for a

325

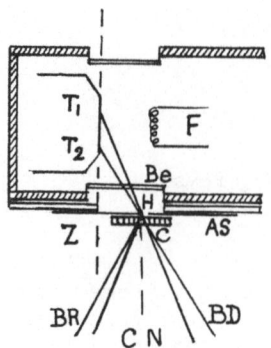

Fig. 1: Bragg-Borrmann spectroscopy
using convergent rays from a
line-focus target T_1T_2 to the
crystal C set immediately
after the pin-hole H. The
reflected Borrmann spectra BR
and the direct mirror image
Borrmann spectra BD appear
symmetrically on either side
of the crystal normal CN.

fixed position of the crystal. The spectral lines appear on either
side of the crystal normal as usual in Borrmann spectra.

We have measured the wavelengths of W- L-α, -β, and -γ spectral
lines recorded on a film placed at a distance of 1 meter from the
crystal and the values agreed very well with Bearden's table. With
a demountable target T_1T_2, the experimental setup as shown in Fig. 1
turns out to be a convenient method of Bragg-Borrmann spectroscopy.
The position of the pin-hole at H on a movable tungsten plate is
adjusted to cover the spectral range under investigation.

In our experiments we have used (a) a new x-ray tube with Ag
target and (b) an old Ag target tube coated with tungsten. We have
recorded in both cases the hot spots of low divergence superposed
on Ag- K- series lines channelled via (440) reflection. The tungsten L- spectra obtained with the old Ag target tube channelled via
(220) planes do not show any hot spot on any of L-α, -β, or -γ lines
of tungsten. The hot spots appear superposed on characteristic Ag-
K- series lines only when rays from the line focus target are channelled via (440) reflection that is in 2nd order of (220) planes.

Borrmann and Hartwig[1] observed a significant enhancement of the
intensity of the (111) reflection, anomalously transmitted through
thick germanium plates when either (11$\bar{1}$) or (1$\bar{1}\bar{1}$) planes are brought
to diffracting position simultaneously with the (111) planes. Ben
Post et al.[2,3] reported the results of a simultaneous four-beam Borrmann diffraction through germanium crystal. Ben Post et al., in de-
scribing their experimental setup remarked: "The use of a divergent
incident beam enormously facilitates the alignment of the crystal.
It is only necessary to adjust the crystal to within 2 to 3° of the
optimum setting: the divergent beams will then seek out the effects
sought".

I have recently repeated the investigation of the low divergent

 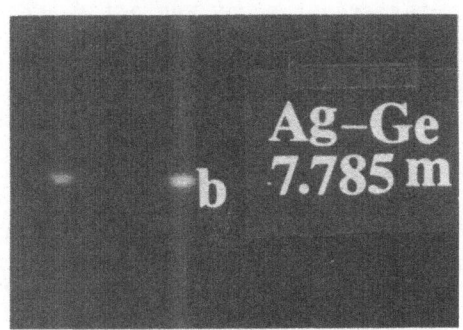

Fig. 2 (a): Hot spots are supersposed on Ag Kα₁ and Kα₂ lines channelled via (440) planes of Ge crystal. The film was
kept at a distance of 2 m from the crystal.

2 (b): The same spots appeared at a distance of 7.785 m from
the crystal. The vertical divergence of the hot spot
measured by a micro-comparator is 2.5 sec of arc.

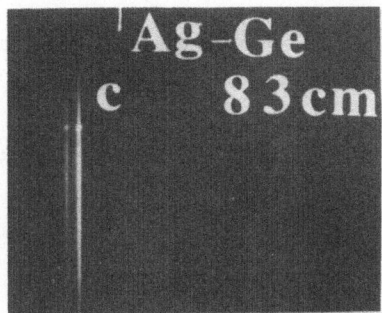

2 (c): In a new experimental setup, we have used a collimated
beam of x-rays with horizontal divergence of 25 min of
arc and mounted the Ge crystal onto a Bragg spectrometer.
Prominent hot spots appeared superposed on AgKα₁ and Kα₂
lines via (440) planes in Borrmann transmission.

hot spots by using a highly collimated x-ray beam of horizontal divergence of 25 min of arc with a 2° take-off angle from target T_1T_2
and mounted the germanium crystal onto a Bragg spectrometer. Using
a collimated beam of less than 0.5° striking the germanium crystal,
I have reduced, enormously, the possibility of multi-diffraction effects yet observed the hot spots in Fig. 2 (c).

    Both of our experimental methods (a) using a convergent set of
rays to a pin-hole and (b) a highly collimated beam of x-rays striking
the crystal mounted onto a Bragg spectrometer are, in essence, methods

of x-ray spectroscopy where for a definite lattice spacing of the germanium crystal different wavelengths are dispersed according to $n\lambda = 2d \sin \theta$. The experimental setup for the study of the enhanced Borrmann diffraction by Borrmann and Hartwig,[1] Ben Post et al.,[2,3] and many others is essentially a method to obtain a multiple diffraction for a number of lattice planes using a divergent beam that strikes a finite surface area of the monocrystal.

We tested a silicon crystal with an orientation cut similar to that of germanium. The spectral pattern is similar to that of germanium in every respect but we found no hot spot via (440) channelling of $AgK\alpha_1$, $K\alpha_2$, etc.

Our preliminary measurements of the hot spots in Fig. 2 (a), 2 (b), and 2 (c) by pulse height spectroscopy using a multi-channel analyzer revealed strong peaks at 8.6 and 9.5 keV. These results are now being verified by using a two-germanium-crystal spectrometer and the results will be reported in the near future.

ACKNOWLEDGEMENTS

This research has been supported by the Air Force Office of Scientific Research Grant Number 76-3098 and the Robert A. Welch Foundation Grant Number D-243.

REFERENCES

1. G. Borrmann and W. Hartwig, Z. Kristallogr. 121:401 (1965).
2. Ben Post, S. L. Chang, and T. C. Huang, Simultaneous Four-Beam Borrmann Diffraction, Acta Cryst. A33:90 (1977).
3. T. C. Huang and Ben Post, Experimental Methods for the Study of Multiple Borrmann Diffraction, Acta Cryst. A29:35 (1973).

X-RAY RESIDUAL STRESS MAPPING IN INDUSTRIAL

MATERIALS BY ENERGY DISPERSIVE DIFFRACTOMETRY

C.J. Bechtoldt, R.C. Placious, W.J. Boettinger and
M. Kuriyama

National Bureau of Standards
Washington, D.C.   20234

ABSTRACT

An application of energy dispersive diffractometry to the measurement of residual strains (stresses) in the interior of industrial materials is described with particular emphasis on the use of high energy (up to 250 keV) x-ray photons. The use of high energy photons permits better penetration into materials. Hence diffraction data for evaluating bulk residual strains can be obtained in the transmission geometry in contrast with the conventional angular dispersive diffractometry, which uses Bragg reflections from the surface of materials. The reliability and sensitivity (detectability of small strains) of the energy dispersive method are demonstrated through its application to mapping of residual stress distributions across weld zones in Alaskan pipe line segments (API5LX65). The detectability of strain variations within materials depends on x-ray optical resolution and statistics.

The energy dispersive system is simple and compact, and involves no moving parts. Through the present demonstration, this energy dispersive method shows great promise for providing a powerful nondestructive tool for the evaluation or mapping of residual stress distributions within bulk materials. This method is particularly suitable for inspection and monitoring of industrial materials.

329

INTRODUCTION

For monitoring the structural stability of industrial materials, stress distributions within materials should be measured quantitatively in a nondestructive manner. The most accurate method for the determination of residual stresses is x-ray diffraction from the surface of materials[1]. Unfortunately such measurements are incapable of detecting stresses in the interior of bulk materials (>1mm thick). When the incident x-ray beam is polychromatic (white radiation), the change in the atomic interplanar spacings can be detected as energy changes in the peak positions of Bragg diffraction instead of the changes in the Bragg angles. This mode of measurement is called energy dispersive diffractometry[2]. A particular advantage of this mode is found in the availability of high energy x-rays which satisfy Bragg's law in the materials. The high energy x-ray photons permit better penetration into the materials, and hence strains deep inside the materials can be detected. The use of an energy dispersive system for residual stress evaluation was tested in 1973 by Leonard[3], who concluded, however, that this system did not have sufficient accuracy for this purpose. In the past several years, more effort has been made on the application of energy dispersive diffractometry to the detection of residual stresses within materials[4-6]. In this paper, some results are presented using bulk industrial materials such as a commercial aluminum block and a welded portion of Alaskan pipe line to demonstrate the potential of this technique as an industrial tool for inspection of materials in the field.

Energy dispersive diffractometry involves the use of solid state detectors and white radiation. To measure residual stresses, higher resolution and accuracy are required than for numerous other applications of solid state detectors, as for example, scanning microscopy (microanalysis) and x-ray fluorescent analysis. If one could improve the resolution sufficiently and overcome other important industrial requirements, energy dispersive diffractometry becomes a practical method for residual stress measurement.

Measurement of strain tensors by energy dispersive diffractometry

In energy dispersive diffractometry, one is concerned with energy spectral profiles and peak positions at different energy values. For the present measurement of strains, only the peak positions are of interest since these Bragg energy peaks represent lattice constants in a local volume, and hence macroscopic strains. Bragg's law can be written

$$d(\text{Å}) = \frac{6.119}{\sin \theta} \frac{1}{E(\text{keV})} \quad (1)$$

to give lattice constants in terms of the energies of diffracted photons, where d is an atomic interplanar spacing related to lattice

constants, E is the energy of diffracted photons that is a peak position in the spectrum; the coefficient, including half the scattering
angle θ, is constant for a given scattering geometry. Entrance and
exit collimators guarantee the energy resolution as well as the spa-
tial resolution to the desired degree. A solid state detector is
placed after the exit collimator at a fixed scattering angle. For
the present purpose, the scattering angle is small to ensure complete
transmission geometry for high energy diffraction peaks.

Let us denote the geometrical orientation of a sample by the unit
vector, $\underset{\sim}{n}$, which is perpendicular to the bisector of the scattering
angle. For a given sample geometry, an energy spectrum consists of
many Bragg peaks associated with $h\ k\ \ell$ diffracting planes which are
perpendicular to $\underset{\sim}{n}$. From a set of interplanar spacings for different
$(h\ k\ \ell)$, one can define the "stretch" in the $\underset{\sim}{n}$ direction as

$$\lambda_n = \left(\frac{\Delta d}{d}\right)_n = \frac{d^n_{hk\ell} - d^{\circ}_{hk\ell}}{d^{\circ}_{hk\ell}} , \quad (2)$$

where $d^{\circ}_{hk\ell}$ is the interplanar spacing for $hk\ell$ diffraction obtained
either from a reference unstrained material, or from an arbitrary
reference part of the sample. This quantity is related to the strain
tensor by

$$\lambda_{\underset{\sim}{n}} = \sum_{ij} \varepsilon_{ij} n_i n_j , \quad (3)$$

where ε_{ij} is the i, j component of the strain tensor and n_i is the
$i\underline{\text{th}}$ component of the vector $\underset{\sim}{n}$ in a frame imbedded in the sample.
Since the strain tensor has, in general, six independent components,
ε_{11}, ε_{12}, ε_{13}, ε_{22}, ε_{23} and ε_{33}, one must have a set of six simultaneous equations, which can be obtained using (3) for independent mea-
surements performed in six different sample orientations.

There are many ways to perform six independent experiments.
Here we mention one, in which the same predetermined volume in the
interior of the sample is viewed by the detector for all six measurements. As shown in figure 1, the sample is rotated around the incident x-ray beam three times using two detectors (only one detector is
shown in Fig. 1) to obtain six independent measurements. (When the
incident beam direction is used as a rotational axis, one cannot obtain six independent sets of data for one scattering angle.) As the
sample is translated up, down and sideways, one can map any desired
volume in the sample. The incident beam must be prepared using a
slit system so that it has a very small cross section, and also it
should be extremely parallel (no angular divergence). In addition,
the detector must receive the scattered beam within an extremely narrow angular range so that the scattered beam should originate from a

well determined volume inside the sample. This simple geometrical
argument concerning the scatterer is justified so long as the scattering (Bragg diffraction) is kinematical.

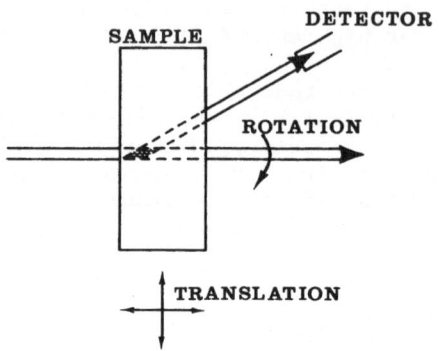

Fig. 1. An example of the geometrical arrangements for the evaluation
of residual strains in a predetermined volume inside materials. The
sample is rotated three times around an x-ray beam to give six independent sets of measurement with two detectors. The mapping of
strains throughout the volume of the sample is achieved by translations.

Gaussian curve fitting

A measure of the resolution is given by

$$\frac{\Delta d}{d} = \frac{\Delta E}{E} + \cot \theta \, \Delta \theta \;, \quad (4)$$

where ΔE is the energy resolution of a detector and $\Delta \theta$ is the angular
divergence created by the incident beam divergence and the receiving
angle of the detector. For intermediate energy photons, $\Delta E/E$ is of
order 10^{-2}. This term becomes smaller as the photon energy increases.
The second term is controlled by the geometrical arrangement. In the
application to residual stress measurements, a strain of less than
10^{-4} (or 10^{-2} percent) should be detectable. One may claim that
the energy resolution of current solid state detectors is not small
enough to reduce the value in (4) for the evaluation of residual
strains[7,8]. It is known, however, that each diffraction profile in
the energy spectrum obtained from a solid state detector is very close
to a Gaussian. If the mathematical shape is known for a spectral profile, then the claim mentioned above can be circumvented. Mathemati-
cally fitting the observed peaks with Gaussian functions, one can

determine the peaks of the profiles far more accurately than the $\Delta E/E$ value in (4) would indicate. Indeed, it has been experimentally proved that the accuracy of determining the peak positions can be improved by a factor of 100 for photons of intermediate energies (up to 30 keV)[5].

The objective of this paper is to study the same technique applied to the transmission geometry using more penetrating high energy photons through much thicker industrial materials. In order to ensure high energy diffraction peaks with low index diffracting planes to be present in a spectrum, the scattering angle should be reduced to about 5°. This obviously increases the second term in (4), even if the incident beam divergence and the receiving divergence are well controlled by the entrance and exit collimators. The effect from the relatively large second term generally results in broadened line profiles. However, in most industrial materials microscopic inhomogeneous strains may be much smaller than macroscopic strains. Thus the profiles of diffraction peaks may have only symmetrical broadening. As long as the profiles are Gaussian, the line broadening due to inhomogeneous strains does not hinder the determination of the peak positions.

The energy dispersive diffractometry technique using high energy photons must meet the following three major conditions before it becomes useful in industrial applications. (a) Well defined diffraction energy peaks should be obtained in the energy range of about 100 keV. High energy photons can penetrate through rather thick materials, as, for example, a steel plate 2.5cm thick or an aluminum plate 5cm thick. Hence information on stresses within bulk materials can be obtained from the diffraction peaks: (b) These diffraction peaks should have a Gaussian profile so that the curve fitting technique[4,5] can be used for the determination of the peak positions: And (c) the measurement accuracy of lattice constants and their local variations (hence, residual stresses) should be increased by a factor of 100 as compared with the resolution limit of solid state detectors after the profiles are fitted by Gaussian curves. The improvement factor (that is, the accuracy of determined strains) should be studied as a function of counting statistics.

An earlier experiment, using low energy photons and thin materials, has proved that all of the three conditions are satisfied[5]. The use of high energy photons normally creates more difficult situations in controlling scattering problems and background problems which directly affect the accuracy of the stress determination. It is therefore necessary again to test these conditions using high energy photons with primary interest in the industrial application of this technique as a possible inspection and monitoring tool for material integrity.

Mapping of residual stress distribution in industrial materials

An energy dispersive system consists of two simple collimators,

a solid state detector, an industrial radiographic x-ray source and a
sample stage with a translator. The x-ray source was operated at
270 kV and 7 mA with a tungsten target and a Cu window. The source
size was estimated to be 4mm x 4mm. Each of the two collimator pipes
had two holes which were separated by a distance of 36cm. In the
majority of experiments, the hole diameter was 1.5mm or less. The
incident beam divergence and the receiving divergence were thus effectively much less than 0.5°. The area irradiated on the sample surface was a circle of less than 2mm in diameter. The scattering angle
was set near 5°. An intrinsic germanium planar detector having an
area of 100mm^2 and a thickness of 5mm was employed.

As shown in Figure 2, diffraction energy peaks of significant
intensity are clearly separated. This spectrum was obtained from a
commercial steel plate (AISI-C1015-SAE 1015) 3/8 inch (=9.53mm) thick.
The scattering angle was set at $2\theta = 5.7°$ with the incident divergence
3 m rad. In this geometry, the diffraction peaks were obtained through
the entire thickness of the sample at a given irradiated point. Note
that the well defined (200), (211), (220), (310) and (321) diffraction
lines appear in the energy range between 90 keV and 170 keV in this
transmission geometry. This result confirms that condition (a) is
completely satisfied. Figure 3 shows an example of an observed (110)
diffraction profile which has been fitted by a Gaussian curve plus a
linear background. This profile was obtained from a cold worked steel
plate 1/2 inch (12.7mm) thick using a scattering angle of 3.8°. The
excellent Gaussian fit shown here suggests a high precision in the
determination of peak positions for the observed diffraction peaks.
Hence, condition (b) has also been satisfied in experiments using high
energy photons.

The accuracy of determining the peak positions depends simply
upon the counting statistics, as shown in Table I. This result agrees
with previous results obtained from thin samples using low energy photons[5]. Table I also shows the reproducibility of the peak positions
in a series of experiments where the incident beam views the same volume within the sample; the peak positions remain identical within the
statistical error. As a reference to possible instrumental instabil-
ity, W K_β spectra along with nuclear radiation line spectra from an
isotope in each run and between runs were fitted with Gaussian curves,
as shown in Table I. These results indicate that instabilities, if
any, are not significant so as to jeopardize the accuracy in the determination of diffraction peak positions. Table I demonstrates that
the measurement accuracy can be increased by a factor of 100 and more
over the resolution limit of solid state detectors with reliable reproducibility if sufficient counting statistics are established. This
implies that strains of the order of 10^{-4} can be reliably detected.
Condition (c) has thus been shown to be satisfied as well.

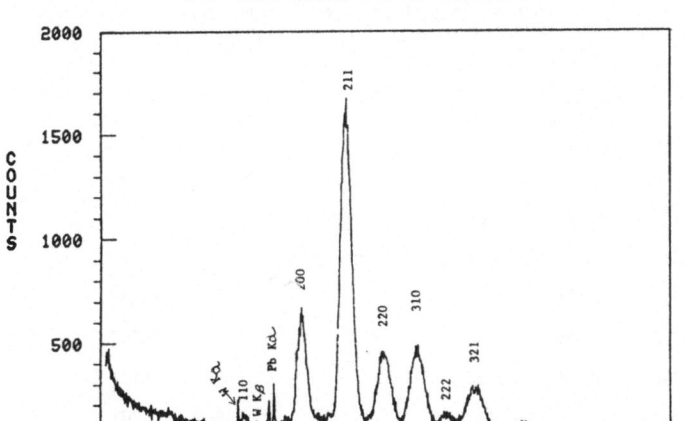

Fig. 2.   A Bragg diffraction energy spectrum obtained from a commercial AISI-C1015-SAE 1015 steel plate 3/8 inch (9.53mm) thick.

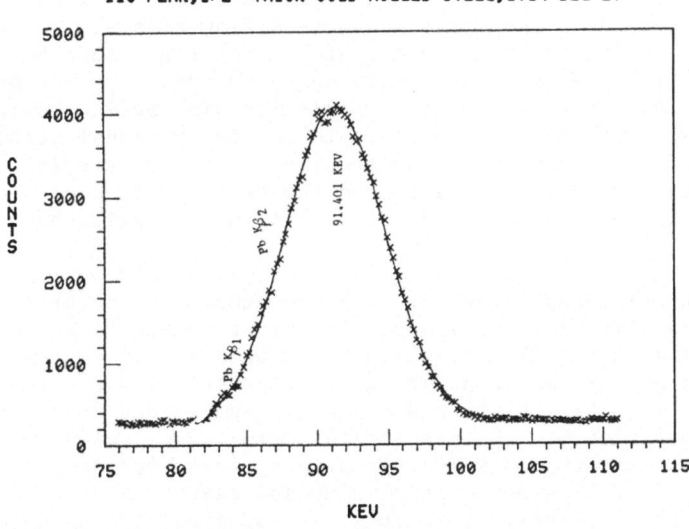

Fig. 3.   An example of the observed (110) diffraction profiles from a
steel plate 1/2 inch (18mm) thick fitted with a Gaussian curve with
linear background.

TABLE I

Relation between the accuracy of peak position determination and counting statistics.

Run	211			W Kβ₂		
	Peak Height (counts)	Peak Position (kev)	2σ	Peak Position* (kev)		2σ
1.	846	119.928	± 0.066	69.031	±	0.040
2.	1733	119.994	± 0.040	"	±	0.031
3.	2578	119.967	± 0.040	"	±	0.023
4.	3465	119.843	± 0.032	"	±	0.018
5.	4311	119.853	± 0.030	"	±	0.017
6.	5196	119.894	± 0.024	"	±	0.015
7.	6042	119.894	± 0.024	"	±	0.015

*The peak position of the tungsten $K\beta_2$ line is assigned to be 69.031 kev, after curve fitting.

Finally, we come to the last, but most important question, as to whether this technique is workable in practice as a nondestructive tool to monitor the material integrity for industrial purposes. For industrial monitoring, the inspection in each small volume within a material may not be necessary. It is more important to locate the approximate section of the material where a flaw exists. One can always pinpoint the exact location inside the material later by repeating the test with greater sensitivity and accuracy. (In the present method, this means better counting statistics with refined beam collimation.) We have therefore decided to map the residual stress distribution inside a material across the length of the material. We do not simultaneously have information on the depth profile of residual stress distributions in this mode of operation. We will show here one example.

The scattering angle was set at 5° to guarantee the transmission geometry. The slit size was 1.5mm. The first sample is a section of one of the Alaskan pipe lines, which has a weld zone. Figure 4 shows the variation of strains (and, in turn, stresses) across the weld zone. The material is 14.6mm thick, and has a minimum welded zone 6mm wide centered at position 0. The statistical error (2 σ- value) in this measurement is less than 3×10^{-4} in strain, which corresponds to 3000 psi in stress in this steel. This value indicates the detectability of stress in the present measurement. Reproducibility was also confirmed. In this measurement the counting statistics were only moderate. With longer counting times or an increased photon flux, the counting statistics can be improved. It is interesting to note that the observed stresses in the sample are compressive, and the stress

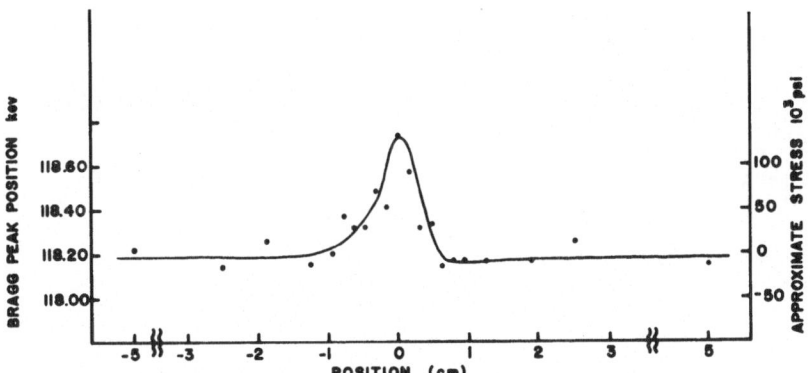

Fig. 4. Residual stress distribution across a welded zone in an Alaskan pipe line steel (API5LX65) 14.6 mm thick. This mapping obtained
at a scattering angle 5° in the transmission geometry using (211) diffraction. The beam divergence and receiving divergence are much less
than 3 m rad (0.17°). The beam size on the sample is less than 1.5mm
x 1.5 mm². Positive values of stress represent compression.

value near position 0 has reached the yield stress of this material.
We have indeed confirmed by radiography that this sample has a crack
in the weld zone. During welding, alloy compositions might vary lo-
cally and different metallurgical phases might be produced, and hence
might create a local lattice constant variation. Such a variation
would provide a profile similar to Figure 4. We have taken x-ray
diffraction pattern from both weld bases and weld metal and similarly
from sections cut parallel to the surface in the $\theta - 2\theta$ scan mode and
hence $\psi = 0°$. No variations in lattice constants between weld base
areas and weld area were observed. Hence the result observed in
Figure 4 is the manifestation of residual stress variations in the
sample.

References

1.   J.B. Cohen, H. Dölle and M.E. James, Proceedings of Symposium
      on Accuracy in Powder Diffraction in: "NBS Special Publ. 567,"
      p. 453 (1980).
2.   B.C. Giessen and G.E. Gordon, Science 159 973 (1968).
3.   L. Leonard, "Franklin Institute Research Lab Report F-C3454,"
      Philadelphia, July (1973).
4.   M. Kuriyama, W.J. Boettinger and H.E. Burdette, ASNT National
      Fall Conference, p. 49 (1978), October, Denver, CO.
5.   M. Kuriyama, W.J. Boettinger and H.E. Burdette, Proceedings of
      Symposium on Accuracy in Powder Diffraction, "NBS Special Publ.
      567," p. 479 (1980).

6.   M. Kuriyama, Proc. of Synposium on Nondestructive Measure of
      Wheel/Axle Residual Stress, Dept. of Transportation, (1981).
      In print.
7.   E. Laine, I. Lahteenmaki and M. Kantola, X-Ray Spectrometry,
      1 93 (1972).
8.   T. Kukamachi, S. Hosoya and D. Terasaki, J. Appl. Cryst. 6 117
      (1973).

STRESS MEASUREMENT AND PRECISION DIFFRACTION ANGLES

ON LARGE GRAINED SPECIMENS

Charles S. Barrett

University of Denver Research Institute

Denver, Colorado   80208

This note describes a diffractometer technique that is successful on specimens that yield sharp diffraction peaks but that have grain sizes somewhat too large to give reliable results with the usual powder diffraction methods, owing to the peak profiles being distorted or displaced by having too few grains reflecting. Such specimens are not uncommon in stress analysis and the errors encountered are sometimes reduced by oscillating the specimen through an angle range, $\Delta\beta$, of a degree or so about the diffractometer axis, thereby increasing the number of recorded reflections. The technique discussed here is basically a single crystal technique rather than a powder technique, and with a diffractometer capable of measuring reflections at high angles on both sides of the initial beam this technique would amount to W. L. Bond's precision method.[1] For the usual diffractometer, however, high diffraction angles must be measured on one side only, even though some instruments permit the lower angle reflections to be measured on both sides for accurately locating the zero of the 2θ scale. We are discussing a technique as applied to these usual types of diffractometers.

The initial beam is collimated so as to strike a narrow spot on the specimen. The spot is centered on the specimen surface at the diffractometer axis. To reduce errors that arise from lengthwide divergence of the initial beam along the axis, the divergence slit at the x-ray tube is shortened (blocked off at both ends, leaving a short opening at the center). The slit at the detector is also shortened and centered so that any beam passing it lies very close to the plane that contains the primary beam and is normal to the diffractometer axis. The diffractometer arm is set at the assumed 2θ for the high angle reflection that is to be used,

and while the diffractometer arm remains fixed at this position,
the specimen is manually rotated slowly through an angle range,
$\Delta\beta$, of a degree or two centered on the desired angle β between the
primary beam and the normal to the specimen surface. For reflection from planes that are parallel to the surface (the $\psi = 0$ condition in stress analysis), $\beta = (180° - 2\theta)/2$, but other values of
β can also be used. The individual reflections that occur during
this rotation are noted; if many are seen, the divergence slit is
narrowed, and if none are found the specimen is shifted in its own
plane for additional tries, and if necessary the divergence slit
is widened. The aim is to find a strong reflection that is well
isolated from all neighboring ones as β is altered, so that its
peak profile is undistorted by any overlapping peaks. A strip-
chart record is used during this search in order to see better
the possible overlaps, distortions and relative intensities.

Fig. 1A is a strip chart record that illustrates a condition
in which too many reflections occur per degree change of β (at
constant 2θ). In this example the divergence slit at the x-ray
tube was too wide, causing obvious overlapping of peaks, and also
increasing the danger that some overlaps are undetected. In this
example the divergence slit was narrowed until very few reflections
occurred during the rotation through range $\Delta\beta$; a strong one that
was completely separated from all neighbors is illustrated in Fig.
1B.

When a strong isolated reflection is found, its intensity is
maximized by manually adjusting alternately the diffractometer arm
and the specimen angle β in an iteration procedure. The high intensity is used to indicate a reflection that is not clipped at
the sides or ends of the receiving slit. Since the chosen reflection is strong this iteration of 2θ and β adjustment can be done
fairly rapidly; moderate precision in the results obtained have
been obtained with adjustments that took less than a minute, but
higher precision can be obtained with step counting. Narrow slits
are required for good accuracy--and a 2θ scale that is accurately
zeroed on the direct beam that has been centered on the diffractometer axis. The usual care to avoid specimen displacement errors
is needed. When the maximum intensity adjustment of 2θ and β has
been found with narrow enough slits for the desired precision,
the 2θ and β rotations are then coupled together and the peak profile is scanned with the usual $2\theta : \beta$ ratio of $2 : 1$. The reflection angle is then determined by one of the common methods, such
as step counting and parabola fitting. A single peak measured in
this way may give sufficiently precise results if the geometrical
conditions in the diffractometer are close enough to ideal, and
if the different grains in the specimen reflect identically, but
in general it is advisable to repeat the procedure with additional
grains and determine a mean 2θ value.

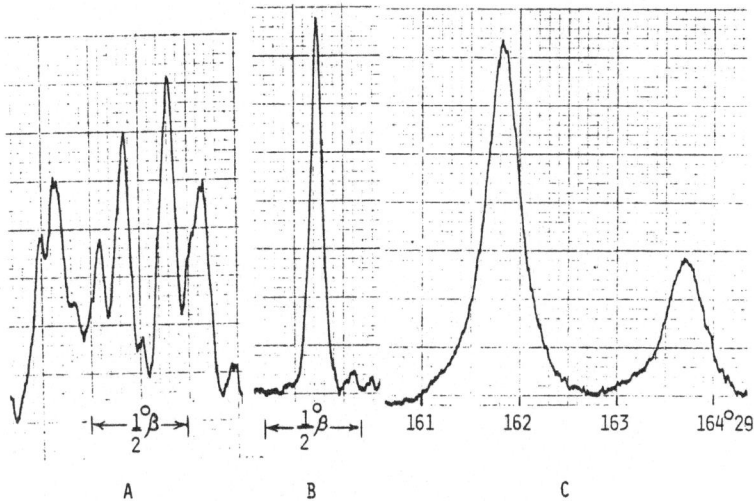

Fig. 1. CuKα 511 and 333 peaks from individual grains in a com-
         mercial aluminum sheet.
         (A) Kα₁ overlapping peaks with β varying (i.e. specimen
         rotated), 2θ fixed. Illustrating condition to be avoided
         in using this individual grain method.
         (B) Isolated Kα₁ peak with β varying, 2θ fixed.
         (C) Isolated reflections of Kα₁ and Kα₂ with 2θ coupled to
             β after maximizing the peak intensity, with 2θ scanned
             at twice the rate of β.

     The following is an example from work that was done using a
Siemens Krystalloflex II diffractometer that had been modified[2] so
as to carry a curved graphite crystal monochromator at the counter
and to permit reflections up to 2θ = 164.4°. A commercial aluminum
sheet that gave the sharp reflections of Fig. 1A and 1B for CuKα₁
with indices 511 or 333 gave the profiles of Fig. 1C, with well
resolved Kα₁ and Kα₂ peaks when the 2θ and β rotations were coupled.
The peak from each individual grain in these tests was measureable
with a precision (standard deviation) of ± 0.002° 2θ, but the average of reflections from several grains in these samples had a considerably larger uncertainty than this, but was suitable for stress
analysis. The divergence slit was 1/8° or 1/4° (0.16 or 0.33 mm
wide), 5 mm long in different runs, at 105 mm from the specimen;
the counter slit, of similar dimensions, was 177 mm from the
specimen.

REFERENCES

1. W. L. Bond, Acta Cryst. 13:814 (1960).
2. Paul Predecki and Charles S. Barrett, J. Compos. Mater. 13:61
    (1979).

DETERMINATION OF RESIDUAL STRESSES IN AUSTENITE AND MARTENSITE IN CASE-HARDENED STEELS BY THE $\sin^2\psi$ METHOD

Chongmin Kim

Climax Molybdenum Company of Michigan
A Subsidiary of AMAX Inc.
Ann Arbor, Michigan

INTRODUCTION

Compressive residual stress is one of the essential elements which contributes to the exceptional fracture resistance of case-hardened components. The sign and magnitude of residual stress are determined by transformation characteristics, thermal and mechanical histories of the steel. Accurate determination of residual stresses is essential in research on case-hardened steels.

In this investigation, a modern x-ray diffraction method was used to analyze residual stresses in two groups of steel samples; one group comprised carburized specimens of SAE 4028 steel (0.80Mn-0.24Mo-0.27C) with and without a shot-peening treatment, and the other consisted of carbonitrided specimens of SAE EX55 (0.99Mn-0.74Mo-1.78Ni-0.65Cr-0.19C) with and without a sub-zero treatment. The details of the steel compositions, heat treatments and post-case-hardening treatments are presented elsewhere.[1] Reference 1 contains data on fracture properties of the steels studied in this paper, thus providing necessary information for discussing the importance of residual stresses on fracture of case-hardened steels.

X-RAY DIFFRACTION ANALYSES

The profiles of residual stress in the cases of the two case-hardened steels were determined by the parallel-beam x-ray $\sin^2\psi$ method. The x-ray intensity data were also used for calculating retained austenite contents. A Rigaku x-ray diffraction system was used with a chromium-target x-ray source, a vanadium filter and a scintillation detector. The ψ-angles were chosen so that $\sin^2\psi$

values would vary by 0.1 for the austenite 220 peak from 0 to 0.5, and by 0.15 for the martensite 211 peak to 0.6. Both positive and negative ranges of ψ were used. Therefore, the total number of exposures was nine for martensite stress determination and eleven for austenite.

The measured diffracted beam intensities were first corrected for absorption due to ψ-tilt according to a previously determined dependence of intensity versus tilt angle,[2] and then normalized with respect to the Lorentz polarization factor (L.P.F.). The background intensities were measured at locations at which no martensite and austenite peaks were present but near the peaks of interest, normalized with the L.P.F., and then subtracted from the normalized diffraction peak intensities. Presently popular procedures of x-ray stress measurements seldom specify such background correction, apparently based on assumptions that the material is essentially martensitic/ferritic and that the peak-to-background ratio is high. However, case-hardened steels usually contain significant quantities of austenite in the case, so that the background correction becomes necessary.

The diffraction peak position was determined by least-squares curve fitting a parabola to five to seven evenly spaced data points in the top 15 percent of the peak. In addition to the diffraction peak angle, the area under the parabola and the width of the parabola at half the peak height, the half-peak width, were also computed. Comparison of the areas at various ψ-angles gives a measure of preferred orientation in the material, and the average of the areas was later used to determine the volume fraction of retained austenite. The half-peak width has been shown to indicate the level of hardness in hardened steels.[3]

The residual stress was determined by the $\sin^2\psi$ method. A plot of $(d_\psi - d_\perp)/d_\perp$ versus $\sin^2\psi$ was constructed, and a straight line through the origin was fit by the least-squares method. The residual stress value was obtained by dividing the slope of the line by the constant $\frac{1}{2}S_2$, which was taken from Reference 4 as $5.8 \times 10^{-6} MPa^{-1}$ for martensite 211 and $5.5 \times 10^{-6} MPa^{-1}$ for austenite 220.

The retained austenite content was determined from the ratio of martensite 211 peak and austenite 220 peak intensities, each of which was the average of the corrected intensities at all the ψ-angles employed. This procedure has to be superior in terms of accuracy to the conventional one-exposure method, even if it may be expected that at times the method would be inferior to the rotating-tilting method.[5] To verify the accuracy and validity of the new method, austenite contents in two hardened steel samples from another investigation were determined by both the new method and the rotating-tilting method. The results agreed excellently; they were 68.5% and 10.5% with the former method and 70.0% and 11.0% with the latter.

The residual stress and retained austenite content in the hardened cases were determined up to 0.2 mm (0.008 in.) in depth for SAE 4028 and to 1.1 mm (0.043 in.) in depth for EX55. The specimens were prepared for x-ray analyses by sequentially electropolishing with a perchloric-ethanol electrolyte (78 cc perchloric acid, 100 cc butylcellosolve, 120 cc water and 700 cc ethanol) at a 40 V applied potential.

The bulk residual stress (macrostress) was computed by weighting the stresses in austenite and martensite with respective volume fractions and summing. The bulk residual stress profiles were further corrected for the effect of consecutive removal of stressed layers according to the mathematical method given in Reference 6. The same amounts of correction were also given to measured stresses in austenite and martensite phases.

The microhardness profiles of the hardened cases were determined by measuring Vickers Hardness Numbers with a Tukon microhardness tester and a 100 gram load. The objective was to measure changes of hardness due to the thermal and mechanical treatments and to observe how the x-ray peak half-peak width varies with hardness.

RESULTS AND DISCUSSION

Profiles of retained austenite content in the cases of carburized SAE 4028 and carbonitrided EX55 steels are presented in Fig. 1.

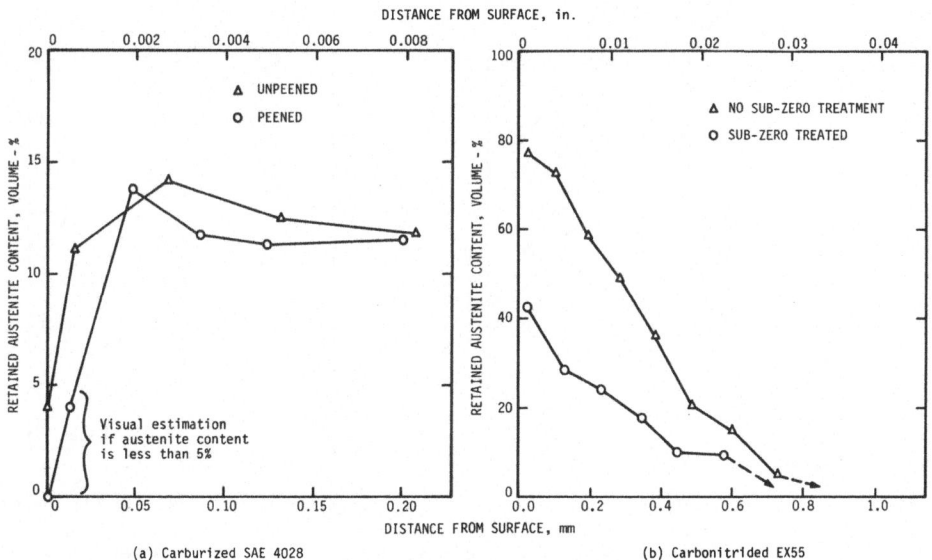

(a) Carburized SAE 4028   (b) Carbonitrided EX55

Fig. 1.   Retained Austenite Contents at Various Depths from Surface

Shot-peening of the former sample resulted in the transformation of austenite to martensite very near the surface; the depth to which the transformation occurred was less than 0.05 mm (0.002 in.). The sub-zero treatment given to carbonitrided EX55 resulted in a marked reduction in austenite content, Fig. 1b.

Examples of $(d_\psi - d_\perp)/d_\perp$ vs. $\sin^2\psi$ plots for martensite and austenite phases are shown in Fig. 2. They were chosen to demonstrate the need for employing both positive and negative ranges of ψ-tilts. These examples show the so-called ψ-splitting, which is explained theoretically through the triaxiality of the stress state at the measured sample surface.[7] If only one of the positive and negative ψ-ranges had been used, the resultant stress value would have differed from the value obtained using both ψ-ranges by as much as 70 MPa (10 ksi).

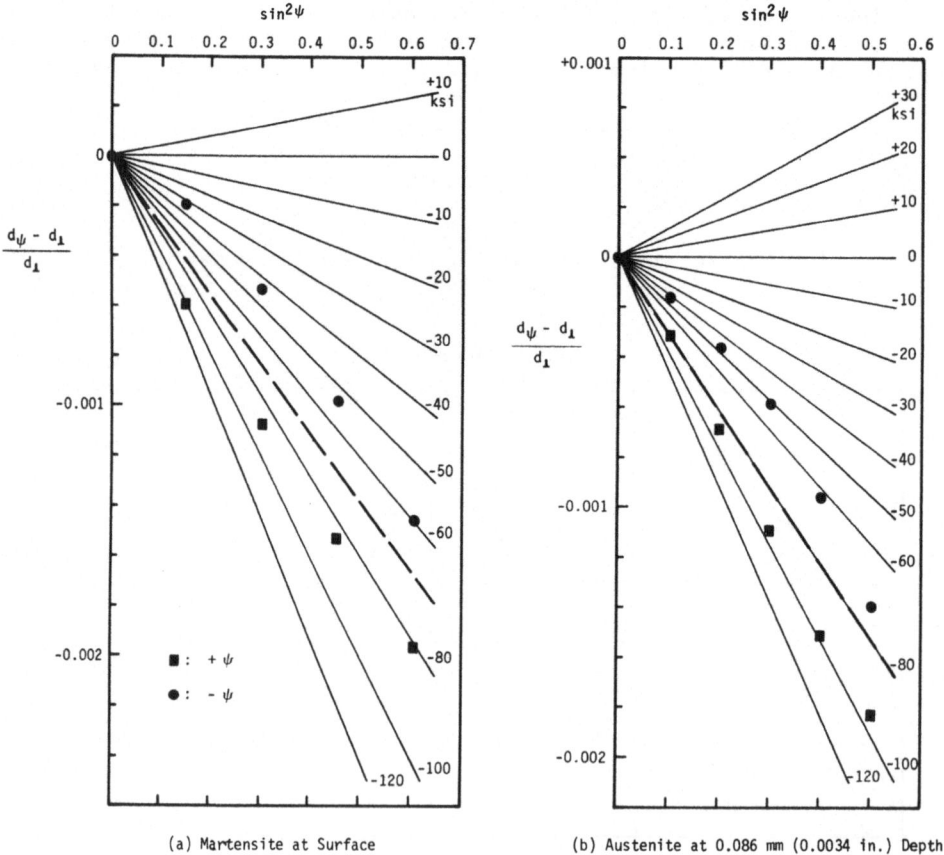

(a) Martensite at Surface   (b) Austenite at 0.086 mm (0.0034 in.) Depth

Fig. 2. $(d_\psi - d_\perp)/d_\perp$ versus $\sin^2\psi$ for Martensite and Austenite at Indicated Depths in Shot-Peened, Carburized SAE 4028

The residual stress patterns are shown in Figs. 3 and 4. The
residual stress pattern in the case of carburized SAE 4028 was influenced markedly by the shot-peening treatment. Differences existed
and the effect persisted even beyond the depth where austenite transformation due to the treatment had occurred. The compressive stress
in the martensite phase in the unpeened sample remained at about
165 MPa (24 ksi) to a depth of 0.2 mm (0.008 in.); the residual
stress in the austenite was low in this sample. In marked contrast
to the unpeened sample, the residual stresses in both martensite and
austenite phases in the shot-peened sample were highly compressive,
the observed maxima being 650 MPa (94 ksi) at a depth of 0.013 mm
(0.0005 in.) and 600 MPa (87 ksi) at 0.048 mm (0.0019 in.), respectively. The residual stress in martensite in the shot-peened sample
decreased to the same level found in the unpeened sample at a depth
of about 0.2 mm (0.008 in.). However, the stress in the austenite
phase at the same depth in the shot-peened sample remained highly
compressive at 250 MPa (36 ksi), compared to 20 MPa (3 ksi) for the
unpeened sample.

The residual stress pattern in the carbonitrided case of EX55
was greatly affected by the sub-zero treatment, Fig. 4. The stress
in the austenite of the steel without the treatment was compressive
at depths up to 0.2 mm (0.008 in.) from the surface, slightly tensile
between about 0.2 to 0.6 mm (0.008 to 0.025 in.) depth, and finally

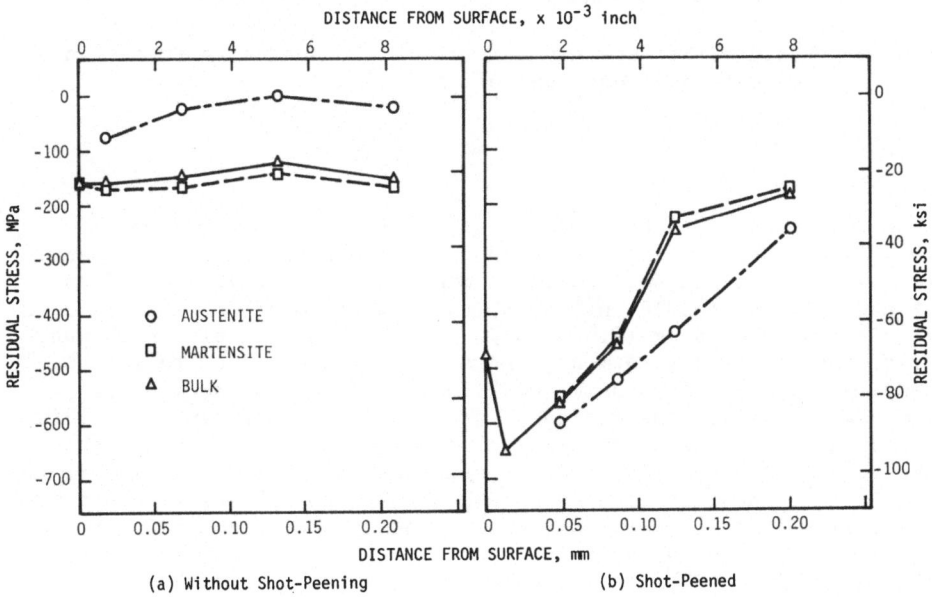

Fig. 3.   Residual Stress Patterns in the Case of Carburized SAE 4028

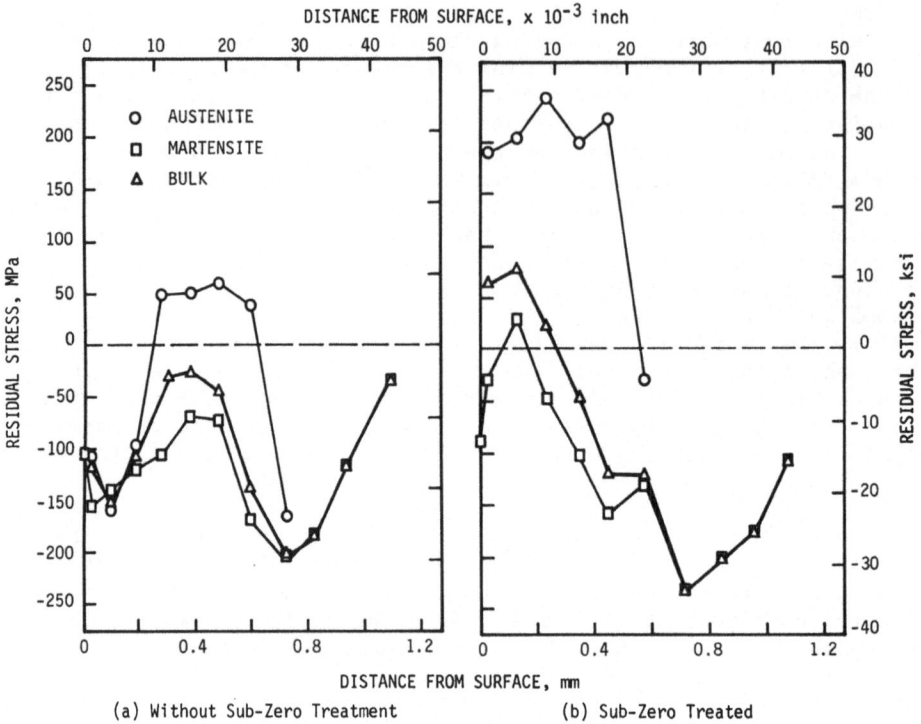

Fig. 4. Residual Stresses in Martensite and Austenite Phases in
         the Carbonitrided Case of EX55

changed rapidly to a high compressive value at about 0.7 mm
(0.028 in.) depth. The stress in the martensite of the same sample
was always compressive, showing a minimum value of 70 MPa (10 ksi)
at about 0.4 mm (0.016 in.) depth and a maximum of 205 MPa (29.5 ksi)
at a depth of 0.75 mm (0.029 in.). In the steel sample which was
given the sub-zero treatment, the stress in the austenite was highly
tensile in the first 0.5 mm (0.02 in.) from the surface, about
210 MPa (30 ksi). The residual stresses in the martensite in this
region also were less compressive than in the sample without the
sub-zero treatment. The results seem contradictory to an intuitive
guess that transformation of more austenite to martensite would make
the residual stress more compressive due to the volume expansion associated with the martensitic transformation. It is speculated that
the martensite, with its larger specific volume, stretches neighboring austenite and therefore leaves the untransformed austenite
under considerable tensile stress.

The normalized intensities of the martensite 211 peak did not vary significantly with ψ-angles. However, the austenite 220 peak showed a significant degree of preferred orientation in all steel samples. Examples are shown in Fig. 5, where the intensity changes with ψ are illustrated for carbonitrided EX55 before and after the sub-zero treatment. It is clearly observed that an intensity determination of a peak by a single exposure may lead to a considerable under- or overestimation of the peak intensity thereby resulting in an erroneous value of austenite content.

The half-peak widths of the diffraction peaks were compared together with microhardness data of the steel specimens across the case, keeping in mind that for martensitically hardened steels the half-peak width increases with the hardness.[3] However, shot-peened SAE 4028 had quite low half-peak widths near the peened surface even though the hardness level was similar to that of the unpeened one, Fig. 6. Therefore, it was concluded that the relationship between the hardness and half-peak width found in Reference 3 applies to martensitically hardened steels, but does not necessarily apply to mechanically worked martensitic structures.

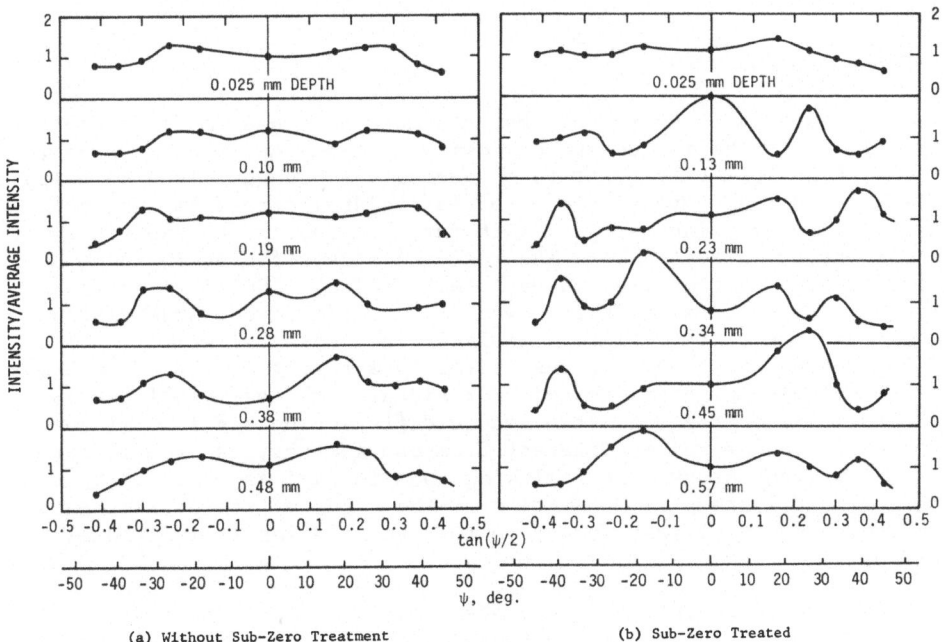

(a) Without Sub-Zero Treatment   (b) Sub-Zero Treated

Fig. 5. Intensity of Austenite 220 Peak at Various ψ-Tilt Angles, Showing Preferred Orientation at Indicated Depths in Carbonitrided EX55 Steel

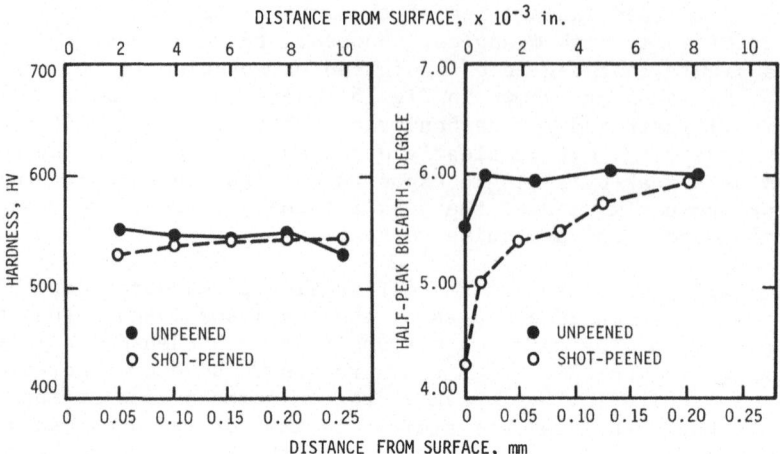

Fig. 6.   Profiles of Hardness and Half-Peak Width of Martensite 211
          Peak of Carburized SAE 4028 Steel

GENERAL DISCUSSION

     It is relatively recent that attempts have been made to quanti-
tatively describe influences of residual stress on fracture properties. Linear elastic fracture mechanics has been utilized to compute
crack tip stress intensity due to the residual stress fields, and
the results were incorporated with fatigue theories to explain residual stress effects.[8-11] The residual stress fields in these
studies are in principle "macrostress" fields. (Cullity[12] defines
the term macrostress as stress which is uniform over large distances.)
In order to perform a similar analysis on a material which consists
of more than one phase, it must first be agreed that the macrostress
is the volume-weighted average of residual stresses in all phases
present. Since residual stresses in different phases in a microstructure can be quite different from one another, as exemplified
in the present study, the following question arises: According to
the present systems of definition of the kinds of residual stress,
what is the classification scheme for the stresses in individual
phases?

     In the U.S., residual stresses are classified into two cate-
gories, macrostress and microstress.[12,13] In the German literature,
residual stresses are categorized in three groups: residual stresses
of the first kind, of the second kind, and of the third kind.[14] The

residual stress of the first kind is identical to the macrostress.
The residual stress of the second kind is the difference between the
average residual stress within a grain and the residual stress of
the first kind, and the third kind pertains to stress variations
within a single grain.

Both U.S. and German systems of residual stress definitions
fail to acknowledge the important facts that most engineering materials, especially steels, contain more than one phase in the microstructure and that differences in residual stress exist among phases.
This difference in residual stress among phases in a microstructure
is more important for theoretical as well as practical treatments
than stress variations from one grain to another.

The author proposes a new alternative set of definitions for
three categories of residual stresses. The first one would be the
same as the macrostress or the German residual stress of the first
kind. The second kind of residual stress would be defined as the
average residual stress in a particular phase in a microstructure.
The weighted average of the residual stresses of the second kind
would then be identical to the residual stress of the first kind.
The third kind of residual stress pertains to variations of stresses
of the first and the second kinds over small distances. Specific
terminology for the above three kinds of residual stresses remains
to be determined by general agreement.

The importance of knowing residual stresses in individual phases
for understanding fracture properties of case-hardened steels was
demonstrated in Reference 1, which contains fracture property data
on the very steels studied in the present paper. The residual stress
in austenite not only contributed to determining the macrostress,
but also was shown to affect fracture path. Case-hardened steels
usually fracture along grain boundaries, but the hardened cases of
the steels studied exhibited transgranular fractures at locations
where the residual stress in austenite was highly tensile. It appeared that the austenite, high in carbon and often as hard as martensite, was the "weakest link" and therefore the preferred fracture
path in the structure because of the high tensile residual stress to
which it was subjected.

SUMMARY AND CONCLUSIONS

The profiles of residual stress and retained austenite contents
were determined by the parallel-beam x-ray $\sin^2\psi$ method for two case-
hardened steels with and without sub-zero or shot-peening treatments.
The following are conclusions derived from this study and a proposal
for new definitions for residual stresses:

1. The parallel-beam x-ray $\sin^2\psi$ method is a viable technique for determining residual stresses in martensite and austenite phases in case-hardened steels. It is necessary to employ both positive and negative ranges of ψ-tilt angles.

2. Accurate determination of retained austenite contents is possible by processing the intensity data of diffraction peaks used for stress measurements. The method is superior to the single-exposure method, which is prone to errors arising from neglecting the preferred orientation. Such a preferred orientation was present to a significant extent in the austenite in steels studied.

3. The residual stress in austenite can be significantly different from that in martensite in both sign and magnitude. The stress in austenite not only contributes to the macrostress but also influences fracture properties significantly.

4. The relationship that diffraction peak width increases with material hardness did not hold for the shot-peened steel. The shot-peened sample retained very high hardness, but the diffraction peak width decreased markedly compared to that for the un-peened sample.

5. A proposal is made for defining three kinds of residual stresses as follows:

    The first kind: average residual stress, equivalent to the macrostress or the residual stress of the first kind as defined in the German literature.

    The second kind: average residual stress in a particular phase in a microstructure, for example, the residual stress in austenite. The weighted average of the residual stresses of the second kind is the residual stress of the first kind.

    The third kind: residual stress at a specific location in the microstructure.

REFERENCES

1. Chongmin Kim, D. E. Diesburg and R. M. Buck, submitted to J. Heat Treating, 1981.
2. J. Fukura and H. Fujiwara, Soc. of Mat. Sci. Japan, Vol. 15, No. 159, 1966, p. 825.
3. R. E. Marburger and D. P. Koistinen, Trans. ASM, Vol. 53, 1961, p. 743.

4. V. Hauk and H. Kockelmann, Arch. Eisenhüttenwes., Vol. 50,
    No. 8, 1979, p. 347.
5. Chongmin Kim, J. Heat Treating, Vol. 1, No. 2, 1979, p. 43.
6. M. G. Moore and W. P. Evans, SAE Trans., Vol. 66, 1958, p. 341.
7. W. Lode and A. Peiter, Härterei-Tech. Mitt., Vol. 35, No. 3,
    1980, p. 148.
8. Grzegorz Glinka, ASTM STP 677, 1978, p. 198.
9. J. F. Throop, "The Effects of Residual Stress in Fatigue," a
    summary report from ASTM Task Group No. 3 of E09.90 sub-
    mitted to ASTM Committee E09 on Fatigue, Nov. 9, 1980, p. 51.
10. Chongmin Kim, D. E. Diesburg and G. T. Eldis, presented at ASTM
    symposium "Residual Stress Effects in Fatigue," May 11, 1981,
    Phoenix, AZ.
11. A. P. Parker, presented at ASTM symposium "Residual Stress Ef-
    fects in Fatigue," May 11, 1981, Phoenix, AZ.
12. B. D. Cullity, Elements of X-Ray Diffraction, 2nd ed., Addison-
    Wesley, 1978.
13. Leonard Mordfin, presented at ASTM symposium "Residual Stress
    Effects in Fatigue," May 11, 1981, Phoenix, AZ.
14. U. Wolfstieg and E. Macherauch, Härterei-Tech. Mitt., Vol. 31,
    Nos. 1 and 2, 1976, p. 2.

X-RAY CHARACTERISTICS AND APPLICATIONS OF LAYERED SYNTHETIC

MICROSTRUCTURES

J. V. Gilfrich, D.J. Nagel, N.G. Loter*
Naval Research Laboratory, Washington, DC 20375

T. W. Barbee, Jr.
Stanford University, Stanford, CA 94305

INTRODUCTION

Practitioners of wavelength-dispersive x-ray spectroscopy are always seeking better dispersing devices. The wavelength-dispersive instrument is called a "crystal spectrometer" because natural or synthetically-grown crystals are most often used. Occasionally, other "manufactured" dispersers are suggested for specific applications: highly oriented polycrystalline graphite (1) provides much higher intensities than the crystals usually used for the K-lines of P, S and Cl; Langmuir-Blodgett films (heavy metal salts of fatty acids) provide 2d-spacings over the range of 70 to 130 A, making soft x-ray spectroscopy practical in a wavelength range for which natural crystals are not available (2).

Vacuum deposition techniques have been used in the last few years to produce layered synthetic microstructures (LSMs) which have interesting x-ray diffraction characteristics and the potential for valuable applications in x-ray spectroscopy. The preparation of these LSMs will be discussed briefly in the next section, followed by a summary of their x-ray character-ization. Finally, some applications to different areas of x-ray spectroscopy will be examined.

* Work done while a National Research Council Postdoctoral
  Fellow at NRL. Present address:
    American Science and Engineering
    Arlington, MA 02174

MULTILAYER PREPARATION

LSMs are produced by evaporating or sputtering alternate
layers of two components onto ultrasmooth substrates. The specific
devices being reported here have been fabricated at a facility
at Stanford University by sputtering layers of carbon and a
transition metal (TM) onto smooth silicon wafers(3). Figure 1 shows
a schematic cross-section of such a microstructure. The layers
may be as thin as about five angstroms but the two components
need not be the same thickness. LSMs can be produced with bi-
layer thicknesses from about 10 to over 100 A giving 2d-spacings of
20 to beyond 200 A. The compositions and 2d-spacings of multilayers
used in this investigation are listed in Table 1.

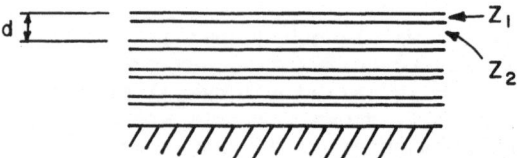

Figure 1. Schematic cross-section of a LSM. Z_1 and Z_2
represent the two components. The d-spacing is
the sum of the two dissimilar layer thicknesses.

Table 1. Components and 2d-Spacings of LSMs

Ti-C	Nb-C	Mo-C	W-C
			42 A
			61 A
113 A	104 A	109 A	120 A

X-RAY CHARACTERIZATION

Measurement of the x-ray characteristics of these LSMs over the range of about 5 to 15 A was carried out in a conventional x-ray fluorescence wavelength-dispersive spectrometer. Minor modifications to improve the instrument for this purpose included removing one of the two crystal holders so that the primary beam could be measured, installing an aperture ahead of the primary collimator so that the entire beam could be intercepted by the LSM even at the low Bragg angles sometimes required, and equipping the detector with a pressure-controlling apparatus so the the counting gas could be used at less than atmospheric pressure for the longer wavelengths. The details of the experimental measurements are given elsewhere(4).

Resolution

The use of a single crystal spectrometer precludes the determination of the rocking curve widths but the resolution for these LSMs can be compared with the more usual crystal analyzers run under the same conditions. Figure 2 compares the resolution (FWHM) of a typical LSM with energy-dispersion (solid-state detector and proportional counter) and with wavelength-dispersion (RAP and LiF spectrometers using fine collimation). The dashed curves indicate the resolution that might be achieved due only to the divergence allowed by the collimator, i.e. assuming the crystal had no contribution to the peak width. This collimator contribution is sensitive to theta and hence 2d, so individual curves are shown for each dispersing element. The solid lines for the three wavelength-dispersion spectrometers represent actual measured values showing the degradation due to the rocking curve of the dispersing device. At the shorter wavelengths within its range, the LSM can be seen to behave much like the crystals, while at longer wavelengths it diverges from the collimator-determined curve, illustrating the significantly poorer resolution. It is expected that improved fabrication techniques will provide LSMs with better resolution in the near future.

Diffraction Efficiency

Absolute measurements of the diffraction efficiency (R - the integral reflection coefficient) have been made for the LSMs listed in Table 1. For the W-C LSMs of varying 2d-spacing, Figure 3 compares the R-values with those for a Lead Stearate Langmuir-Blodgett film (2) and a KAP crystal (5). The increased diffraction efficiency of the LSMs is easily seen. Compare the 42 A LSM with KAP and the 120 A LSM with Lead Stearate. Similar measurements as a function of the atomic number of the TM show that the R-value of W-C at 120 A 2d-spacing is about twice as large as the 109 A Mo-C, which is about twice as large as the 113 A Ti-C.

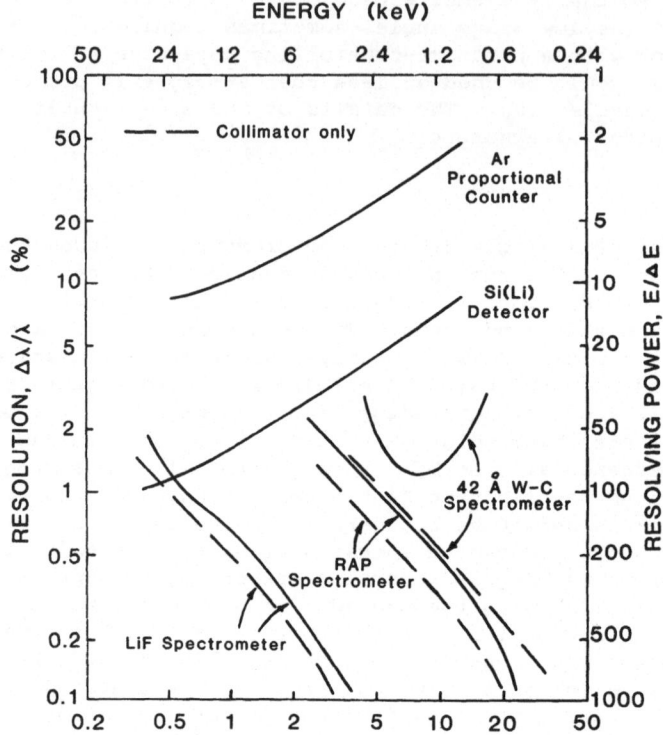

Figure 2.   Measured resolution (resolving power) of various x-ray
             spectrometers. The dashed lines give the d-dependent
             resolution minimum set by a soller collimator with a pass
             angle of 1.3 mrad. as employed with the flat crystals LiF
             (200) and RAP (001) and a W-C multilayer with 42A 2d
             spacing.

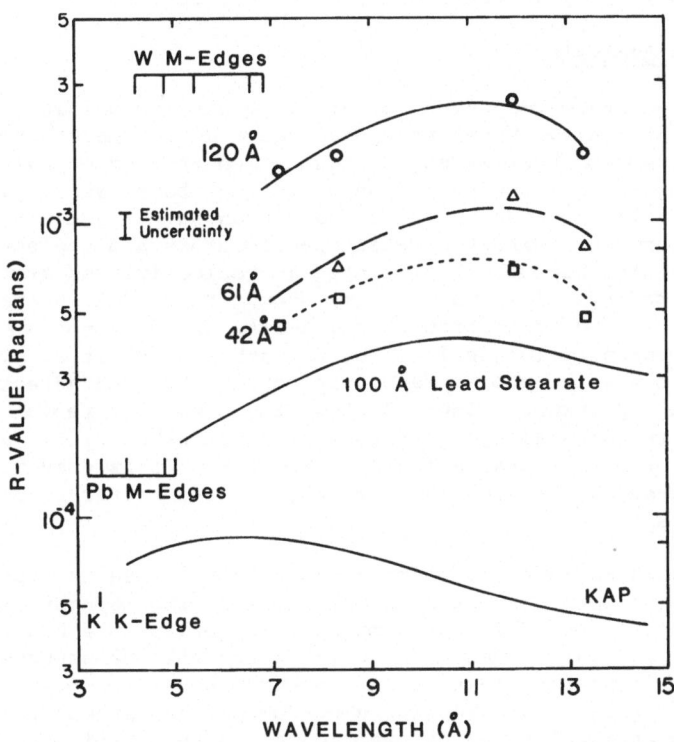

Figure 3. Measured diffraction efficiency for a KAP crystal, a lead stearate Langmuir-Blodgett film and 3 sputter-deposited W-C multilayers of the indicated 2d-spacings.

APPLICATIONS

Multilayers have potential applications wherever diffraction
or reflection is employed to analyze x-rays with energies below
about 10 keV. They can be used in a wide variety of x-ray
instruments; x-ray monochromators, analyzers for fluorescence and
scattered spectra, and x-ray optics are all candidates for their
use. The many disciplines in which x-rays are employed can also
exploit LSMs. Illustrations of their use in three areas will be
discussed briefly in the remainder of this section.

Materials Analysis

X-rays are widely used to determine the composition and structure of materials. Wavelength-dispersive x-ray spectrochemical
analysis depends heavily on the resolution and reflectivity of the
dispersing element employed. As discussed above, spectrometers
using multilayers made by vapor deposition have resolution intermediate between ordinary crystal spectrometers and the energy-
dispersive instruments; their integral reflectivities are large
compared to crystals. Hence, they should be useful in analytical
situations where interference from nearby lines is not a problem
but the concentration is low. On the other hand, it is possible to
illustrate how the poorer resolution of the LSM can affect the
analytical spectrum. Figure 4 shows the 8 to 10 A region of
the spectrum of a Al_2O_3-MgO ceramic, contaminated by As, as recorded by a RAP crystal and a 42 A W-C LSM. It is easy to see
the increased intensity with the LSM, and the better resolution
with the RAP.

Multilayer x-ray monochromators will find use in materials
studies with both conventional and storage ring (synchrotron
radiation) sources. If the composition profile in a LSM is nearly
sinusoidal, higher order diffraction peaks will be suppressed or
absent, an important feature in monochromators. Chemical determinations by x-ray photoelectron spectroscopy and structural studies
using conventional x-ray diffraction can both benefit from the
nearly monochromatic, high intensity beams focussed by LSMs; high
reflectivity devices can be prepared on curved surfaces or on
deformable materials such as mica.

Plasma Diagnostics

The natural x-ray emission from a multimillion degree plasma
provides rich information on conditions within the plasma and on
its radiative cooling. X-ray spectral measurements are especially
useful in providing temperature and density data. Multilayers
have several applications in plasma diagnostics. Their high peak

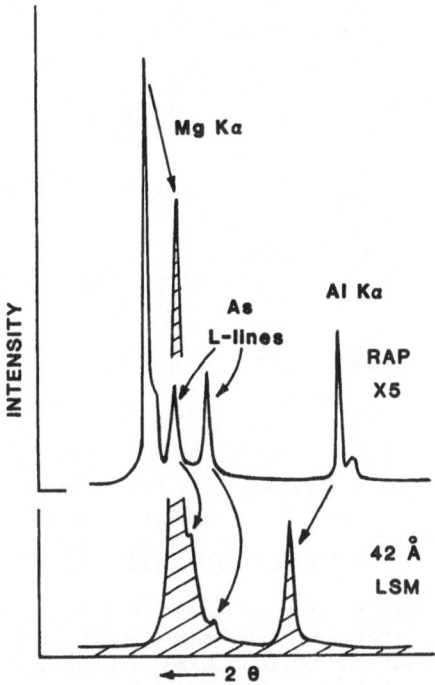

Figure 4.   A portion of the spectrum of Al$_2$O$_3$-MgO ceramic
            contaminated with As, measured with a RAP crystal and
            with a 42 A W-C LSM in a spectrometer with a colli-
            mator divergence of 1.3 milliradians.

reflectivities are needed for work with relatively cold or otherwise faint plasmas if the emission lines do not overlap, as is
the case for low atomic number ions. Also, the arbitrary 2d-
spacings are useful in fixed-angle instruments including
polarimeters and the long 2d-spacings find use where ineffi-
cient grazing-incidence gratings would be employed otherwise.
Further, mirrors coated with multilayers can be used to obtain
images of plasmas

     In order to demonstrate the applicability of a multilayer
diffraction element to plasma diagnostics, a W-C LSM and a RAP
crystal were used to record simultaneously the plasma spectrum
from highly ionized fluorine. An 8 J, 20 nsec. pulse from a ruby
laser was focussed onto a LiF target in vacuum. X-rays from
one- and two-electron fluorine ions were diffracted onto Kodak
101 film. Figure 5 shows densitometer traces of the resulting

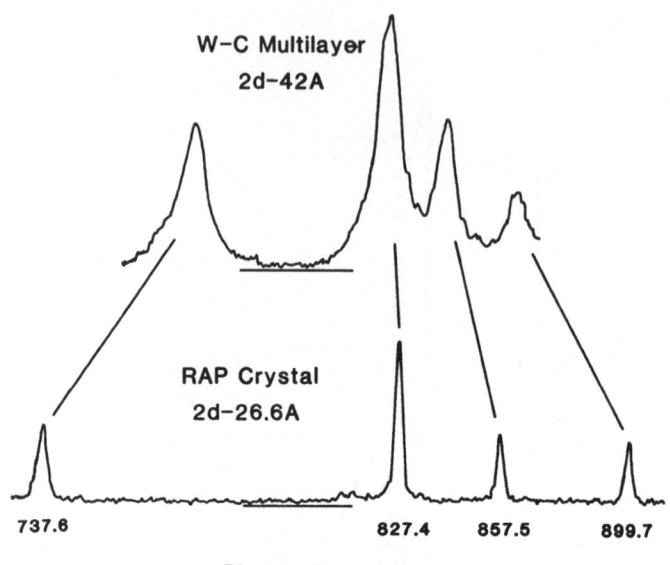

W-C Multilayer
2d-42A

RAP Crystal
2d-26.6A

737.6   827.4   857.5   899.7

Photon Energy (eV)

Figure 5.   Densitometer traces of spectra from a laser-heated
plasma as diffracted by a W-C LSM (top) and a RAP
crystal (bottom). Approximate baselines are indicated
for each spectrum. Lines from fluorine ions containing
only one or two electrons appear at the indicated
energies with different dispersion, peak widths and
heights due to differences in the diffraction characteristics of the two dispersion elements.

spectra. As in Figure 4, the LSM gave poorer (but adequate)
resolution with significantly higher peak intensity. The integrated intensity from the multilayer is about 11 times that from
the RAP crystal. The small size (about 100 μm) of the laser-plasma
makes it useful for determination of both resolution and reflectivity of the LSMs, relative to crystals of known characteristics.
Details of this work will be available elsewhere(6).

X-ray Astronomy

Analysis of astrophysical, as well as laboratory, plasmas by
x-ray techniques is valuable. Rocket and satellite measurements
of solar x-ray spectra have provided detailed data on the compo-

sition, temperature, density, equilibrium and time variation of
solar plasmas. Until recently, dispersive x-ray spectroscopy of
faint stellar sources was not possible. However, crystal-disper-
sed spectra from stars have now been obtained because of the long
counting times offered by the Einstein Observatory. Just as
multilayers are making new and better x-ray measurements possible
in other fields, it is expected that they will have a significant
impact on x-ray astronomy. For example, multilayer-coated elements
which will focus radiation onto an x-ray detector have been sug-
gested(7). Such normal-incidence x-ray mirrors were recently
demostrated at 44A (8). Two such devices, each taking a different
spectral cut, are indicated in Figure 6. Fast response detectors
with such multilayer collectors might make possible spectral
studies of x-ray flares on remote stars with relatively good
spectral resolution.

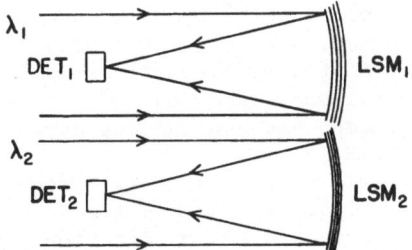

Figure 6. Two channels, each using a LSM with different 2d-spacing,
           for measuring two different wavelengths in a stellar
           spectrum.

DISCUSSION

     This paper has concentrated on the x-ray characteristics and
areas of application of multilayers produced on flat, curved or
curveable surfaces. The major value of such multilayers lie in
their large d-spacings and in their high reflectivities. The
ability to suppress higher-order reflections by use of a proper
composition profile is also important in some applications.
Multilayer structures with lower absorption which will produce
sharper diffraction peaks (that is, higher resolution) might be
possible. Hence, one of the primary values of multilayers is the
ability to trade off resolution and intensity with greater freedom
than is possible with crystals or Langmuir-Blodgett films.

Multilayers offer a unique opportunity which is simply not possible with ordinary diffraction elements. It appears possible to make multilayers with graded layer thickness, both along the surface (laterally) and in depth. This opens unique opportunities for the design of x-ray instrumentation. For example. an x-ray spectrum could be scanned at fixed theta angle (i.e. with a fixed detector) simply by moving a laterally-graded multilayer parallel to its surface. Further, a depth-graded multilayer would provide a wide pass band (i.e. poor resolution) which might be useful for x-ray lithography. The characteristics and uses of graded-thickness multilayer structures are discussed elsewhere (9). It is anticipated that such structures will lead to the design of many new and improved x-ray optical devices.

REFERENCES

1. R. W. Gould, S. R. Bates and C. J. Sparks, Appl. Spectrosc., 22:549 (1968).
2. B. L. Henke and M. A. Tester, Techniques of Low Energy X-Ray Spectroscopy, in: "Advances in X-Ray Analysis, "Vol. 18, W. L. Pickles, C. S. Barrett, J. B. Newkirk and C. O. Ruud, eds., Plenum Press, New York, NY (1975).
3. T.W. Barbee, Jr., Sputtered Synthetic Multilayer Structures, in: "Proc. of Topical Conference on Low Energy X-ray Diagnostics," D.T. Attwood and B.L. Henke, eds., Am. Inst. Physics, New York, NY (1981).
4. J. V. Gilfrich, D. J. Nagel and T. W. Barbee, Jr., Layered Synthetic Microstructures as Dispersing Devices in X-Ray Spectrometers, accepted for publication, Appl. Spectrosc., 36: (Jan.-Feb. 1982).
5. J. V. Gilfrich, D. B. Brown and P. G. Burkhalter, Appl. Spectrosc., 29:322 (1975).
6. N. G. Loter, D. J. Nagel and T. W. Barbee, Jr., submitted to Review of Scientific Instruments.
7. J. H. Underwood, T. W. Barbee, Jr. and D. C. Keith, SPIE Vol. 184 Space Optics-Imaging X-Ray Optics Workshop (1979) p. 123.
8. J.H. Underwood and T.W Barbee, Jr., submitted to Nature.
9. D. J. Nagel, J. V. Gilfrich and T. W. Barbee, Jr., submitted to X-Ray Spectrometry.

THE USE OF ENERGY DISPERSIVE DIFFRACTOMETRY TO MEASURE THE

THICKNESS OF METAL AND GLASS THIN FILMS

Glen A. Stone

South Dakota School of Mines and Technology
Rapid City, South Dakota

INTRODUCTION

This paper presents a new method to measure the thickness of very thin films on a substrate material using energy dispersive x-ray diffractometry. The method can be used for many film-substrate combinations. The specific application to be presented is the measurement of phosphosilicate glass films on single crystal silicon wafers.

Phosphosilicate glass (PSG) films on silicon wafers have found a variety of applications in semiconductor device fabrication (1,2,3). For example:

    (1)   Solid-state diffusion sources in preparation of p-n junctions in bipolar and MOS devices;

    (2)   Electrically stable MOS gate oxides;

    (3)   Stabilizing effect on the electrical properties of the SiO_2-Si interface and reducing the drift of positive ions in the oxide;

    (4)   Gathering sources for cations or metal impurities; and

    (5)   Over-coat mechanical protection.

The measurement of the PSG film thickness on silicon wafers and the phosphorous concentration in the PSG film represent a major quality control problem in the semiconductor industry.

ENERGY DISPERSIVE X-RAY DIFFRACTOMETRY

Energy dispersive x-ray diffraction is not new to science. The Laue photograph of a crystalling material is an excellent example. Max Laue in 1912 established that a crystal is a three-dimensional grating for x-rays and can be used, therefore, to disperse the continuous x-ray spectrum by energy (or wavelength).

Within months of Laue's publication, W. L. Bragg explained Laue's results in terms of x-ray (waves) reflecting from atomic planes. Thus, Bragg's law, (Equation 1) written below in terms of x-ray energy, was discovered:

$$E = \frac{nhc}{2dSin\ \theta} \qquad\qquad\qquad \text{Equation 1}$$

where h is Planck's constant; c the speed of light; d the inter-planer spacing; and θ the diffraction angle.

With the development of solid-state x-ray detectors, it became possible to observe diffraction from many crystal planes simulta-neously in real time. The procedure requires the use of the unfiltered continuous x-ray spectrum from an x-ray tube while holding the diffraction angle θ constant.

EXPERIMENTAL CONDITIONS FOR THIN FILM MEASUREMENTS

The experimental conditions for successful application of the method follow:

(1)   the substrate material must be a different crystalline material than the surface film;

(2)   the substrate should be an accurately cut single crystal; and

(3)   two diffraction lines from the substrate must be present such that the difference in the mass absorption co-efficients for the thin film material is as large as possible.

In order to optimize the third condition for different thin film-substrate configurations, it is necessary that the Bragg angle be adjustable. This can be accomplished by mounting the energy dispersive x-ray detector on a diffractometer. All conditions are met for PSG thickness on (111) silicon wafers.

For this study of PSG film on silicon single crystal wafers,
a diffraction angle of 60 degrees was selected. The silicon wafers
were cut with the (111) orientation. Therefore, Bragg's law is
satisfied for the (111) reflection at 2.38 Kev and the (333) reflection at 6.85 Kev. The mass absorption coefficients for the PSG film
at the energies of the (111) and (333) reflections are 20 m^2/kg and
1 m^2/kg respectively. This is shown in Figure 1.

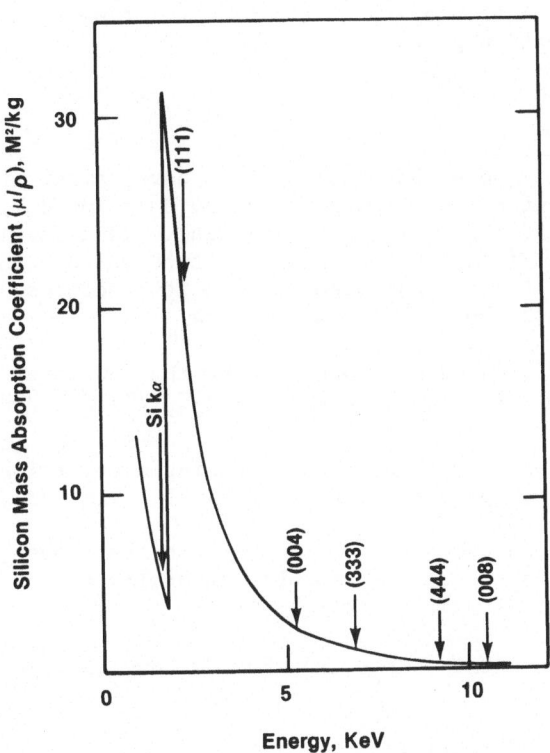

Figure 1. Energy dependence of the mass absorption coefficient for
           Silicon. Position of (111) and (00ℓ) diffraction peaks
           for a Bragg Angle of 60 degrees is shown.

The x-ray source was an unfiltered rhodium target tube, operated at 10 Kv and 0.60 ma. All data collection and data reduction was conducted using a XERTEX/Dohrmann EXAC 5000 x-ray spectrometer. See Figure 2 for (1) examples of the x-ray spectrum from a silicon wafer with 2.354 μm (23540Å) PSG and (2) a polished silicon wafer with no PSG.

Inspection of Figure 2 will reveal that the ratio of the (111) and (333) diffraction peak intensities for the two specimens are quite different. Note in particular this condition: that the (111) peak generated in A (the sample with 2.354 μm PSG) is significantly reduced in intensity when compared to the (111) peak in B where no PSG is present. The reason for this is the large difference in the mass adsorption coefficients for the PSG film for the (111) and (333) reflections. These observations show that PSG thickness can be expressed as a function of the ratio of the intensities of the (111) and (333) diffraction peaks.

THEORY

Based on the above observations, the development of a model for the measurement starts by fixing the x-ray beam intensity from the x-ray tube for the two energies of interest. That is, $I_o(111)/I_o(333)$ is a constant and is a function of x-ray tube voltage and current only. There are three x-ray beam, sample interactions, shown in Figure 3, that must be considered, that is:

1. The incident beam is absorbed by the PSG layer of thickness X, producing the ratio $I_x(111)/I_x(333)$;

2. this beam is diffracted by the single crystal silicon substrate, producing the ratio $I_x^D(111)/I_x^D(333)$;   and

3. this is absorbed by the PSG layer of thickness X, producing the only measurable intensity ratio $I_X^D(111)/I_X^D(333)$.

The measured diffraction intensity should obey, therefore, the following equation:

$$X = A - B\ln\{I_X^D(111)/I_X^D(333)\} \qquad\qquad \text{Equation 2}$$

The constants A and B must be obtained through the use of standards by regression analysis.

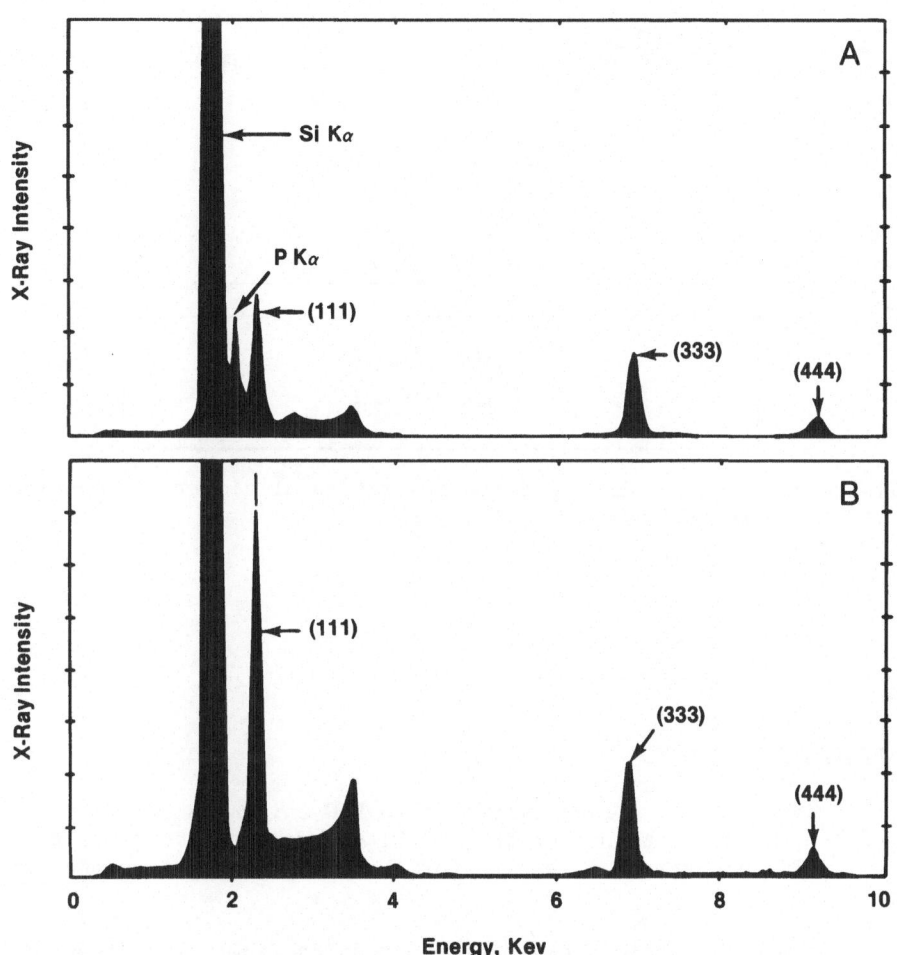

Figure 2.   Spectrum A was generated from a silicon wafer
with 2.354 μm (23540Å) PSG and 5.45% phosphorus.
Spectrum B was generated from a polished silicon
wafer, no PSG and no phosphorus.   The x-ray
fluorescence lines for Si and P and the diffraction
peaks are identified.

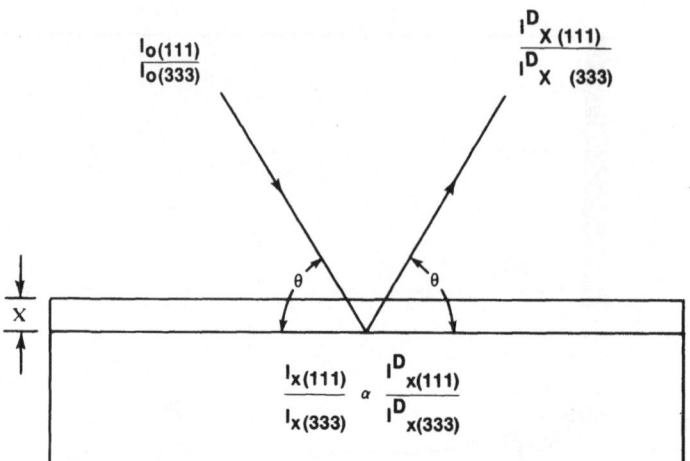

Figure 3. Diagram showing intensity ratios at different locations
in the sample.

EXPERIMENTAL RESULTS

A set of (111) silicon wafers with PSG thickness from 0 to
2.454 μm, accurate to better than ±0.01 μm, were obtained for this
study. The experimental results for the analysis are presented in
Table I.

The final equation, given below, results from the regression
analysis and has a correlation coefficient (R) = 0.999993:

$$X = 1.6092 - 1.6920 \ln \{ I_X^D(111)/I_X^D(333) \} \qquad \text{Equation 3}$$

The above data was collected from the center of each wafer. Sample
C was left in the spectrometer for several days with a measurement
being made periodically. The value of 3σ for ten measurements was
±0.005 μm (50Å).

Table I.   Evaluation of PSG Thickness Using the
Diffraction-Absorption Model

Sample	Given Thickness of PSG	Calculated Thickness of PSG	Difference
	μm	μm	μm
C	1.012	1.014	−0.002
D	0.334	0.329	0.005
F	1.346	1.346	0.000
X	2.454	2.454	0.000
Z	0.000	0.004	−0.004

CONCLUSION

A method using x-ray energy dispersive diffractometry has
been developed where: under specific experimental conditions,
a total instrument error of ±0.005 μm (50Å) was obtained. This
was accomplished for the measurement of PSG thickness on (111)
Si wafers in the thickness range of 0 to 2.454 μm (24540Å).
This method should work for any thin film-substrate pair where
the three experimental conditions discussed can be met.

ACKNOWLEDGMENT

The author wishes to thank Stephen A. Walton, General Manager,
XRF Business Center, Dohrmann Division, XERTEX Corporation for
supporting this research.

REFERENCES

1. R. DuRant, Proceedings of the 5th Conference on Chemical Vapor
    Depositions, p. 421 (1974).
2. J. Wong, J. Non-Crystalline Solids, 20, 83 (1976).
3. D. R. Kerr, J. S. Logan, P. J. Burkhardt and W. A. Pliskin,
    IBM J. Res. Develop., 8, 376 (1964).

APPLICATION OF AUTOMATED X-RAY DIFFRACTION

TO ALTERATION MINERAL ZONING STUDIES

H. Salek, H. A. Vincent, and L. Thorpe

Anaconda Copper Company
P. O. Box 27007
Tucson, Arizona 85726

ABSTRACT

The purpose of this study was to develop a quick and valid X-ray method to determine the alteration models for acidic and intermediate type of altered rocks. Four alteration minerals-- alunite, kaolinite, montmorillonite, and sericite--and total quartz in certain altered porphyritic rocks were determined and evaluated by using the APD (Automatic Powder Diffractometer) system with external standard methods.

The findings indicate that the APD system can successfully be used in the quantitative analysis of these five mineral phases. The best results are obtained from samples with low iron minerals and of a grain size not coarser than 325 mesh.

INTRODUCTION

The action of hydrothermal solution and high-temperature gases and vapors on rocks near developing ore deposits produces appreciable alterations in their rock composition. These changes can result in the supply or removal of substances. Hydrothermal alteration is the conversion of an initial mineral assemblage to a new set of minerals more stable under the hydrothermal conditions of temperature, pressure, and fluid composition. Hydrothermal alteration forms a readily visible empirical guide to hydrothermal activity and, thus, the alteration zoning pattern is considered a pathfinder for discovering a new orebody.

Rock alterations cover a greater area than the deposit itself. In a horizontal section, the rock alterations appear as a concentric target generally ellipsoid in form. Alterations of rocks around deposits vary according to the activity of ore-forming processes and the characteristics of the rock. The types of alteration are classified by rock types, such as acid and intermediate, basic and ultrabasic and carbonates.

Each of the alteration types is sometimes associated with certain types of ore.

To determine the minerals in an alteration zoning study, two acceptable methods can be used: (1) optical, a time-consuming method, and (2) X-ray diffraction, a faster method but still time-consuming if not automated. When hundreds to thousands of samples must be processed and evaluated for each project area study, the automated X-ray diffractometer is the method of choice.

The objective of this work was to determine how well the automated X-ray diffractometer could be applied to obtaining the mineral phase information required for a mineral alteration study. At question is whether quantitative phase measurements are adequate to describe mineral zone trends and whether relatively unattended machine interpretations of complex diffraction patterns are satisfactory for that description. Also investigated were the influences of sample particle size and iron content on the quality of the measurements.

EXPERIMENTAL

The external standard method was used for this study. Intensities of selected X-ray lines for each phase present in the unknown samples were compared with intensities of calibration lines obtained standard mixtures. The Phillips APD 3600 system with computer Nova 3/12 and copper radiation was used in the evaluation.

As part of the laboratory procedure for each project, several random samples are studied optically and by X-ray diffraction to determine the general rock composition for the area under study. This is used to set instrumental conditions for the project diffraction work.

A Phillips model 3600 automated powder diffractometer, equipped with a 35 position sample changer and a copper target narrow-focus tube, was used throughout this study. The measurement parameters for the five phases studied are listed in Table 1.

Table 1 - Scanning ranges for five phases (alunite, kaolinite, quartz, montmorillonite, and sericite)

PHASE	LINE USED (hkl)	2θ START	2θ END	ANGLE INCR.	TIME INCR. (SEC.)	TIME/ SCAN (MINS.)	REMARKS
ALUNITE	(220)	51.50	53.00	0.02	3.00	3.75	
QUARTZ	(211)	59.60	60.60	0.02	3.00	2.50	
KAOLINITE	(001)	11.30	13.30	0.02	3.00	5.00	if chlorite is present should be corrected by light optics if gypsum and epidote are present use (020) line
MONTMORILLONITE	(001)	3.70	7.70	0.03	3.00	6.70	if chlorite is present should be glycollated
GLYCOLLATED MONTMORILLONITE		4.00	5.85	0.03	3.00	3.08	
SERICITE	(002)	8.00	9.70	0.02	3.00	4.25	if pyrophyllite is present use (202) line if biotite is present should be corrected by light optics

scanning time 25.28 minutes
Approximate delay time 1.00 minutes
per scan

Total time/single run/sample 26.28 minutes

Synthetic Mixtures

Mixtures of pure mineral phases were prepared in order to assess the influence of contained iron, the effect of particle size upon the diffracted X-ray intensities, and for calibration of the instrument. Mixtures were made with cell particles in each of the range of -100 +200, -200 +325, or -325 mesh, respectively. Blending was done by agitation of the samples in lucite vials for eight minutes with a Spex Mill.

Samples were packed in holders from the back side on a glass plate for orientation and smoothness. The same volume was maintained for all samples.

The results of the diffraction scans for these mixtures are shown in Figure 1 for montmorillonite. Regression curves of diffracted intensities versus the content of montmorillonite show increased measurement precision with small particle size and a better slope and intercept as iron content is decreased. Similar curves resulted from measurements of the other phases. A summary of standard deviations for determinations from the various mixtures is given in Table 2.

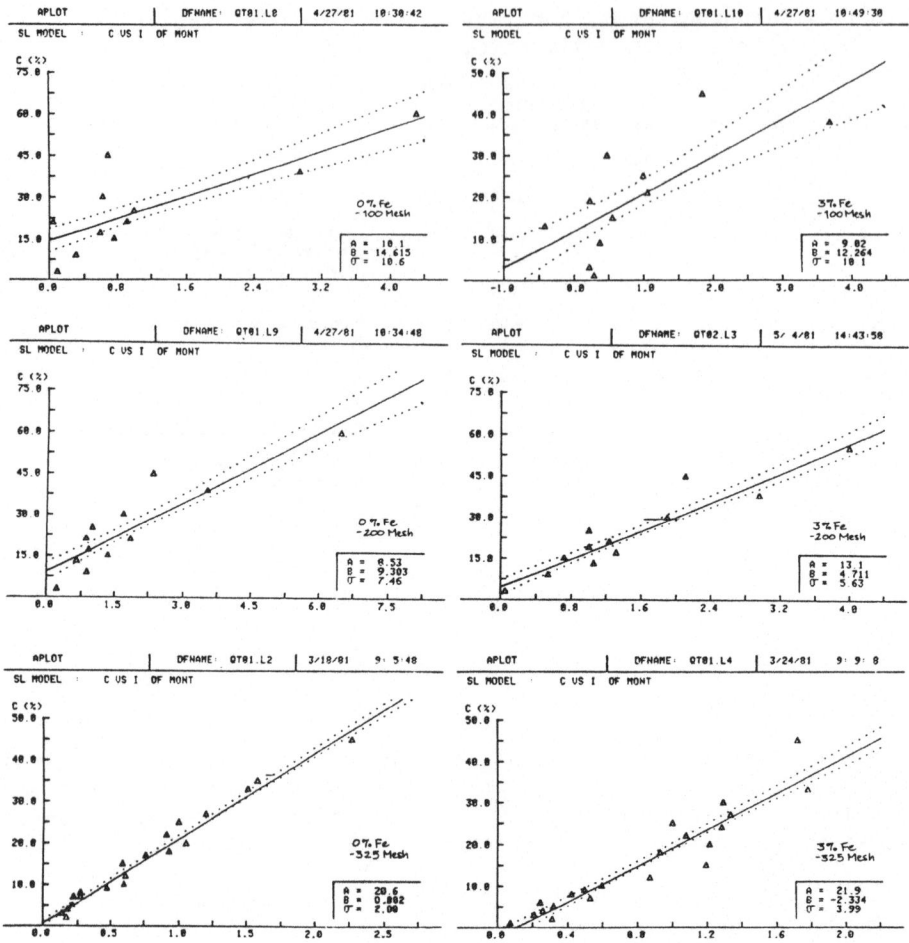

Fig. 1 - Standard curves for montmorillonite (-100, -200, and -325
         mesh ± 3% Fe)

This work indicates that unknown samples to be measured for
these mineral phases should be pulverized to -325 mesh and either a
correction made for the influence of iron or quantitative comparison
made to reference standards containing similar quantities of iron.
Use of the comparison to references similar to the rocks under study
aids in accounting for differences in crystallinity often encountered
in materials from different localities.

Natural Samples

Field samples are usually received as assay rejects weighing
about one kilogram and passing a 10 mesh screen. Approximately one
gram is collected from a splitter and further crushed to -325 mesh

Table 2 - Sigma for four phases (kaolinite, quartz, montmorillonite, and sericite) with various grain sizes and iron content

Mineral	Grain Size (mesh)	SIGMA				
		0% Fe	1% Fe	3% Fe	6% Fe	14% Fe
Kaolinite	-100	4.22		4.01	2.10	2.82
	-200	3.62		3.48	4.03	4.96
	-325	2.79	2.85	4.27	3.28	6.31
Quartz	-100	22.7		14.4	22.1	11.3
	-200	18.9		11.3	9.06	11.8
	-325	2.47	5.35	7.17	7.21	21.4
Montmorillonite	-100	10.60		10.10	8.28	9.80
	-200	7.46		5.63	4.84	3.55
	-325	2.00	3.70	3.99	5.71	7.22
Sericite	-100	7.00		9.68	8.65	7.38
	-200	3.22		10.40	4.73	2.93
	-325	4.43	2.61	4.31	3.71	8.11

using the ball mill canisters for ten minutes on the Spex Mill. The samples are packed into the X-ray holders as described earlier.

Samples for this study were handled in this manner, loaded into the APD cassette system and X-ray data accumulated in the same manner as for the synthetic mixtures. Quantitative phase determinations were automatically calculated using appropriate synthetic mixtures for calibration.

A typical contour plot of these data for one mineral versus location is shown in Figure 2 indicating successful description of the alteration trend for that locality. Series of individual mineral phase and phase ratio plots for a location are used collectively and with other geological information in defining alteration trends. These in turn should give direction to where the geochemical environment was appropriate for metal to be deposited to form ore. On the basis of the defined trends, further drilling, sampling, and diffraction analysis may allow for the discovery of an ore deposit.

The relative percent standard deviations (RSD) determined for the values of the five phases present in a group of samples are: Alunite, RSD = 2.4; Kaolinite, RSD = 6.3; Quartz, RSD = 3.5; Montmorillonite, RSD = 8.6; and Sericite, RSD = 7.5. These values for RSD are lower than those determined in the past by using the microscopic point-counting method.

CONCLUSIONS

Mineral phase measurements by automated X-ray diffraction have been shown to be satisfactory for application to description of an alteration mineral zoning study. The measurement precision is

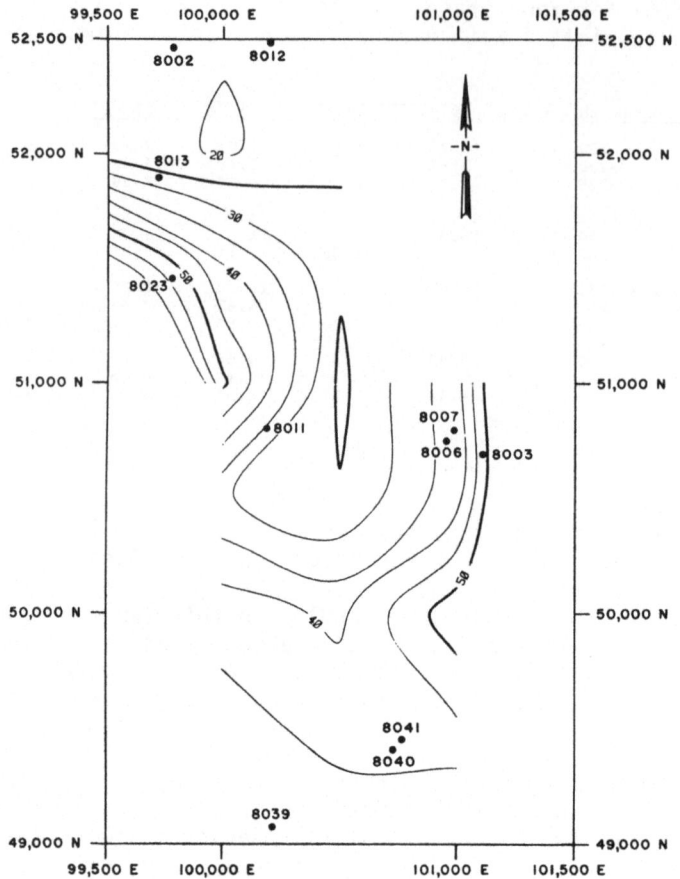

Fig. 2 - Statistical contour plot of alunite content versus location with 10% phase interval

adequate to differentiate zones at an approximate 10 percent relative mineral phase change level.

     Further validation of this approach for exploration will be
necessary from correlation with other mineralogical, geochemical,
geophysical, and geological evidence. The mineral alteration
information may be the strongest of these in the directing of
prospecting for an ore deposit.

     Extension of this work to the measurement of other suites of
altered rocks is continuing in this laboratory with other mineral
phases added, such as chlorite, tourmaline, pyrophyllite, and albite.

THE APPLICATION OF X-RAY DIFFRACTION

FOR GLASS BATCH HOMOGENEITY DETERMINATION

H. S. Kim and C. I. Cohen

Owens-Corning Fiberglas Corporation
Technical Center
Granville, OH 43023

ABSTRACT

An X-ray diffraction (XRD) technique is employed to determine the presence or absence of crystalline phases in glass batch and to determine its homogeneity qualitatively and quantitatively.

Sample preparation problems are discussed, and promising techniques for sample preparation are presented. Qualitative batch homogeneity determination is accomplished by comparing the integrated intensity of a particular reflection of the reference sample to that of the unknown sample. Quantitative batch determination is accomplished by using the internal standard method. Three replicated data sets indicate that the standard deviation of kaolinite and colemanite are higher than those of quartz and calcite. However, the overall data from the quantitative analyses lie within an acceptable range of precision and accuracy.

INTRODUCTION

A missing batch component or a gross error in batch proportioning will seriously affect the glass melting process and resultant glass properties. The XRD technique is a relatively quick and simple procedure for identifying the batch components and determining the homogeneity of the batch.

Quartz, calcite, kaolinite, and colemanite are used as the basic components for this study, because those components constitute important ingredients of the batch. Sample preparation problems accompanying qualitative and quantitative analyses are investigated.

SAMPLE PREPARATION

Grinding and Particle Size: Each ingredient (quartz, calcite, kaolinite, and colemanite) is ground separately with a Pulverisetto-0-Electromagnetic micropulverizer and subsequently passed through a 325 mesh U.S. sieve (approximately 37 um). Integrated intensity data is obtained by using a sample spinner which gives a greater area of reflection and randomizes the sample during the running procedure.

The integrated intensity data from different size fractions demonstrates strikingly the effect of variations in the crystalline sizes of quartz powder upon both the intensity and the reproducibility of the intensity of the 3.34A° reflection (Klug and Alexander (1)). For the 15um to 50um powder fraction, the mean deviation in intensity from Klug and Alexander's values is 18.3 percent, whereas for the 5um powder, it is 1.2 percent. Our mean deviation in intensity measurement is 4.7 percent, which is higher than their data at the 5um particle size, but lower than would be anticipated from their other values.

Mixing: Mixing of the powder samples is performed in either a polyethylene bottle or a tungsten carbide container. Sample A is mixed in a polyethylene bottle with leucite balls for 4 minutes in the wiggle bug mixer. Sample B is mixed in a tungsten carbide container with a tungsten carbide rod for 4 minutes in the wiggle bug mixer. A constant amount of sample and the dense packing method are used for both samples A and B. The data obtained from fifteen measured values for each sample demonstrate the different effects of the mixing procedure on the intensity of the 3.34A° quartz reflection. For Sample A, the mean deviation in the integrated intensity is 2.5 percent, whereas for Sample B, it is 4.7.

Packing: A novel method for reproducible sample preparation has been devised which utilizes the Philips fiber holder accessory. The powdered samples are packed from the back of the sample holder against a rough glass surface. This method provides for a positive rear seal, control of compaction through the plastic properties of sealing wax as well as the ability to modify and use the same packing surface for all samples in a given series. In addition, new sample holders can be easily machined from aluminum tubing. Four packing methods using the back-filling technique in the specimen spinner are selected as the basic methods to perform this study. Those methods are: constant amounts of sample with dense packing (Method 1), variable amounts of sample with dense packing (Method 2), constant amounts of sample with loose packing (Method 3) and variable amounts of the sample with loose packing (Method 4).

The data show the marked difference in the reproducibility of the integrated intensity measurements generated by the different packing methods. The mean deviation in intensity from Method 1 is 4.7 percent, whereas the mean deviations of intensity from Methods 2, 3, and 4 are 11.5, 19.7, and 19 percent respectively. Method 1 is selected as a promising technique for the packing of the sample into the specimen holder.

Preferred Orientation: In this study, we pack the sample from the back of the specimen holder against a flat rough surface which is later removed. The flat rough surface is prepared by grinding the glass plate to a 30um finish. This possibly disorients the surface of the sample minimizing the preferred orientation.

QUALITATIVE DETERMINATION OF THE BATCH HOMOGENEITY

This study is carried out with pelletized glass batch components which are subsequently ground to 37 um. A Philips Automatic Powder Diffractometer (APD 3500) is used to accumulate integrated intensity measurements. A preliminary diffraction pattern of an unknown batch sample is obtained in order to determine the presence or absence of phases in the batch sample and to show the presence of interfering reflections.

Qualitative batch homogeneity determination is accomplished by comparing the integrated intensity of a particular reflection from the reference sample to the rest of the unknown samples. Sample number 2200, an arbitrarily selected single batch pellet against which other batch pellets will be compared, is chosen as the refererce sample. Quartz, kaolinite, and calcite are chosen as the indicating minerals in the samples. The integrated peak intensity ratio of the indicating minerals of each unknown sample versus sample Number 2200 indicates that calcite appears to be generally consistent from one sample to another, while quartz and kaolinite are quite variable. Only qualitative variations can be determined in this technique.

QUANTITATIVE DETERMINATION OF THE BATCH HOMOGENEITY

Quantitative analysis of the amount of each phase is based on the fact that the concentration of the component is proportional to the ratio of the integrated intensity of a selected reflection from the samples to the integrated intensity of an internal standard which is added in a constant proportion. The weight percent of the unknown sample is then calculated by using the I(sample)/I(internal standard) ratio with standard calibration curves.

Samples prepared using the previously discussed techniques produced the following results:

Precision: The ranges of quartz and calcite standard deviations of the observed mean are 1.11 to 1.89 and .50 to 1.04, respectively. The ranges of kaolinite and colemanite standard deviations of the observed mean are 2.21 to 4.10 and 1.50 to 2.52 respectively. The higher standard deviations of kaolinite and colemanite are probably caused by the preferred orientation, the crystallinity, and the absorption of those samples.

Accuracy: If we assume that the actual samples contain well crystallized phases and represented true values, then the standard deviations from the actual percent will represent the degree of accuracy. The standard deviation ranges of the actual percent for quartz and calcite are 1.5 to 2.13 and .37 to 2.85, whereas those for the kaolinite and colemanite are 2.70 to 4.14 and 1.84 to 2.52 respectively. The high standard deviations of both kaolinite and colemanite are probably due to sample mixing and preparation, crystallinity, particle size, degree of nonreproducibility of preferred orientation and the interaction of kaolinite and colemanite with the internal standard.

SUMMARY

In conclusion, our results demonstrate that acceptable quantitative phase analyses can be obtained using the reported sample preparation techniques. These include:

mixing the sample in a tungsten carbide container with a tungsten carbide rod in the wiggle bug mixer to minimize the error due to mixing.

back-filling into the sample holder and compacting the sample through the plastic properties of the sealing compound thus providing uniform control of sample packing densities for uniform sample irradiation.

using a sample spinner to give a greater area of reflection and to randomize the sample during the analytical procedure.

The overall data on the quantitative analyses lie within an acceptable range of precision and accuracy even when a 37um particle size is used. The errors contributed by preferred orientation are not established at this time. However, the standard deviation of four replicate analyses indicates that these errors are minimum.

REFERENCE

H. P. Klug and L. E. Alexander, X-ray diffraction procedures, 2nd Ed., John Wiley and Sons (]974), p. 365-376.

X-RAY DIFFRACTION AND FLUORESCENCE IN

THE ANALYSIS OF PHARMACEUTICAL EXCIPIENTS

A. J. Durbetaki and T. F. Quail

FMC Corporation, Chemical Research and Development
Center, Princeton, New Jersey   08540

INTRODUCTION

The majority of pharmaceutical dosage forms are marketed as
tablets which are formulated to satisfy various basic requirements.
The analysis of the resultant multicomponent pharmaceutical and
product is usually a lengthy task.

This paper describes the complementary use of X-ray diffraction
(XRD) and X-ray fluorescence spectroscopy (XRS) to characterize and
quantify excipients in pharmaceutical tablets.

EXPERIMENTAL

A number of pharmaceutical tablets were ground in a diamonite
mortar and pelletized. These tablets were used to obtain the
concentration of Mg, Si, P, S, Cl and Ca by XRS. The measurements
were made with a wavelength dispersive spectrometer (Philips
Electronic Instruments, Inc.) equipped with a chromium target X-ray
tube, a flow proportional counter and germanium and thallium acid
phthalate crystals.

The XRD measurements were made with an automated powder
diffractometer (APD 3500, Philips Electronic Instruments, Inc.).
Spectral data were obtained using peak-search and profile-scan
mode programs.

To carry out the studies on the crystallinity and differentiation of microfine cellulose, it was necessary to prepare a
standard which was 100% crystalline cellulose type I, a 100%
amorphous standard of cellulose, and a series of crystalline-
amorphous admixtures with a concentration range of 25–85% crystalline
cellulose by weight.

The crystalline and amorphous standards were prepared from
Avicel® PH101 using the procedures reported by Wakelin et al.[1] in
the determination of the crystallinity of cotton cellulose.

The XRD spectra for qualitative analyses and crystallinity
measurements were recorded from $2\theta = 6°-50°C$ (Figure 1).

The cellulosic excipients of the pharmaceutical tablets were
recovered using a ternary solvent system. A weighed, 0.1500 g,
portion of the pharmaceutical was introduced into a glass stoppered
tube containing 25 mL of acetic acid–ethanol–water (1:1:1 V/V).
The tube was immersed into a 100°C bath for 1 hour. The contents
of the tube were then transferred quantitatively into the funnel of
a Millipore® filtration apparatus equipped with a 0.8 μm pore size,
preweighed, silver membrane filter. The residue was washed three
times with 10 mL of hot ternary solvent followed by ethanol and
dried under vacuum at 50°C. The residue was then weighed to obtain
an estimate of the cellulosic excipient content prior to XRD
analyses (Figure 1).

Two methods were used for the measurement of the degree of
crystallinity of the cellulosic excipients. These were the so
called "absolute" and the correlation methods. In the case of the
"absolute" method a line was drawn which connected the minima
between the crystalline reflections of the (101), (101), (021) and
(002) planes. The degree of crystallinity (χ_c) was calculated by
taking the ratio of the integrated crystalline scattering to the
total scattering, both crystalline and amorphous. This is given
by the equation

$$\chi_c = \int_0^\infty s^2 I_c(s)\,ds \Big/ \int_0^\infty s^2 I(s)\,ds$$

where s is the magnitude of the reciprocal-lattice vector =
$(2 \sin \theta)/\lambda$, I(s) is the intensity of coherent X-ray scatter from
the crystalline and amorphous regions of the specimen, $I_c(s)$ is the
intensity of coherent X-ray scatter from the crystalline region
and λ is the Cu Kα wavelength (1.5405 Å).

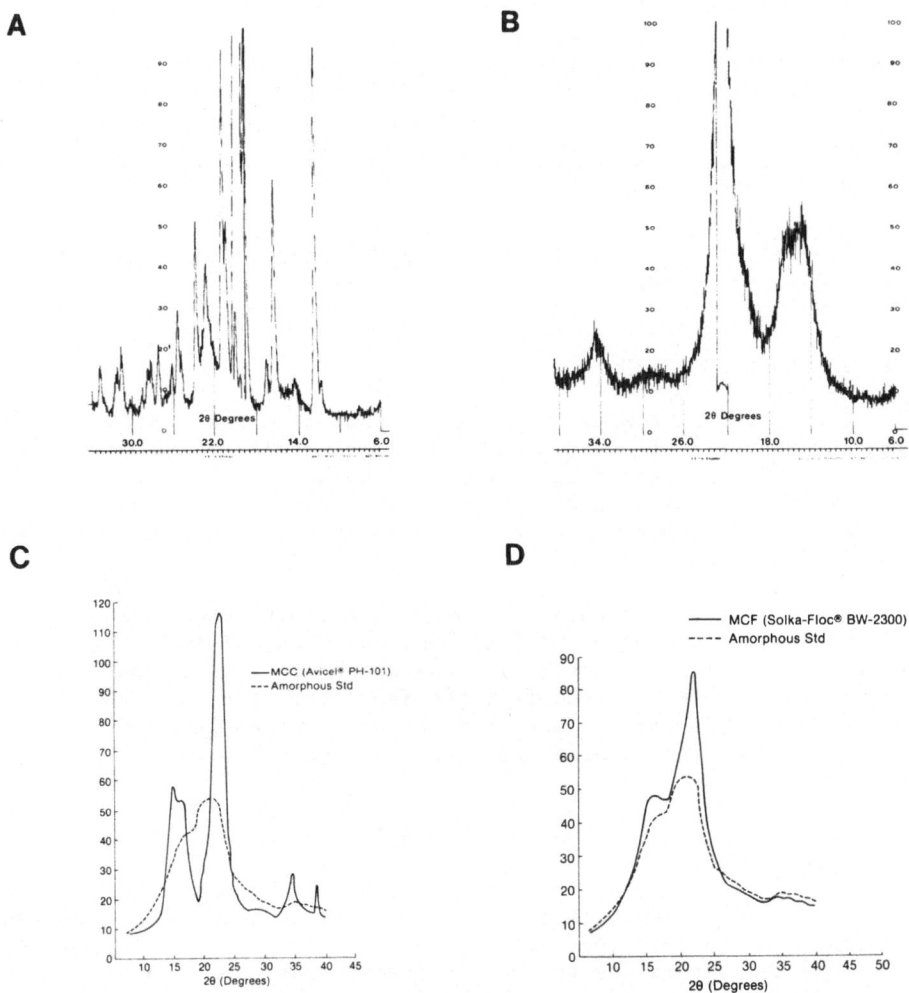

Figure 1.   A – XRD Spectrum of a Pharmaceutical; B – XRD Spectrum of Cellulosic Insolubles (Avicel®); C – X-Ray Scattering Curve of Microcrystalline Cellulose; D – X-Ray Scattering of Microfine Cellulose

In the correlation method, $(I_c-I_a)_{2\theta}$ was taken at every value of scattering angle 2θ from 6° to 50° (I_c = averaged scatter intensity from the crystalline standard specimen, I_a = averaged scatter intensity from the amorphous standard specimen). Average amorphous data were then subtracted at corresponding values of 2θ from data for each cellulose sample (u) to form $(I_u-I_a)_{2\theta}$. Correlation of the values $(I_u-I_a)_{2\theta}$ with $(I_c-I_a)_{2\theta}$ at corresponding scattering angles then gave a regression line whose slope was taken as an index of the crystalline content. This can be expressed as

$$\chi_c = \frac{\Sigma XY-(\Sigma X\Sigma Y/N)}{\Sigma X^2-[(\Sigma X)^2/N]}$$

where $Y = (I_u-I_a)_{2\theta}$, $X = (I_c-I_a)_{2\theta}$ and N is the total number of pairs of observations.

RESULTS AND DISCUSSION

The elemental compositions of two pharmaceuticals were determined by XRS from the net intensities of each element using the fundamental parameters program[2]. The accuracy of the XRS method was evaluated using standard chemical analyses and by comparing the data obtained by both techniques with those calculated from reported compositions (Table 1).

The excipients in these two pharmaceuticals were characterized to be magnesium stearate, $CaHPO_4\cdot 2H_2O$, colloidal silica and microcrystalline cellulose (Avicel®). The concentrations of the non-cellulosics were calculated from the XRS data. The chloride and sulfur in the two samples are from the active constituents chlorpromazine hydrochloride and thiamine hydrochloride, respectively.

The crystallinities on the crystalline-amorphous admixtures are given in Table 2.

The data were subjected to linear regression analyses. As can be seen from the intercept values, the "absolute" method is subject to systematic errors due to the arbitrary separation of the crystalline scatter under the peaks from the scatter of the non-crystalline regions. This can also be observed in the difference between the experimental and actual values.

The average crystalline indices of microcrystalline and microfine celluloses are given in Table 3.

The observed X-ray crystallinity ranges for microcrystalline and microfine celluloses were 77-84% and 39-59%. These values provide a clear distinction between the two types.

Table 1. Elemental Analyses of Pharmaceuticals

Sample	Element	Concentration, %[a]		
		XRS	Chemical	Calculated[b]
A	Mg	0.06±0.01 (4)	0.05±0.03 (2)	0.060
	Si	0.22±0.02 (4)	0.19±0.04 (2)	0.230
	P	8.99±0.06 (3)	8.89±0.11 (2)	8.999
	S	2.52±0.05 (3)	2.50±0.06 (2)	2.526
	Cl	5.59±0.08 (3)	5.52±0.09 (2)	5.587
	Ca	11.65±0.06 (3)	11.59±0.12 (2)	11.645
B	Mg	0.08±0.01 (4)	0.07±0.02 (2)	0.082
	Si	0.22±0.02 (4)	0.17±0.03 (2)	0.230
	S	2.81±0.04 (3)	2.76±0.06 (2)	2.850
	Cl	6.29±0.05 (3)	6.19±0.10 (2)	6.309

[a]Figures in parentheses represent number of analyses
[b]On basis of formulation

Table 2. Crystallinity on Crystalline-Amorphous Admixtures

Mixture (% Crystalline by Wt.)	Crystalline Index (%)	
	Correlation Method	"Absolute" Method
25	26	29
35	34	38
50	49	54
65	67	69
70	72	73
80	79	84
85	84	88
Slope of Regression Line	0.9956	0.9954
Intercept	0.4005	3.8426

Table 3.   Average Crystalline Indices of
Microcrystalline and Microfine Celluloses[a]

Cellulose Type	Crystallinity Index (%)	
	Correlation Method	"Absolute" Method
Avicel® PH101	78.1 (3.2)	77.8 (5.4)
Avicel PH102	80.3 (3.5)	84.0 (6.1)
Avicel PH103	76.9 (4.1)	77.0 (5.5)
Avicel PH105	77.9 (2.9)	76.8 (7.1)
Solka-Floc® BW 60	57.7 (2.7)	58.0 (7.0)
Solka-Floc BW 100	67.2 (3.1)	59.0 (6.9)
Solka-Floc BW 300	56.8 (4.0)	57.9 (7.2)
Solka-Floc BW 2030	45.8 (4.1)	48.1 (7.1)
Elcema® P 100	54.7 (3.0)	50.2 (6.0)
Elcema G 250	58.1 (3.5)	57.3 (5.9)
Sanacel 90	40.5 (4.2)	39.2 (6.6)

[a]Figures in parentheses are the observed standard deviation of
individual replicates; N=5.

In conclusion, XRD and XRS can be used as complementary
techniques to rapidly characterize and quantify excipients amenable
to X-ray analysis. Furthermore, XRD crystallinity measurements can
be used to differentiate microcrystalline from microfine cellulose
powders.

ACKNOWLEDGEMENTS

We wish to express our appreciation to the Food and Pharmaceutical Products Division for permission to publish and to
Drs. T. L. Benney, C. F. Ferraro and H. Stange for their valuable
review of this work.

REFERENCES

[1]Wakelin, J. H.; Virgin, H. S. and Crystal, E., J. Appl.
        Phys. 30, 1654 (1959).
[2]Criss, J. W.; Birks, L. S. and Gilfrich, J. V., Anal.
        Chem. 50, 73 (1978).

CORRECTIONS TO VOLUME 24

R. Bador, M. Romand, M. Charbonnier and A. Roche. "Advances in
Low-Energy Electron-Induced X-Ray Spectroscopy (LEEIXS)."
Advances in X-Ray Analysis 24, 351-361 (1981).

Figures 4 and 5 on p. 355 of the above paper are to be
replaced with the figures on the next page (page 390).

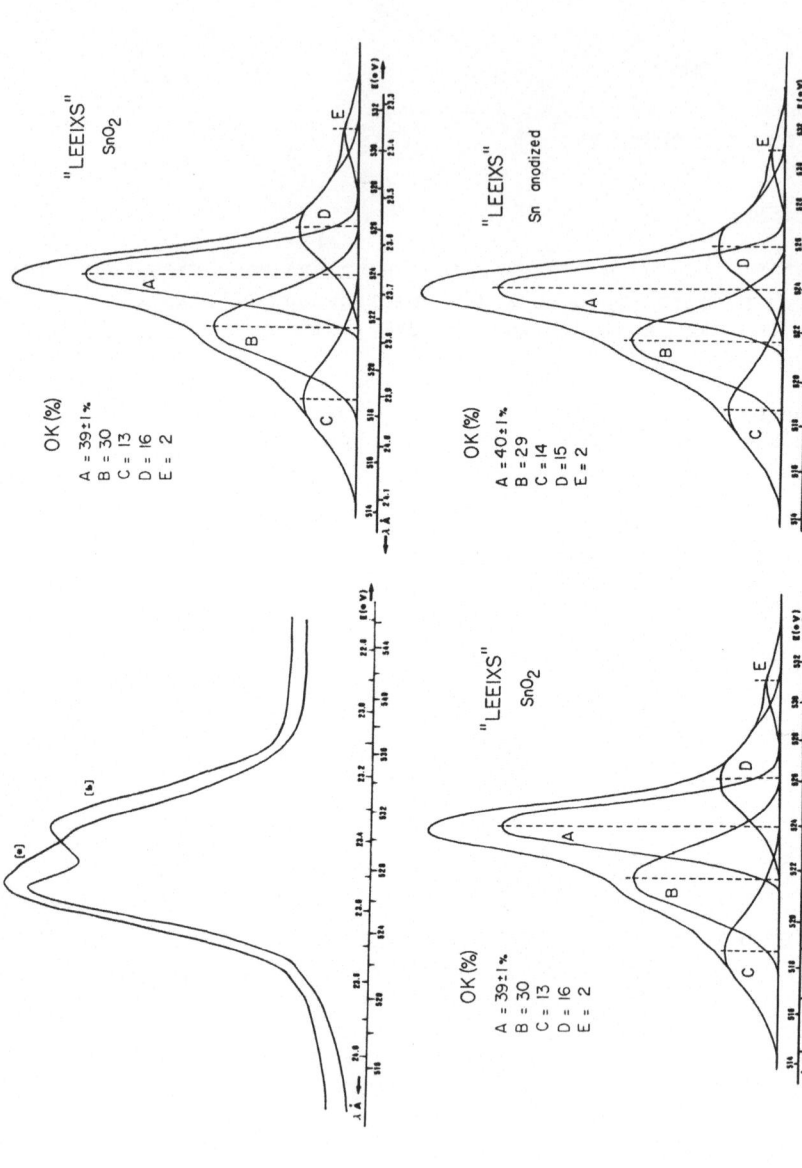

Figs. 4 and 5. Oxygen K emission spectra for CuO (curve (a)) and Cu₂O (curve (b)), SnO, SnO₂, and anodized Sn. All the spectra are obtained at 3 kV operating voltage using a TlAP analyzing crystal. R. Bador et al., Advances in X-Ray Analysis, Vol. 24, pages 351-361 (1981).

AUTHOR INDEX

Adamson, B. W., 169
Aiginger, H., 63
Artz, B. E., 81
Ayers, G. L., 221

Baker, J. W., 91
Ballad, R., 201
Barbee, Jr., T. W., 355
Barrett, C. S., 339
Barten, H., 261
Barton, J. B., 31
Bechtoldt, C. J., 329
Bell, G., 121
Blankenship, D., 85
Boettinger, W. J., 329
Bramlet, H. L., 163
Briden, F. E., 189
Brown, R. H., 181

Carlos, M., 201
Carlson, R. H., 113
Carsey, T. P., 209
Censullo, A. C., 189
Chwaszczewska, J., 23
Cohen, C. I., 379
Conde, C. A. N., 39

Dabrowski, A. J., 1, 23, 31
Das Gupta, K., 325
Davis, B. L., 295
de Campos, A. J., 39
Dow, R. H., 117
Doyle, J. H., 163
Durbetaki, A. J., 113, 383

Entine, G., 31

Ferreira, L. F. R., 39
Furfaro, J., 201
Furnas, R. E., 59
Furnas, Jr., T. C., 59

Garbauskas, M. F., 283
Gilfrich, J. V., 355
Gleason, T. G., 45
Göbel, H. E., 273, 315
Goehner, R. P., 283, 309
Gurvich, Y. M., 139, 145, 169

Hale, D., 85
Hare, T. M., 237
Haynes, B. W., 107
Hom, T., 267, 289
Huang, T. C., 213, 221
Hubbard, C. R., 245
Huth, G. C., 23

Iwanczyk, J. S., 31

Jenkins, R., 231, 267, 289
Jobst, B. A., 273
Johnson, L. R., 295
Johnson, Q., 301
Jurkowski, J., 23

Kim, C., 343
Kim, H. S., 379
Kneip, T. J., 201
Kuntz, G. S., 59
Kuriyama, M., 329
Kusmiss, J. H., 31

LaBrecque, J. J., 151
Ladell, J., 267
Lanzo, M. J., 237
Latuszynski, A., 23
Laurer, G. R., 201
Lei, W., 201
Leyden, D. E., 95
Lis, S., 31
Loter, N. G., 355
Lucas-Tooth, J., 169

Mahan, K. I., 95

Nagel, D. J., 355
Neylan, D. L., 107
Nichols, M. C., 301

Panagiotopoulos, N. C., 245
Parker, W. C., 151
Parrish, W., 213, 221
Patton, M. E., 157
Placious, R. C., 329
Platbrood, G., 261
Purnell, C. J., 181

Quail, T. F., 113, 383
Quitin, J. M., 261

Richter, F.-W., 195
Ricker, G., 31
Rokosz, M. J., 81
Russ, J. C., 85, 237
Ryon, R. W., 63

Salek, H., 373
Sanders, S. C., 121
Satterfield, T., 85
Schorin, H., 127
Schreiner, W. N., 231, 289
Short, M. A., 45
Skillicorn, B., 49
Slapa, M., 23
Smith, D. K., 301
Smith, R. A., 103
Smith, T. K., 133
Snyder, R. L., 245
Squillante, M. R., 31
Stehr, K. C., 173
Stone, G. A., 365
Strittmatter, R. B., 75

Thorpe, L., 373

Vallerga, J., 31
Villamizar, C., 289
Vincent, H. A., 157, 373

Warren, A., 31
Wätjen, U., 195
West, H. M., 177
West, N. G., 181
Withers, E., 181
Wobrauschek, P., 63

Zabronsky, J., 107
Zahrt, J. D., 63, 121

Absorption,
  corrections in XRPD, 296
  macro/micro in XRPD, 301
  use in phase identification,
    302ff
Aerosol filters, 195
Alaskan pipe line weld
    stresses, 336
Algorithms, 267 (see also
    Computer programs)
Alumina, 3 phase mixtures, 311
Am^{241} spectrum with HgI_2
    detector, 13
Amplifier, time variant, 45
Analysis (see specific sub-
    stances, Phase
    analysis)
Archeological artifacts, 121
Automated XRPD systems,
    245, 261 (see also
    Computer programs)
  for mineral zoning, 373
  for phase analysis, 273

Background correction in
    XRPD, 246
Barite, 139
Batholith, California, 298
Bauxites, 127
Beam blanking system, 45
Beryllium X-ray windows, 54
Borate fusion, 95
Borrman diffracted beam, 326

Calibration of diffractometer,
    289
  at low 2θ values, 289
  errors vs 2θ, 251, 269
Calibration standards, funda-
    mental parameters scheme,
    173
Catalysts,
  automobile exhaust, 145
  XRF of, 82, 151
$CaWO_4$ Guinier diffractometer
    pattern, 320
  line widths, 322
Cellulose, crystalline vs.
    amorphous, 384
  scattering curves, 385
Ceramic tubes, 313
Chemical Information System (CIS),
    microcomputer use of,
    239
Clinochlore ferroan, 286
Co^{57} source, 203
Coal, NBS SRM 1635, XRF of, 191
Computer programs (see also
    Automated systems,
    Microcomputer)
  additive phase identification,
    232
  calibrating XRPD patterns, 267,
    289
  error-correcting, XRPD, 231,
    251, 267
  fundamental parameters, 81
  indexing/correcting XRPD pat-
    terns, 267
  minicomputer XRPD, 237

Computer programs (continued)
  NRLXRF, used with XRPD, 285
  peak refinement, 263ff
  profile fitting, 277, 264
  refining lattice parameters,
      268
  SANDMAN, XRPD, 231
  search/match by minicomputer,
      237
  search/match performance, 213
  systematic XRPD errors, 232,
      235
  wavelength dispersive, 81
  XRF minicomputer, 87
Convergent beam spectrum, 325
Copper alloys, XRF, 169
Core samples (XRF), 201
Corrections (see Computer
      programs)
Corrosion product, 284
Count times, optimized, 246,
      252
Crystallinity, methods for
      determination of, 384

Dead time, 47
Detection limits, XRF, 127
Detectors,
  energy resolution, solid
      state, 333
  GaAs, CdTe, HgI_2,
    band gaps, 4
    electron and hole mobili-
        ties, 5, 7
    growth and fabrication, 1
    performance, 3ff
  gas proportional scintilla-
      tion, 39
  germanium, 334
  high purity Ge for ultra-low
      energy, 23
  HPGe, resolution, 23
  ion implanted contacts, 24
  mercuric oxide, 1, 31
Differential X-ray diffraction
      using 2 wavelengths, 301
Diffractometer (see also
      Computer programs)
  automated, 261

Diffractometer, automated
      (Continued)
    in mineral zoning studies,
        373
  automation system, 245
    quantitative analysis for-
        mat, 255
  calibration at low angles, 289
Dissolution, 108

Energy dispersive diffraction,
    at high and low energies, 333
  counting statistics, 334
  energy resolution expected in,
      332
  film thickness measurement, 365
  for residual strains, 329
  profile fitting, Gaussian, 334
  strain sensitivity, 334
Energy dispersive spectrometry
        (EDS, EDXRF), 1, 31
      (see also Detectors)
  fundamental parameters, 81
  gas proportional scintillation
        counter for, 39
  microcomputers in, 85
  room temperature, 39
Environment, occupational, gases
      in, 181
Environmental assessment, appli-
      cations of XRF/XRD to,
      189
Error-correcting XRPD program,
      231, 268
Error windows for XRPD, 215
External standard method, XRPD,
      310

Fe^{55} spectrum,
  with HgI_2 detector, 34
  with HPGe, 23
Figure of merit, XRPD, 281
  search/match, 217
Film thickness measurement,
      XRPD, 365
Filters, XRF membrane, 209
Filtration in XRF of U, 108
Flyash, 296, 299
Fuels, XRF of, 191

Fundamental parameters method, 81
  with environmental samples, 189
Fusion,
  preparation of XRF samples, 95
  use with geological XRF samples, 117
  volatilization of sulfur in, 91

Gaseous contaminants, XRF of, 181
Ge, high purity detectors, 23
Geological samples, XRF, 117
Glass batch homogeneity, 379
  internal standards for, 381
Graphics comparison for XRPD data, 221
Guinier diffractometer with fast collection time, 315

Hardness vs XRPD peak width, 349, 352
Hematite/corundum mixtures, 311
Hollandite, 241

Identification of phases, XRPD, fast interactive system for, 273
Indexing, XRPD automatic, 267
Influence coefficients (XRF), 177
Intensities, XRPD,
  grain size effect, 302
  identifying lines by differential, 302ff
  scaling to αAl_2O_3, 304
Ion exchange
  columns analysis, 157
  resin, 109
Iron content effects on XRPD, 375
Iron sulfide minerals, 113

JCPDS file,
  microfile based, fast, 278
  rearranged for S/M algorithm, 214

JCPDS Round Robin test, 217, 218

$K\alpha_2$ elimination, 263, 277
KAP crystal efficiency in XRF, 359

Langmuir-Blodgett films for XRF, 355
Large grain XRPD technique, 339
Laterites, 127
Layered synthetic microstructures, 355
  applications as monochromators, 360
  focussing onto detectors, 363
Lead, XRF intensities, 79
Lead stearate
  as low 2θ XRPD standard, 292
  crystal efficiency in XRF, 359
Limestone, 286
Lubricating oil additives, by EDXRF, 173

Marcasite/pyrite in ores, 114
Mercuric iodide detector, 1, 13
  energy resolution, 33
  growth and fabrication, 32
Microcomputer, minicomputer
  for EDXRF, 85
  for XRPD, 237
  microfile, XRPD system, 273
  search/match method, 221
  XRPD programs for, 248
Mineral zoning studies, 373
Minerals, mixed, automated XRPD, 373
Monazite, rare earths in, 133
Monochromators,
  Ge(111), 318
  multilayer, for XRF, XRD, 360
  toroidal, 59
Mullite, 286
Multichannel analysis, microcomputer based, XRF, 85
Multiple Borrman diffraction, 326
Muscovite, 286

Ni compounds, identification of XRPD lines of, 306

Noble metals in catalysts, 151
Novaculite as XRPD standard, 290
NRLXRF program, used with XRPD,
        285

Obsidian, identification of
        source by XRF, 121
Occupational environment, deter-
        mination of gases in, 181
Oil additives, 177
On-stream XRF, 139
Optical polarizing microscopy
        with XRPD, 298, 300
Optimizing XRPD count times,
        246, 252

Particle size effects,
    on XRPD, 375
    XRF, 212
Peak search, Kα$_2$ stripping,
        XRPD, 263, 276
Perovskite, 241
Pharmaceuticals,
    cellulose crystallinity in,
        384
    XRD + XRF for, 383
Phase analysis (see also
        individual sample
        listings)
    absorption correction in, 296
    adiabatic equation for, 295
    crystalline/amorphous, 296
    figure of merit, 240
    glass, 379
    multiple amorphous phases,
        297, 299
    quantitative, XRPD + XRF, 283
    quantitative methods com-
        pared, 309, 313
    sources of errors, 311
    thin aerosol layers, 295
    with densitometer data, 238
    with external standards, 309
    with microcomputer, 237
    XRPD vs optical in rocks, 298
Pile up, 245
PIXE, 195
Plasma diagnostics, 360
    multilayers on mirrors for, 361

Plutonium, 163
Polarized X-rays, XRF using, 75
Polarizers for XRF, 63
Position sensitive detector,
    for fast XRPD system, 275
    in Guinier diffractometer, 315
    resolution, 315
Preamplifier noise, 11, 16
Precipitation, 109
Precision 2θ with large grains,
        339
Preconcentration, 107
Preferred orientation in search/
        match algorithm, 216
Preparation of XRF samples by
        fusion, 95
Pressed disks, geological XRF
        samples, 117
Process monitoring by EDXRF, 103
Profile fitting, 229
    in ED diffraction, 334
    XRPD, Marquardt algorithm for,
        264
    XRPD peak search, Kα$_2$ strip-
        ping, 277
Proportional counter, position
        sensitive (see Position
        sensitive detector)
Pulse shaping, 45
Pulsed X-ray tube, 45

Quantitative analysis, advantages
        of XRF with XRD, 283,
        388
Quantitative XRPD program format,
        255
Quartz (alpha) as XRPD standard,
        290
Quartz, Guinier diffractometer
        patterns of, 323

Radioactive waste host rock, 241
Radioisotope XRF for catalysts,
        151
Rare earths, XRF analysis, 133,
        205
Residual austenite, 344
Residual stress, 345
    after subzero cooling, 348

Residual stress (continued)
  by parallel beam, $\sin^2\psi$
      method, 343
  categories and definitions,
      350, 351
  in austenite and martensite,
      343ff, 352
  in case hardened steels, 345ff
  mapping throughout interior,
      329
  method for large grained
      specimens, 339
  relation to fracture, 350, 352
  vs fracture path, 351
  vs psi tilt, 346, 352
Resin loaded paper for XRF, 107
Retained austenite determina-
      tion, 343
Round Robin JCPDS test, 217, 218

Sample preparation, XRPD
  aerosol, 297
  effect of particle size, 380
  mixing, packing, 380
  pharmaceutical, 384
  uranium ores, 113, 114
SANDMAN XRPD program, 231
Savitsky-Golay smoothing
      XRPD data, 246, 264
Scintillation, secondary, 39
Search/match XRPD procedures
  applied to minicomputer, 221
  environmental samples, 192
  fast computerized system, 278
  for electron diffraction data,
      216
  for isostructural phases, 234
  for solid solutions, 235
  subfiles and users files,
      221ff
  systematic error corrected,
      213, 223
  using 2 wavelengths (DVD), 301
  using phase segregation, 301
  with elemental information,
      284
  with modified Snyder's
      program, 261
Sediment samples (XRF), 201

Separation (for XRF), 108
Silicon as XRPD standard, 290,
      292
Silicon wafer film thickness, 365
Slurry, XRF analysis of, 139
Soils (XRF), 81, 201
Solvent extraction of U and V,
      103
Standard background XRF method,
      145
Standard materials for XRD 2θ's,
      152, 289, 322
Steels (see Residual stress)
Strain tensor, in X-ray method,
      331
Stress (see Residual stress)
Sulfide ores, XRF analysis, 91
Surface layers, XRF, 196
SYNROC (for radioactive waste),
      237
Systematic error correction, 215

Thallium acid phthalate as XRD
      standard, 291
Thorium, 205, 207
Tin, XRF intensities, 79
Trace elements, 127
Tungsten-carbon multilayers for
      XRF, 359, 360

Unit cell refinement, 267
Uranium, 107, 203
  ores
      FeS_2 in, by XRPD, 113
      marcasite/pyrite in, 114
  XRF, 103
      analysis, 107
      intensities, 79

Vanadium, XRF, 103

Wavelength-dispersive
      spectrometry, 81
  layered synthetic microstruc-
      tures for, 355
  resolution and efficiency, 357
Welding fumes, XRF of, 209
Widths of XRPD peaks vs hard-
      ness of steel, 349

WKβ_2 diffracted, 336

X-ray
  astronomy, 16, 362
  beam of low divergence, 325
  photoelectron spectroscopy,
        multilayer mono-
        chromator for, 360
  tubes
    for EDXRF, 49
    intensity variation, 55
    processing in manufacture,
        57
    pulsed, 45
    target, electron gun, 51,
        53
    window, 54
X-ray fluorescence (XRF)
  addition method, 127
  aerosol filters, 195
  Americium^{241} source, 158
  analysis (see individual
        substance listings)
  beam blanking, 45
  detection limits, 63, 99, 127,
        206
  fundamental parameter
        program, 189
  hybrid ED/WD system, 59
  influence coefficients, 177
  low power, multichannel
        spectrometer, 169
  methods for Ga, Fe, Ni in
        Pu, 163
  monochromator, 59
  multiple linear regression,
        135
  overlap coefficients, 133
  polarized rays for 25-110
        keV range, 75
  portable systems, 43
  pressed discs, 117
  pulsed X-ray tube, 45
  resin loaded paper, 107
  resolution, 45
  spectra with HgI$_2$ detector,
        13, 15
  standards, fusion, 95
  surface layers, 196

  toroidal crystal, 69
  vs other analyses, 192
  with polarized X-rays, 63
XRPD (see Phase analysis,
        Computer programs,
        Automated systems)

Zeolites, XRF analysis, 157
Zirconolite, 241